Carpet Manufacture

Geoffrey H Crawshaw, BSc, PhD, CText, FTI

Editor: Barbara H Vaile, BA

Designer: Peter J Durrant

Director of Publication: Errol J Wood, PhD

Published by

WRONZ Developments

Private Bag 4749, Christchurch, New Zealand

ISBN 0-908974-25-6

Printed by
Chaucer Press Limited
Christchurch, New Zealand

Dedication

To my wife Anthea

who shares my interest in carpets

Preface

This publication has been produced in response to a need for a reference book covering the technology currently used by the carpet industry and its associated industries. George Robinson's book "Carpets", excellent in its time, was written in 1966 and last revised in 1972. Much has happened in the industry in the interim. Although material relating to individual aspects of the industry has been published and the trade press provides constant updates on new developments, people in the industry itself, its suppliers, its customers, machinery manufacturers, and students, have not been able to have the overall current picture on their shelves for over a quarter of a century.

Chapters containing detailed technical information cover only the processes in common use in the late twentieth century, but the introductory chapter outlines the history of the carpet industry and includes brief descriptions of many processes that are now extinct. A keen student will feel dissatisfied unless s/he has at least some awareness of the background to the subject. Moreover, pathways into the archives should always be maintained, since old technology can sometimes be revived in a different market situation or its feasibility can be enhanced by associating it with newer technology. Indeed, the decline of the carpet weaving industry would not have been resolved without the stimulation provided by modern electronics in the form of electronic jacquards and computer aided design systems. Quick response to orders and very short production runs would not have been practicable without them. Recent history and ancient history are equally important and the introductory chapter leads logically into a description of the situation of the carpet industry at the opening of the third millennium.

The relative importance of the various steps towards making a carpet varies with the style and construction of the carpet. In some cases the style is virtually created by the yarn; in others a basic "substrate" is converted into a virtual work of art by printing; while, often, the carpet making process itself is highly sophisticated and is clearly the creative step. Some carpet manufacturing machines give a product that requires only a little finishing to make it saleable: other machine products would be unusable without major further processing. In this book I have taken a holistic approach, considering that the technologist must have a general understanding of all aspects of manufacture from selection of raw materials to the principles of designing, as well as the detailed knowledge of his/her own speciality that can only be based on experience.

Some perspective on the market situations for tufting, needling and weaving is provided at various points in the book. Comments are also made on the position of weaving carpets by hand. Consumers in many parts of the world have become intensely interested in supporting the living craft of hand knotting, and this historic method of making carpets expanded significantly during the later decades of the twentieth century.

Carpet tests have mostly been standardised, but the requirements for the many facets of carpet performance are highly dependent on the application (some applications have very specific requirements), the culture of the country and the styles of carpet that are favoured there, and national expectations of textile floorcoverings, so I have necessarily taken a broad approach to the subject of carpet testing.

The main purpose of the collection of coloured illustrations of carpet textures and designs is to illustrate the possibilities and limitations of the technologies described in the text. The illustrations also demonstrate that carpet manufacture is undoubtedly an art form.

G H C, September 2002

Acknowledgements

Although references to published material are given at the end of most chapters, and the facilities provided to me by Carol Thomas, the librarian at Wool Research Organisation of New Zealand were invaluable in the preparation of the manuscript, my principal sources of information and illustrations were the machinery, chemical, dye, fibre, yarn and service organisations that support the carpet industry, and the carpet manufacturing industry itself. I am particularly grateful to the people who generously gave their time and used their specialised knowledge to ensure that the information in the various chapters was accurate. The following organisations and individuals provided me with assistance. However, absence of mention of competitive organisations in the text should not be used to infer that their products or services could not equally be applied in the carpet industry.

Allertex Ltd, UK
David Almond Ltd, UK
Amoco Fabrics, Germany
APSO Ltd, UK
Argelich, Termes Y Cia SA, Spain
Asota GmbH, Austria
Asselin, France
Astrograss All Weather Surfaces Ltd, New Zealand
Babcock TextilMaschinen GmbH, Germany
Barmag AG, Germany
BASF
Bekintex NV, Belgium
Ms Katrina Bennett, UK
Bonar Floors Ltd, UK
Bonar Yarns & Fabrics Ltd, UK
Border Textiles Ltd, UK
Boucher (Textile Engineering) Ltd, UK
Samuel Bradley Group, UK
Brintons Ltd, UK
Brintons Christchurch Yarns Ltd, New Zealand
British Carpet Technical Centre
Mr M J Brown, UK

Brückner Apparatebau GmbH, Germany
Burmatex Ltd, UK
Card-Monroe Corporation, USA
Carpet & Rug Institute, USA
The Carpet Tile Company, UK
Carpets International (UK) plc
Cavitec AG, Switzerland
Ciba (UK) Ltd
Clariant (UK) Ltd
Cobble Blackburn Ltd, UK
Colplan Engineering Ltd, UK
Cotinfi NV, Belgium
Courtaulds European Fibres, Germany
Courtaulds European Fibres, UK
David Crabtree & Son Ltd, UK
Crompton & Knowles (UK) Ltd
Oskar Dilo Maschinenfabrik KG, Germany
Don & Low Ltd, UK
DuPont (UK) Ltd
DuPont Speciality Chemicals, UK
DuraLite Carpet Corporation, New Zealand
ETF Machinefabriek BV, Netherlands
Europa Corporation SC, Poland
Textilmaschinenfabrik Dr Ernst Fehrer AG, Austria
Feltex Carpets Ltd, New Zealand
Firth Carpets Ltd, UK
Fleissner GmbH, Germany
Freudenberg Vliesstoffe KG, Germany
Dr K Gandhi, UK
Garnett Controls Ltd, UK
Gilbos NV, Belgium
Mr A van Goethem, Belgium
Griffith Textile Machines Ltd, UK
GT Supplies & Engineering (Acc) Ltd, UK
John T Hardaker, UK
Hofmann Handtuft-Technik GmbH, Germany
Dr I Holme, UK
Hometex, UK
Maschinenfabrik Michael Hörauf GmbH & Co KG, Germany
Houget Duesberg Bosson SA, Belgium
Dr J Ince, UK
Interface Europe Ltd, UK
Ishikawa Seisakusho Ltd, Japan
Kibby-Modra Automation Pty Ltd, Australia
Kirstol Ltd, UK
Eduard Küsters Maschinenfabrik GmbH & Co KG, Germany
Michele Lamperti SpA, Italy
Richard Latham, Textile Agent, UK
B H McCleery & Co Ltd, UK

Mackie International Ltd, UK
Mahlo GmbH & Co KG, Germany
Robert S Maynard Ltd, UK
MCS Officina Meccanica SpA, Italy
Milliken Carpet, USA
Officine Minnetti Italia srl, Italy
Mitter GmbH & Co, Germany
Montell Polyolefins, Netherlands
MP Engineering srl, Italy
m-tec Maschinenbaugesellschaft mbH, Germany
NedGraphics Print Ltd, UK
New Zealand Sports Turf Institute
Novalis Fibres Ltd, UK
Plantex SpA, Italy
Plasticers Ltd, UK
Pneumatic Conveyors (Huddersfield) Ltd, UK
Priest (Lindley) Ltd, UK
RGS Chromatone Ltd, UK
Rieter Synthetic Fiber Machinery, Switzerland
Ryalux Group, UK
Sallee Carpet Looms (NZ) Ltd
N Schlumberger & Co (UK) Ltd
Sellers Engineers Ltd, UK
Shell Chemicals, New Business Development Unit, Belgium
Sillaford Ltd, UK
Mr J H E Spencer, Australia
Spindelfabrik Suessen, Germany
Stephenson Thompson Textile Chemicals Ltd, UK
Superba SA, France
Symtex, UK
Tapistron International Inc, USA
Teppich Forschung International, Germany
Textile Laboratory Services Ltd, UK
Thibeau, France
Tuftco Corporation, USA
Tufting Machine Company Inc, USA
Ulster Carpet Mills Ltd, UK
Michel Van de Wiele NV, Belgium
Vanwyk Textile Systems BV, Netherlands
The Wilton Carpet Factory, UK
WIRA Instrumentation, UK
Wool Research Organisation of New Zealand
Wools of New Zealand, New Zealand
Wools of New Zealand, UK
Yarn Engineering Services, UK
J Zimmer Maschinenbau Gesellschaft GmbH, Austria

The production team, consisting of word processor, editor, designer, printer and publisher, performed magnificently.

The TWO-in-ONE® carpet backing system

For contract carpet tiles and broadloom

Environmentally friendly

The TWO-in-ONE® system is based on environmentally friendly and non-toxic materials. The construction contains no poisonous substances which effect the indoor climate. Furthermore, at the end of its life time it can be incinerated with energy recovery in mind.

Easy installation

TWO-in-ONE® carpets do not require permanent glueing. Installations are performed quickly and effectively by using a tackifier. Cutting of the tiles are easier and faster compared to conventionally constructed carpet tiles. An additional advantage is the low weight of TWO-in-ONE® carpets compared to conventional tiles and the elimination of wall marks.

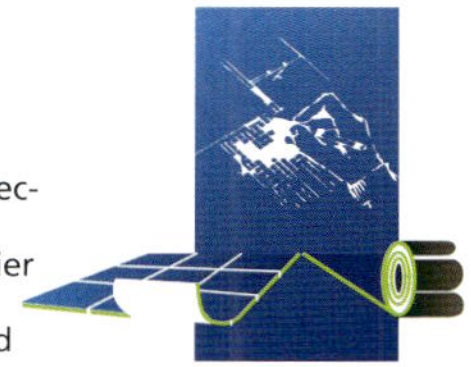

Design flexibility

The TWO-in-ONE® system enables architects, interior designers and specifiers to create a new dimension in textile floor covering without limitations. Combining broadloom and tiles in carpet designs opens up new opportunities which is a bonus to the end user both practically and aesthetically.

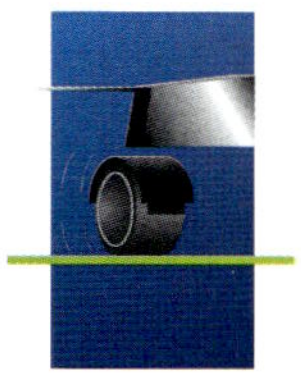

High wear resistance

Edges are not damaged even after extensive use of castor chairs and consequently joints remain intact and do not become visible.

High comfort

The TWO-in-ONE® carpet has a high comfort level. Due to an excellent shock absorption the carpet provides an excellent surface to walk on with orthopaedic advantages.

High performance

In addition to the above, the TWO-in-ONE® carpets meet all technical demands for heavy contract flooring in terms of ignition resistance, dimensional stability, antistatic properties and comfort.

The **TWO-in-ONE®** system is jointly developed by:

Fibertex A/S
Box 8029
Svendborgvej 16
DK-9220 Aalborg Øst
Denmark

Tel. +45 96 35 35 35
Fax +45 98 15 85 55
e-mail fibertex@fibertex.com
web www.fibertex.com

DOW Benelux N.V.
P.O. Box 48
NL–4530 AA Terneuzen

Please contact Fibertex A/S for further information and samples.

B10 MARKETING A/S

Contents

1 Historical Background

Details of the development of the British carpet industry are recorded in an excellent book by Bertram Jacobs: *The Story of British Carpets* [1]. Jacobs' book is particularly important for its coverage of the industrialisation of the carpet industry and because of its emphasis on the achievements of influential personalities in the industry, including the critical interactions between American and British inventors and entrepreneurs. George Robinson's book *Carpets* [2] includes a more general history, but also more technical details. Both books cover hand-crafted carpet manufacture well, but there are better illustrated books that focus exclusively on the subject, such as that by Murray Eiland [3]. The serious student of hand-knotted carpets is referred to the German literature. Current German journals (eg, *Heimtex* [4]) often publish articles relating to hand-craft techniques, design influences, geographic locations, historical development and the collecting and exhibiting of historic carpets. Many useful references to broader publications on the subject can be gleaned from such journals. Historical research on the origins of the carpet tufting industry in the USA is recorded in a book by Thomas Deaton [5]. Deaton's account of the American side of the story has been extended to include Europe from the author's own observations.

1.1 HAND-CRAFTED CARPETS

1.1.1 Origins

Undoubtedly, the first carpets were animal skins, crudely tanned by primitive hunters and laid on the floors of their caves, tents or huts. Weaving of mats and rugs came much later, but it is impossible to say exactly when. The animal and vegetable fibres from which they were made are biodegradable so the very early carpets could not be expected to have survived until the present time. Their existence can, however, be deduced with some certainty from archaeological finds, including impressions of them in clay, illustrations of textile processing, writings about them and records of trading in carpets.

The crudest products in evidence were mats woven from rushes. More sophisticated floorcoverings were produced from vegetable fibres, but there was an early aim to produce pile carpets from sheep's wool or goat fibres, probably to simulate the warmth and comfort of animal skins. The making of true pile carpets is believed to have originated about 6000 years ago, possibly in the region of Persia.

The neolithic (new stone age) period began earlier (about 12 000 BC) in the warm and fertile (more fertile than now) areas of North Africa, the Middle East and Western Asia than it did in North Western Europe. Agriculture and tool making developed hand-in-hand, so that settled communities with a semblance of civilisation were created earlier in these areas, as well. Nomads following their flocks of sheep and goats drifted between these communities. Some of the travellers became traders, and technologies and artistic talents were interchanged along with the goods traded.

By the third century BC the remarkable 'Silk Road', an ill-defined route between China and the Mediterranean, had come into existence through the linking of numerous caravan routes. Among the goods traded were carpets.

The oldest existing carpet was excavated in 1953 from a tomb at Pazyryk in Southern Siberia near the border of Outer Mongolia, and was estimated to be 2400 years old. Its existence is due to the fact that it was preserved in ice. The Pazyryk carpet (Figure 1.1) can be seen at the State Hermitage Museum in St Petersburg.

Figure 1.1. The Pazyryk carpet. (Reproduced with the kind permission of the State Hermitage Museum, St Petersburg)

The Pazyryk carpet is a pile rug in a fine construction (about 36 knots/cm^2) and has a sophisticated design with a central area of repeating motifs and an elaborate border. The presence of these elements of rug design, as produced in subsequent centuries, points to the fact that rug making was a well established art about 2500 years ago. Various features of the Pazyryk rug indicate multicultural influences. Although the Turkish or Ghiordes knot (see Section 8.1.1) is used, the outer border is typically Persian. The ground colour is Turcoman red, with blues, yellows and white, and the horse motifs have Persian characteristics, such as the head-dress of the riders, the treatment of the manes and tails, and the type of saddle cloths. In the central motif there is a quatrefoil of Assyrian design, and the inner border of reindeer might be considered Nordic. To add to the confusion, there are residues of silk and swan-like designs in felt appliqué, suggesting a Chinese influence. These heterogeneous aspects of design and construction support the theory that nomadic tribes spread the art of carpet making throughout Asia.

Probably the art of carpet making was known in the Greek and Roman civilisations, but was certainly lost in Rome when that centre fell to the barbarians. The traditions of carpet weaving were conserved in the East and, with the foundation of Islam in AD 622, the Moslems tended to monopolise carpet weaving and Islamic motifs were extensively incorporated in carpet designs. Many of the older centres of weaving in Turkey, Iran, Caucasia, Central Asia and China remain today, and in the last quarter of the twentieth century the hand-knotting industries of India and Pakistan were greatly expanded. A new hand-crafted carpet industry has been formed by Tibetan refugees in Nepal, who produce their own characteristic designs.

Carpets are still made by nomads from hand-spun yarns, but the bulk of hand-knotted carpets are made from machine-spun yarns, either in carpet centres or by individual families in villages. In the latter case, large companies organise the operation, providing yarns and marketing the products. There are a few hand-knotting organisations in industrialised countries, working to reproduce historic carpets and generally operating a custom (or bespoke) manufacturing service. Details of both traditional and custom hand-knotted carpets are given in Chapter 8.

1.1.2 Hand Knotting in Europe

Carpets arrived in Europe from two directions: from the Middle East and from North Africa.

Islam spread westwards through Egypt along North Africa and the craft of carpet making was transferred with it. Moslems from North Africa invaded Spain in AD 711 and controlled virtually the whole country 20 years later. Carpet weaving centres were established in Spain and many of them remained active after the Moors departed in 1492.

Crusaders from Western Europe undoubtedly encountered carpets during their military expeditions to the Christian Holy Land, in the twelfth and thirteenth centuries. A few carpets may have been brought to Britain to impress royalty, but there was no widespread interest in carpets in Britain in the thirteenth century. However, craftsmen were displaced by the Turkish invasion of Byzantium in 1453 and they set up carpet weaving centres in Italy and southern France.

Towards the end of the sixteenth century, centres of carpet weaving were known to exist in France as far apart as Aubusson and Beauvais. Artists and craftsmen alike contributed to the development of carpet making, and French carpets were certainly not copies of oriental carpets. Aubusson carpets were constructed rather like flat tapestries and were woven sideways, from left to right. Designs were unmistakably French and were destined to furnish palaces.

In 1596, Henry IV of France was impressed with a knotted 'Turkey-work' carpet made for him by Pierre DuPont and a few years later set up a workshop for the young man in his Louvre palace. In 1620 the project was expanded into a disused soap works (savonnerie); the factory remained the property of the French Crown and its business was protected by an import ban. Savonnerie carpets became famous and the factory was to stimulate carpet making outside France.

A hundred years later, a visitor reported that two distinct types of carpet were being made at the Savonnerie. Hand-knotted carpets were made on upright looms, but by an original technique. The carpets were woven from bottom to top, as in Asia, but the knots were tied over a sharp edged iron rod extending across the loom. When the rod was withdrawn, the loops were cut. In addition to the hand-knotted carpets, the Savonnerie was also making loop-pile (Brussels) carpets on treadle operated horizontal looms. In 1825, the Savonnerie was closed and its looms were moved to the Gobelin factory at Beauvais, where the National Tapestry Factory was located.

In Britain, it was Cardinal Wolsey who stimulated real interest in carpets among royalty. He solicited gifts of carpets, among other valuables, from Venice and a consignment of 60 carpets was received in 1520. Subsequent demand from the nobility led to the establishment of upright knotting looms in the British Isles. For example, it is recorded that in 1539 Robert Rothe, Earl of Ormond, imported weavers from the East to make Turkish style carpets in Kilkenny, Ireland. Floral and heraldic designs were developed and quality was high, as demonstrated by the Verulam carpet made for Queen Elizabeth I in 1570.

An alternative British trend was to make floorcoverings for the gentry, which were cheaper than the sophisticated carpets wanted by the nobility. In the early seventeenth century, apparel weavers in the English town of Kidderminster converted their coarser wools into simple 'fote cloths' that were applied to cover the floors of castles and possibly country mansions. Under the Puritan influence during the Civil War, weaving of fote cloths prospered at the expense of hand knotting.

During the eighteenth century a period of gracious living led to a growing demand for high quality hand-knotted carpets, concurrent with the industrialisation of carpet making.

In 1749, two master weavers from the Royal Savonnerie fled to London and Peter Parisot, a Huguenot refugee, secured finance to start a knotting workshop based around their skills. A weaver from Axminster, Thomas Whitty, had ambitions to make carpets and in 1754 indulged in industrial espionage in Parisot's now prospering factory. He taught his family of daughters to weave the Turkish knot, and they became highly skilled. Whitty's Axminster Carpet Manufactory prospered for 75 years. Later, the term 'Axminster' was taken to mean machine-made carpets simulating the hand-knotted carpets as made in the town of Axminster in Devon. Around the same time as Whitty's initiative, further knotting factories were established in Exeter, London and Frome.

The prosperity of the British hand-knotting industry continued through the Regency period, but the Napoleonic wars claimed huge resources and led to a period of economic decline, so that even Whitty's factory closed in 1835 and the looms were consolidated in the Wilton factory (see Section 1.2.3). A few hand-knotting centres survived into the twentieth century.

1.2 INDUSTRIALISATION OF CARPET WEAVING

Steam power was first applied in the textile industry as early as 1785, when James Watt with his partner Matthew Boulton installed one of their engines in a cotton mill in Britain, then the world's leading industrial nation. However, it was 54 years later before a powered carpet loom was made.

Power looms for various carpet manufacturing techniques were initially developed in America, although they were at first more widely used in Britain. In fact, the transatlantic carpet trade prompted the movement of entrepreneurs and inventors in both directions, and cross-stimulation of ideas between Britain and America contributed strongly to the technical and commercial growth of the world's carpet manufacturing industry. Steam railways, introduced around the same time as powered looms, facilitated trade in carpets.

The prime objective of industrialisation was to produce, at lower cost, products similar to hand-knotted carpets. In other words, to extend the market for carpets outside palaces and castles and into the houses of the middle and even lower classes. With the exception of the 'fote cloths', where the targets were comfort and low price, a key objective was to be able to create designs as elaborate as those available from hand knotting: the aesthetic importance of carpets at that time cannot be underestimated.

Installation and transmission of steam power was a major undertaking for a single carpet mill and an interesting event in the history of Kidderminster was the building of the Lord Ward shed. The basic premises included an engine house, a rope race running in a space down one side of the shed to transmit the power, and a dyehouse. Space was leased to about ten carpet manufacturers who were able to take advantage of the common facilities.

Before power was applied, treadle looms had been developed for making many technically different types of carpet. These could theoretically have been driven by water or steam power, so mechanisation of the carpet industry may be thought of as beginning with any loom where the pile was not inserted by hand.

1.2.1 Brussels Weaving

In the latter part of the fifteenth century, pile carpets were being woven on treadle operated looms in the area of Brussels in Belgium. It is not clear when the technique spread further, but certainly, the technique was well established at the Royal Savonnerie factory in Paris in 1720. It survives today.

Although the Brussels technique produces a pile carpet, its structure is very different from that of a knotted carpet. The pile is looped, not cut, and a great deal of pile is incorporated in the backing fabric ('dead pile'), representing the colours not required on the surface at any particular point in the design. Details of the mechanism of modern Brussels weaving are given in Chapter 10.

Originally, pile warps were selected and raised manually for looping over 'wires'. Similarly, the wires used to create the loops were inserted and removed by the weaver's assistant. In modern looms, the pile colours are selected electronically, but for well over a century punched cards (jacquards) controlled the design.

In Tournai, also in the Low Countries, the technique was modified at an early date to create cut pile, probably by running a knife along a groove in the wire.

1.2.2 Ingrain Weaving

Ingrain carpet is a reversible double cloth without a pile, but with a slightly ribbed surface similar to that of a hand-woven tapestry carpet. Pearsall and Broom set up a factory in Kidderminster in 1735 especially for weaving ingrain carpets (see Section 12.2). In 1778, weaving of similar carpets commenced in Kilmarnock in Scotland and a triple cloth was developed there. The products were first named Kidderminster or Scotch carpet, but the term ingrain became established in Continental Europe and the USA.

In 1801, Joseph Jacquard exhibited his punched card patterning mechanism in Paris. His work was highly respected in France and elsewhere and his invention was rapidly adopted by the fabric weaving industry in France. Broom installed jacquard systems on his ingrain looms in 1825 and many firms in Britain quickly followed his example, not only on ingrain looms but also on Brussels looms.

In 1839, an American, Erasmus Bigelow of Massachusetts, built the first powered ingrain loom and it was bought by Scottish weavers and some English companies.

Production of ingrain carpets declined rapidly in the 1930s, but the technique has been proposed in more recent years for producing carpets from 100% natural fibres (wool and/or cotton) without using adhesives to meet the demand for 'environmentally friendly' products.

1.2.3 Wilton Weaving

Brussels weaving was introduced to Britain as a result of Lord Pembroke's visit to the Savonnerie in 1720. He persuaded two master weavers, Pierre Jemaule and Antoine Dufossy, to come to his home town of Wilton in England and set up an establishment for producing Brussels loop-pile carpet.

Their stroke of genius was not well recorded, but some time in the 1740s it seems that they applied the Savonnerie technique of knotting to produce cut-pile carpets on treadle looms. The rounded wires over which the loops were formed in Brussels weaving were simply replaced by wires having knives at one end, so that when the wires were withdrawn the loops were cut.

The cut-pile products attracted acclaim and for a few years the Wilton factory enjoyed a monopoly on their manufacture. Then, in 1749, Broom, who was already manufacturing ingrain carpets in Kidderminster, acquired the skills through recruiting weavers from Belgium or Wilton (or both). Quickly, Brussels looms were installed in other companies and weaving of Brussels or Wilton style carpets became more important than ingrain weaving. By the end of the eighteenth century there were more than 1000 carpet looms in Kidderminster.

Ten years after producing the first powered loom to weave ingrain carpets, Erasmus Bigelow produced his first steam powered Brussels loom, which he exhibited in London and Kidderminster in 1851. It was bought by Frank Crossley of Yorkshire who had already installed steam power in his Halifax mill to drive both tapestry and Brussels looms. Crossley's staff developed power looms further and for some time virtually controlled their development in Britain.

In the first half of the twentieth century, Wilton looms were widely used for weaving high quality patterned squares for residential applications. Their higher cost (because of the cost of dead pile) led to their progressively being displaced by Axminster and face-to-face squares, and their main residential application became for plain carpets. Narrow-width Wilton weaving was used extensively in the middle of the century for producing hard wearing contract carpets, usually in 2- or 3-frames (colours) to customers' specifications (and therefore short runs). When Axminster looms more suitable for short runs became available in the last decade of the century, this application began to decline. A unique facility of the wireloom that remains is for producing textured carpets in loop pile, cut pile and cut/loop constructions.

Wilton weaving survives today for the production of contract carpets and special products and its technology is described further in Chapter 10.

1.2.4 Tapestry (Warp-printed) Carpets

Early nineteenth century inventors who were interested in carpets were conscious of the valuable pile buried in the base (dead pile) of Brussels, Wilton and ingrain carpets alike. A clear research target was to produce mechanically carpets with most of the pile on the surface, as in knotted carpets.

It was a multi-talented Scot, Richard Whytock, who conceived the idea of printing the pile yarn and weaving it on what was essentially a Brussels loom with only one frame of pile yarn. His process gave a good simulation of Brussels carpets, although the colours were not as clear; and tapestry velvet carpets could be made by using bladed wires. His patent was filed in 1832 and he went into production in Edinburgh in Scotland in 1833. The patent was licensed first to Crossleys in Yorkshire and later to other carpet companies in Britain and the USA.

The tapestry process economised on yarn but was tedious to operate. Pile yarn, usually worsted yarn, was wound parallel on a large drum (drums could be up to 12.2 m in circumference) and strips of colour were printed across it. The huge hank that resulted was removed, steamed batchwise to fix the dye, and loose colour rinsed off. This represented one end of pile in a warp bearing repeats of one sequence of part of the design. Other batches of yarn were printed in other sections of the design and each warp end was carefully labelled to mark its position in the design.

All the required printed warps to make up the total design were assembled in order and wound on to a warp beam. The beam was mounted at the back of the loom and, after threading the yarns through the heald, the carpet could be woven. Of course, the design as printed on the yarn was attenuated in a warpwise direction as the yarn was folded over the wires and the wefts. Careful allowance for the structure of the carpet had to be made when preparing the design.

Tapestry carpets were made for many years, but the system declined in importance after the gripper Axminster (see Section 1.2.7) process came into widespread use.

1.2.5 Chenille Axminster

James Templeton had a business commissioning the manufacture of shawls in Paisley, Scotland, which included French-style chenille products (chenille means hairy caterpillar in French). One of his weavers, William Quiglay, had an idea for confining all the pile to one surface of the shawl fabric. This was not commercially successful for shawls but Templeton adapted the principle for carpets and in 1839 set up a carpet factory in Glasgow to exploit the joint invention.

Templeton's process allowed unrestricted use of colour and there was no dead pile in the backing, as in Brussels and Wilton carpets. For the first time, it was possible to produce mechanically carpets very much like the hand-knotted carpets made in Axminster, so the process became known as chenille Axminster. The process was widely used for more than a century and was the leading system of carpet manufacture for many years.

Chenille Axminster carpets were made in three basic stages: weaving the pile yarns into chenille fabric; converting the fabric into chenille fur; and weaving the carpet on a 'setting loom'.

Pile yarn was first woven as weft across widely spaced groups of interlacing cotton warps. The colour sequence corresponded to the colour sequence in a weft of the ultimate carpet and was selected first by hand changing of shuttles and, about 100 years later, by a gripper needle controlled by punched tape (Verdol paper).

After weaving, the chenille fabric was slit into warpwise strips, as shown in Figure 1.2(a), to give chenille fur (Figure 1.2(b)). Immediately after cutting, the strips of fur were steamed and pressed into grooves in a heated drum to set them into a V-shape, as illustrated in Figures 1.2(c) and (d). One length of chenille fabric contained only a limited number (perhaps 12) of the colour sequences required warpwise, so a whole design repeat might require several different fabrics and many (eg, 144 in a 36 in. repeat) strips of fur that had to be carefully marked and used in the correct sequence.

The base structure of the carpet was similar to that of a two-shot Wilton, except that additional warps (catcher threads) were introduced to pass between the tufts of pile and lock the chenille fur to the base. The fur was introduced by a shuttle or, later, by a travelling arm. The weaving sequence was to insert four picks of jute weft and then one of fur. At this stage the loom was stopped and the weaver erected the pile with the aid of a steel comb. The fur was beaten up as the start of another sequence of loom operation.

Because of the labour-intensive nature of chenille Axminster production, the process declined in the mid-twentieth century. The last production unit, which was at Templeton Carpets Ltd, was closed in 1971. In the later years, Templetons used chenille yarns for producing plain and semi-plain bonded carpets.

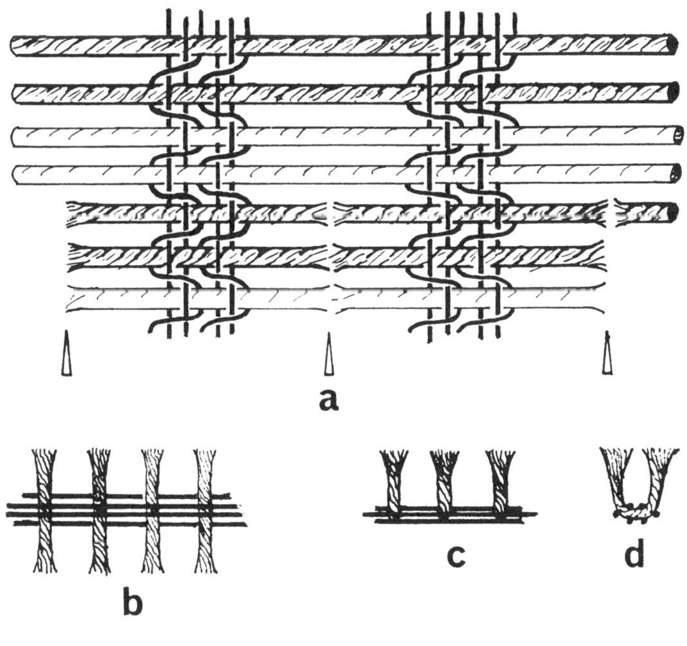

Figure 1.2. (a) Section of chenille fabric, showing how it is cut into strips; (b) fur after cutting; (c) and (d) section of folded fur [2].

1.2.6 Spool Axminster

An inventive employee of Alexander Smith's carpet mill in New York, Halcyon Skinner, was given the task of devising a power loom for producing carpets similar to Axminster knotted carpets and in 1856 Skinner and Smith patented "the Axminster Tufting Loom". The principle was to convert the design element of each row of the carpet into a spool of the required colours of yarn wound parallel. Spools representing consecutive rows of pile were mounted in order on a chain and 'nippers' took a row of tufts from a spool and wound them around the warps.

One of the early spool looms was exhibited in London in 1862 and was bought by a company in Brussels. However, the first spool looms were only moderately successful and Skinner worked for several years to create an improved loom.

Commercial success was achieved in 1877 with a loom that wrapped the tufts around the weft instead of around the warp, taking them directly from a spool that was transferred temporarily from the chain to the weaving position. To distinguish the new material from Axminster, the Americans called it moquette, the French name for hand-knotted pile carpet. As a result of Skinners' invention, Alexander Smiths grew from a small tapestry and ingrain operation to dominate the American carpet industry by the turn of the century.

As early as 1878, Michael Tomkinson of Kidderminster negotiated the British patent rights for Skinner's invention and renamed the product Royal Axminster. As well as setting up production in the Tomkinson and Adam factory, Tomkinson sub-licensed the process to other British manufacturers. Spool Axminster manufacture became widespread.

Erasmus Bigelow produced a variant on the Skinner invention that was introduced to Britain by James Templeton. The spool remained on the chain and grippers took the yarn ends, pulled out tuft lengths that were cut off and transferred them to the weaving position. This spool-gripper Axminster system, detailed in Section 9.2, is used today for producing cut-pile carpets with unlimited design possibilities.

1.2.7 Gripper Axminster

Although Halcyon Skinner had patented the gripper, it was Brintons, a manufacturer of Brussels and Wilton carpets in Kidderminster, who encouraged their technologists to apply the principle to a jacquard loom. Their invention was patented in 1890.

Essentially, the gripper Axminster loom took the creel of a Wilton loom, with its frames of different coloured yarns, and coupled it to the weaving mechanism of a spool loom. Between the two, the jacquard mechanism was used to select the required colours of pile yarn and align them to be taken by a row of grippers operating like birds' beaks. Tuft lengths were cut off and transferred to the weaving position by the grippers, where they were folded over weft yarns to create a row of U-tufts.

The gripper Axminster system obviates the need for spool setting and, although the number of available colours is not as great as in spool Axminster, the system is more versatile. Designs can be changed simply by installing a different set of jacquards or, in modern looms, by switching to alternative designs held on computer disk.

The gripper Axminster system has been progressively developed over the years by Brintons for their own use and by machinery manufacturers, principally David Crabtree and Sons, Bradford, UK, so that it has remained an important carpet manufacturing process over a long period of time. Improved methods of weft insertion have further stimulated interest in the system.

Gripper Axminster carpets have the outstanding advantage of having a high proportion of the pile above the backing and alternative weave structures are available for squares (with the pattern showing on the back, as in knotted rugs) and wall-to-wall carpets. Less wastage of pile yarn occurs than in spool-gripper manufacture. Production rates have been progressively improved and now exceed 10 m^2/h. Looms are now available for weaving a wide range of tuft densities, and a wide range of yarn counts and constructions can be woven in the same carpet to produce patterns of texture as well as patterns of colour. In recent years, the technique has received a major boost from the widespread use of electronic jacquards, CAD systems and rapid weft insertion techniques.

Gripper Axminster weaving is an important current system of carpet manufacture and the technology is described in Section 9.1.

1.2.8 Face-to-Face Weaving

In 1857, Jacinto Barrau y Cortes of Spain patented the first face-to-face weaving system for power looms, but this was for velvets rather than for carpets. Figure 1.3, taken from Barrau's patent, illustrates the principle of creating two fabrics from one weaving operation.

Because of financial difficulties, Barrau was compelled to sell his patent to Samuel Cunliffe Lister of Bradford, who commercialised the concept over the next 16 years to become the most successful producer of velvets in Europe. Prior to World War I, attempts were made to adapt the principle to the weaving of carpets, but success was not achieved until 1921. It was the Belgian engineer Michel Van de Wiele who introduced first, single-shuttle face-to-face carpet looms and, a few years later, double-shuttle looms.

Van de Wiele's company has continued to be innovative. During the period 1969–1975, they introduced shuttleless carpet looms to the Belgian weaving industry. Their modern looms utilise rapiers for weft insertion: single or double rapiers may be used, depending on the carpet construction, and in the mid-1990s a triple-rapier system was introduced to provide a high rate of output (up to 2000 m^2/day) in selected constructions.

In 1993, Van de Wiele applied the principle of the Bonas electronic jacquard to their looms and later developed their own electronic jacquard systems, providing the weaving industry with a major advance in flexibility of operation.

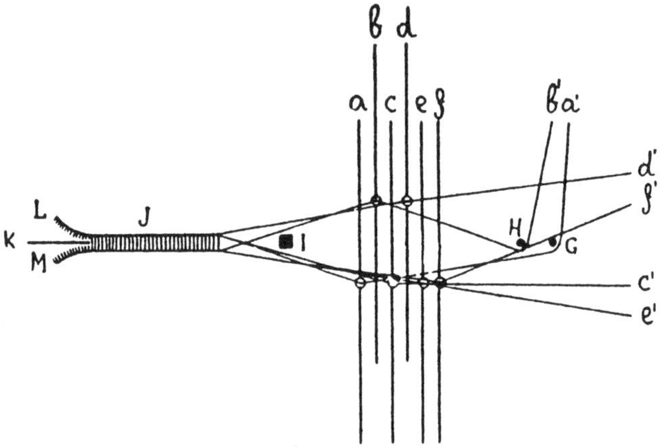

Figure 1.3. The principle of face-to-face weaving taken from Barrau's original patent: a'-f' warp yarns, a-f corresponding heald frames; G and H guides for pile yarns; I shuttle; J woven sandwich; K knife; L, M pile fabrics.

Face-to-face weaving for many years has been the most popular system for weaving carpet squares and is increasingly used for wall-to-wall constructions. It is discussed further in Chapter 11.

1.2.9 Broadloom Carpets

The early manufactured carpets were mostly woven 27 in. wide, coinciding with the Flemish ell (68 cm), to give 'body carpet'. Large squares were made by sewing together several sections of body carpet and four strips of border carpet, mitred at the corners. The borders were woven separately 18 or $22^1/_2$ in. (46–57 cm) wide. An advance that required skilled designing and additional jacquard cards was to make up the squares from 27 in. wide body carpet. The inner strips had body design with border design at each end; and the outer strips had partly body design and border design along their lengths, with corner design at the ends.

In 1896, Alexander Smith in the USA patented a wide spool loom for weaving seamless Axminster carpet and once again Michael Tomkinson took the idea to Britain. Templetons also adapted their spool-gripper looms to weave seamless squares. Four 27 in. spools clipped together gave carpets 9 ft (2.7 m) wide. At an early stage in the development of the jacquard-gripper loom a target of weaving seamless squares was set and Brintons had a 15 ft (4.5 m) wide machine working by 1905. Seamless Brussels squares, up to 12 ft (3.7 m) wide, were being made early in the twentieth century. Considerable advances in engineering were required to scale up from 27 in. looms to giant machines that operated reliably.

When the fashion for carpeting rooms wall-to-wall arrived, the industry was well placed to provide the consumer with non-bordered broadloom carpet. For the hospitality industry and commercial applications it was common for many years, however, to provide wall-to-wall installation from body carpets, since short runs in the customers' colours and designs could be provided more conveniently in this way.

1.3 TUFTING

Carpet tufting (detailed in Chapters 6 and 7) developed initially in a somewhat random way from the bedspread tufting industry to become by far the most important method of carpet manufacture. It is relevant to consider the origins of the tufting process in Georgia, USA.

1.3.1 Bedspreads

Catherine Evans laid the foundations of the tufting industry [5]. She did not originate the basic concept of tufted fabrics, nor did she invent tufting machinery, but she was responsible for the industrialisation of the craft of tufting. Her activities initiated a chain of events that led to the multibillion dollar American carpet tufting industry being centred around Georgia, an industry that has influenced carpet manufacturing throughout the world. There is no logistical reason for North Georgia to be the hub of the American carpet industry; it is simply that Catherine Evans was brought up in the countryside close to Dalton.

The exact origins of tufted fabrics are unclear. Historians believe that the art of tufting was introduced to America during the colonial period by immigrants from France and England. It is said that the trimmed wicks from home-made candles were sometimes used for the tufts. Hand tufting was developed in New England and became sophisticated in the South in the eighteenth century, but the art declined and almost disappeared after the American Civil War.

Catherine Evans was impressed with a tufted bedspread from pre-Civil War days that she had seen as a child and in 1895, when she was 15, decided to make one like it. She plied 12 strands of number 8 cotton yarn together and, with a large needle, formed a pattern of squares on a cotton sheeting fabric, skipping half an inch or so between stitches. When the pattern was finished, she cut between the stitches to create a bedspread having a pattern of fluffy tufts.

She soon started making tufted bedspreads on a regular basis for friends and family contacts. When the family moved to a farm near Dalton in 1909, she developed a good market there, too, and started to employ neighbours to help her.

The general process was to stamp the pattern on the base fabric using tin lids or kitchen dishes tipped with meat fat (later, sheet lead printing blocks carrying wax containing fugitive tints were used), stitch the fabric, cut the tufts, boil the cloth to shrink it so gripping the tufts, and fluff the tufts by beating with a broom or cane.

Catherine Evans (Catherine Evans Whitener after her marriage) became a skilled designer, commissioning others to produce the bedspreads. Her sister-in-law, Mrs Eugene Evans, became active in expanding the business, looking for additional home workers, distributing the raw materials and collecting the finished products. The Evans Manufacturing Company was formed in 1917 and, despite the depression following World War I, was selling throughout the Eastern States by the early 1920s.

Tufting companies were formed by others and the manufacture of bedspreads became a significant industry in Georgia. It remained essentially a home craft industry, but the cotton mills and dyehouses also benefited. After the financial crisis of 1929, home tufting enabled thousands of poor farmers in Georgia to survive.

The tufted bedspreads were often called candlewick and sometimes, incorrectly, chenille. True chenille fabric is actually woven from chenille yarns that are prepared by weaving the pile yarns as weft in a temporary fabric that is subsequently slit into yarns or, nowadays, spun on a special chenille machine.

1.3.2 Mechanisation of Tufting

Franklin D Roosevelt's 'New Deal' of 1933, designed to pull America out of recession, caused the bedspread industry to change dramatically. The new legislation included labour codes for industry, setting minimum wages and maximum hours of work. Although the position of home workers was not clear, the industry faced the prospect of wage rates being quadrupled and the possibility that tufted bedspreads would be priced off the market. People in the industry began to talk of a need for tufting machinery.

Practically every mechanically minded person in the bedspread industry tried to create a tufting machine, usually by modifying an old sewing machine, so there may be a couple of hundred people who claim to have invented 'the tufting machine'. Patents for tufting machinery had actually been filed as early as 1882 and 1889, but do not seem to have reached commercial fruition. Some names of people who developed practical single-needle tufting machines are August Carter, Ernest Moench, Erskine Boyce and O C Moore. Glenn Looper, a graduate of the Georgia Institute of Technology, is credited by the Georgia State Legislature as the inventor of "the first single-needle chenille machine" and the machine is exhibited in the state museum. He created a variety of items of equipment used in the tufting industry (and other industries), including a single-needle machine that sewed heavy count yarn and cut the tufts by a scissors mechanism.

Vast quantities of bedspreads were produced in the mid-1930s on modified gooseneck sewing machines, often Singer model 3115. The operator guided the base fabric under the needle, following the design marked on the fabric. Tufting moved from the farms to workshops and factories where rows of machines were installed. Several famous carpet companies and tufting machine manufacturers were founded in this era.

An alternative type of machine developed and widely used in the 1930s was the needlepunch machine. The base fabric was mounted on an oversized (1.2 m) embroidery hoop which the operator could move under a reciprocating hollow needle. Yarn was fed through the needle from overhead bins. Inventors working on this type of machine were Glenn Looper, Bob McCamy, Lamar Westcott, Peek Smith and Joe McCutchen.

1.3.3 First Tufted Carpets

Carpet tufting started as a minor offshoot of the bedspread industry. There were two quite different motivations for the move towards making floorcoverings.

A problem for the bedspread tufting industry, as it became channelled into larger companies, was to match supply with demand. Peak demand for bedspreads from the retailers was in the autumn, which coincided with the cotton picking season. Delivery times became undesirably extended, while business was slow at other times of the year. It was expensive to build up stocks during periods of low demand and a more satisfactory way of levelling out production over the year was to introduce product variety. Tufters started to produce tufted housecoats, bath sets and small rugs.

With the outbreak of World War II the industries around Dalton were diverted to producing military supplies: tents, parachutes, aeroplane wing covers, sleeping bags, mosquito nets and so on. Demand for household textiles existed, but supply of raw materials was restricted. Occasionally, tufting companies were able to acquire coarse materials, such as cotton yarns intended for mops and military duck fabric of reject quality, and made them into crude tufted rugs. Distributors flocked to the mills for these scarce rugs. Pent-up demand for the new tufted carpets had been created and this was unleashed when the war ended.

1.3.4 Machines for Tufting Carpets

Although bath sets and small rugs were produced on single-needle machines designed for producing bedspreads and goose-neck machines having two or more needles were made, it was the advent of the 'yardage machine' that gave tufting a highly significant role in the carpet industry.

The brothers Albert and Joe Cobble were among the few with experience in textile machinery to enter the industry at this stage, and arguably made the contributions that were most required at the time. Brought up in their parents' knitting businesses, they started an engineering company to repair, rebuild and modify knitting machines. In the late 1930s, they were occasionally asked to make parts for single-needle tufting machines, and in 1940 made a 50 in. (1.3 m) wide machine for producing candlewick bedspreads.

It is not within the scope of this book to record company structures; and the progression of Cobble, its subsidiary companies and its one-time ownership by Singer is complicated. However, companies created by the Cobble brothers, their relatives (Lewis Card was their nephew), and their partners and employees, were highly innovative. Cobble set up a machinery plant in Blackburn, UK, in 1955 to compete with the young British Tufting Machinery Ltd (formed in 1953) and took them over in 1961. In the same year, Singer-Cobble was formed. As had happened in the USA, Singer-Cobble in the UK spawned other inventive tufting machine companies. Much later, tufting machine companies were founded in Germany and Japan. Working contemporaneously with the Cobble brothers in the USA, George Hanson, Mose Painter and others also contributed positively to the creation of broadloom tufting machinery.

In 1945, Albert Cobble built a 3/8 gauge tufting machine having 300 needles. It was used principally to produce fabric for 'chenille' housecoats, but cotton rugs were also made on it. The following year, Joe Cobble and George Muse, in a different company, made a 5/8 gauge cut-pile machine with 185 needles which was designed specifically for manufacturing rugs and carpets.

Such relatively narrow machines were used to produce larger rugs and carpets by sewing strips together. Later, a wide (eg, 12 ft) base fabric was covered towards one edge first with, say, a 36 in. wide strip of tufts and in three subsequent passes through the machine the remainder of the backing fabric was covered in pile. 'Pass machines' are still used today to fulfil orders for custom carpets (see Section 8.2.1). In 1952, Cobble made a 12 ft wide machine to tuft 5/32 gauge loop pile and shortly afterwards made a broadloom machine to tuft 3/16 gauge cut pile.

An important objective was to produce finer gauges (defined in the USA as the spacing in inches between needle centres). It was essentially fineness of gauge that distinguished between a chenille machine and a carpet machine. Much later, more sophisticated engineering techniques, including the introduction of modular tufting elements (needle blocks, loopers, knives), improved the accuracy of their interaction, enabling gauges as fine as 1/20 to be produced.

Cobble has been credited with many improvements to tufting machinery, including yarn creels, yarn feed systems, various types of patterning attachments and the individually controlled tufting needle. Mose Painter introduced tufting from beams, which occupied less space than creels, and was the first to use a tenter to control the dimensions of carpet after piece dyeing.

Important developments arose from within the carpet industry itself. For example, the Honesty machine was developed by Callaway Mills Company and was later licensed for manufacture by machinery companies [6]. The Honesty machine had hollow needles but no loopers, and yarn was fed through the needles by compressed air. The height of the individual loops (the machine could not produce cut pile) was determined by a mechanical let-off system so that a pattern of high and low loops could be created to the required dimensions without robbing previously formed loops, as in the conventional eyed needle/looper system.

1.3.5 Formative Years

Whereas the carpet weaving industry was located principally in the north-eastern USA, the tufting industry had the advantage of cheaper labour in the south. There was a scarcity of carpet wools in the post-war years, so the tufting industry (with a few notable exceptions) turned to cheaper fibres. The American public was therefore offered carpets at prices much lower than ever before, owing to a rapid method of production, cheaper labour and cheaper fibres. Total demand for carpets escalated as consumers moved from carpet squares to wall-to-wall carpets in their sitting rooms and dining rooms and carpeted other areas of their homes (bedrooms, bathrooms, kitchens, verandas). Carpets also became more commonly used in offices, schools, hospitals and other public buildings.

Figure 1.4 shows that US production of tufted carpets expanded at an exceptional rate, while carpet weaving slowly declined [7]. Examples of rapid expansion of tufting in other countries are given in Figures 1.5 and 1.6: note that the trend towards tufting came a few years later than in America and that the outcomes were by no means identical.

The early days of the carpet tufting industry in the USA were characterised by energy, improvisation, high financial gearing, leadership by flamboyant characters and a scarcity of sound technical information.

Washable rugs made from cotton pile tufted into cotton duck did not translate readily into broadloom carpets. The cotton pile lacked resilience, the base fabric was not available 12 ft wide (except by circular weaving) and locking of the tufts by washing and shrinking the fabric, as practised in the bedspread industry, was unreliable and dimensions were difficult to control.

Jute yarns had been commonly used in the backings of woven carpets, and the jute industry around Dundee, Scotland, geared itself to weave 12 ft wide base fabrics for the US tufting industry, and later made secondary backing fabrics in a different weight and weave structure. Jute companies on the Indian sub-continent joined the supply chain. Jute offered good strength, stable dimensions and low price. Later, jute as a primary tufting fabric was displaced by synthetic materials partly because the finer tufting gauges required backing materials that did not deflect needles between warps, but also because of unreliability of quality and delivery times.

Cotton was still the principal pile fibre in 1954, but rayon was rapidly displacing it and yarns were being engineered specifically for the carpet industry. Filament nylon was also beginning to be used as a pile fibre – it was hard wearing and could be bulked to cover the backing fabric at low weight and low cost. Heavier staple nylon yarns were used in the higher grades of carpet, and a few tufters used wool.

Locking the tufts into place was a serious problem for the newly created tufted carpet industry. Coating the back of the carpet with rubber latex was the first practical solution, and this proved to be an enduring one. At first, rugs were painted with latex and later they were sprayed. Broadloom carpets were also coated by hand-held spray guns as they were drawn past the operative; then lick-roller application was introduced and used for many years. The idea of foaming the latex to create an integral underlay was introduced, but fell into disrepute in the USA owing to cost cutting and consequent poor quality. In Europe, foam backings are still widely used. In the USA, application of a secondary jute fabric to simulate the appearance of a woven carpet proved more popular; the jute fabrics were later largely displaced by synthetic secondary backing fabrics.

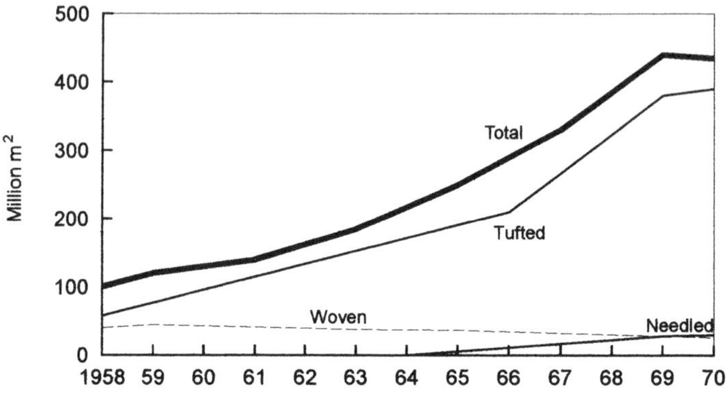

Figure 1.4. Trends in production in the US carpet industry during the period of establishment of the tufting industry [7].

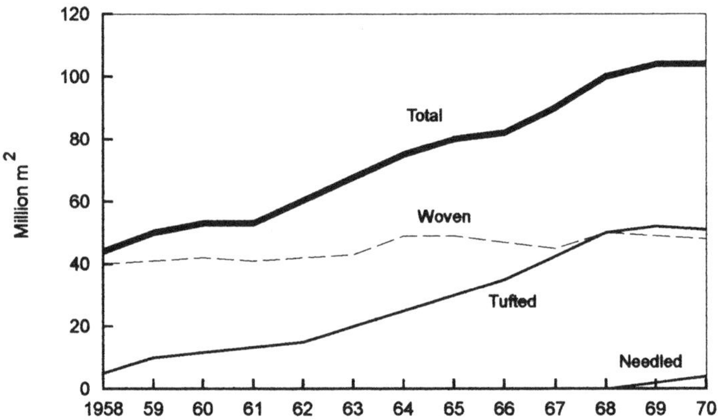

Figure 1.5. Trends in the production of carpets in the UK during the period of establishment of the tufting industry [7].

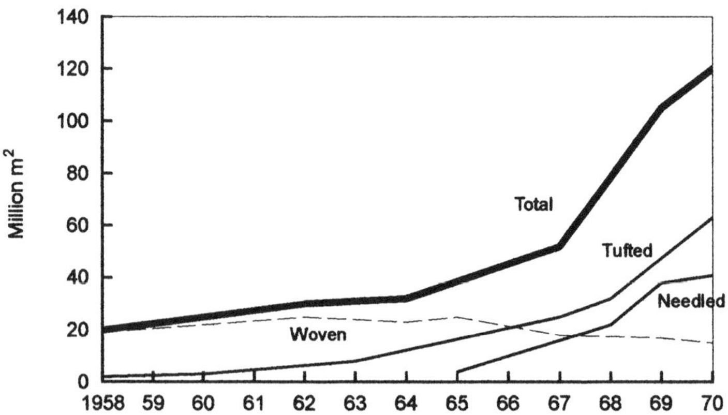

Figure 1.6. Trends in the production of carpets in West Germany during the period of establishment of the tufting industry [7].

To dry and cure the latex or foam, enormous ovens were developed. The early infrared heaters proved to be a fire hazard and hot air ovens replaced them, although infrared heaters have been retained for pre-gelling foams.

1.3.6 Maturity

The contributions of the suppliers to the tufted carpet industry – fibre manufacturers, spinners, dyestuff manufacturers, latex manufacturers and compounders, and machinery companies of various kinds – to the technical maturity of the tufted carpet industry should never be underestimated.

Du Pont was the first synthetic fibre company to offer major support to the carpet tufting industry. In 1955/56 they introduced Type 501 bulked continuous filament (BCF) nylon exclusively to Lees and to Barwick, and the three companies collaborated to solve the many teething problems. Du Pont continued to develop many types of carpet nylon, both filament and staple, over the years, having variations in crimp, fibre cross-section, dyeability, soil hiding properties, soil repellency, stain resistance and antistatic properties. Other fibre manufacturers entered the carpet market. Some market penetration was made with acrylic and polyester fibres, but eventually the situation settled, with nylon the most important pile fibre in volume terms and polypropylene taking the largest share of the lower priced market segments.

Producers of primary and secondary backing fabrics engineered products to suit a variety of carpet product lines and production routes, and latex manufacturers developed synthetic latices that could be used more controllably by the carpet industry. They, and the carpet industry itself, developed filled latex compounds that gave optimum results in terms of performance and cost.

Synthetic tufted carpets presented the dyestuff manufacturers with new problems in terms of performance (principally high fastness to light) and application (physical structure of the carpet substrate, scale of dyeing operation, etc). Of course, pile was also dyed in the form of loose fibre or yarn, as for woven carpets. Yarn printing (space dyeing) was introduced. The dye companies applied the appropriate effort, and highly skilled dyers joined the carpet industry.

The scale of demand led to continuous carpet dyeing (see Section 15.9). Various techniques were used to achieve the necessary uniformity of dye application, but it was Küsters of Germany who became the principal machinery suppliers for this area. Printing became a major technique for creating patterned tufted carpets (see Section 15.10). Roller printing, developed by Stalwart, UK, gave rather simple styles, while flat screen printing, promoted by BDA, UK, was capable of applying more colour effects. The market leader in carpet printing machinery proved to be Peter Zimmer, Austria, who produced both flat bed and rotary machines. Because of high set-up costs, printing was for many years confined to long runs, giving printed carpets a down-market image. Shorter runs became feasible when a US carpet company, Milliken, introduced computer injection dyeing in the late 1970s [8, 9]. Peter Zimmer later developed his own computerised jet printing system and the more flexible computer controlled printing became potentially more important than screen printing in the last decade of the twentieth century.

The foregoing are only a few examples of major contributions made towards the evolution of a technically sound tufted carpet industry, outside of the tufting process itself.

Financial maturity of the tufting industry was necessitated by the first oil shock of 1973/74. Prior to that, the industry had thrived on rapid expansion. Profits were often less important than capital gain – and the latter could be considerable during a period when the carpet industry was classified by stock markets as a high-growth industry. The woven carpet industry had at first looked askance at the tufting industry with its emphasis on cheap products, but increasingly the weavers were impressed with its scale and many felt obliged to become involved with tufting. Their technical and marketing expertise combined with the vitality associated with tufting was undoubtedly beneficial. It was the first oil shock, however, that prompted a complete reappraisal of the direction of the industry. Escalating prices of raw materials, a subdued market and a more cautious approach to investment clearly meant that rates of expansion would be much reduced.

The International Wool Secretariat (IWS), anxious to introduce wool into the tufting industry to compensate for the anticipated decline of the wool minded weaving industry, set out to exploit this situation, pointing out that profit per square metre could be increased by making high value wool carpets. They introduced a wool carpet styling programme, initially based around tweed (berber style), friezé and shag pile carpets, to stimulate the industry. With the consumer mood at that time against disposable products and in favour of natural materials, their ideas were quickly taken up in Benelux, Denmark, Switzerland and the UK and, a little later, Japan. Wool tufted carpets were already popular in Australasia, where wool is grown on a large scale. Thus, wool interests led the carpet tufting industry upmarket, particularly in Europe. Synthetic fibre manufacturers willingly followed this lead – it was in their interests to increase the face-fibre weight of tufted carpets. Now, tufted carpets encompass the whole range of qualities.

1.4 NEEDLEFELTS

A system of creating textile fabrics by consolidating fibrous webs by the action of barbed needles was developed in the second half of the nineteenth century in Great Britain. Originally, coarse fibres, such as jute, sisal and animal hairs, were mechanically bonded in this way to produce pillow fillings, horse blankets, saddle felts, mattress components and carpet underlays. Later, more sophisticated needling machinery was used to produce a wide variety of industrial fabrics, blankets and even apparel fabrics. During the depressing years following World War II, attempts were made to upgrade needled underfelts to serve as substitutes for carpets.

The earlier needlefelt floorcoverings were made from fibrous wastes and had inferior performance in terms of fibre shedding and durability. Two major technical innovations transformed this situation to one where needled floorcoverings rank among the best performing products: mechanical bonding of the fibres was augmented by resin bonding; and synthetic fibres were developed specifically for use by the needlefelt industry (or fibre bonding industry as it is sometimes called).

Manufacturers of needling machinery also made key contributions to the development of the needlefelt sector of the carpet industry. Incremental advances in machinery provided improvements in quality in terms of uniformity and density of product; and the introduction of new techniques allowed surface textures, simulating both cut- and loop-pile carpets, to be produced. Modifications to the texturing processes allowed surface patterns to be created.

The availability of good quality needled floorcoverings at competitive prices led to their rapid acceptance in many countries and in different sectors of the market. Market development was particularly rapid in the German speaking countries (see Figure 1.6) where they proved particularly suitable as a wall-to-wall base over which the popular oriental rugs were laid.

In public buildings and offices, their superior acoustic properties gave needlefelts an important competitive advantage over resilient and hard non-textile floorcoverings. The automobile industry quickly accepted needlefelt products for carpeting cars and lining shelves and luggage compartments.

Needled products participated in the development of textile sports surfaces. Particularly coarse fibred products were developed for entry mats. Needled textile floorcoverings were associated with the development of carpet tiles: indeed the first carpet tiles seem to have been made from needlefelts. Self-laying tiles offer particular advantages over broadloom carpets in terms of facilitating access to sub-floor services, simplicity of installation, and ease of transportation, particularly in terms of access to installations in high-rise buildings.

The development of carpet tiles also contributed to overcoming one of the disadvantages of needlefelts – limited facilities for patterning. Designs could be created by installing tiles in patterns, such as simple chequerboard effects, panels of different colours, and room borders. Areas in open-plan offices could be distinguished by tiles in different colours. More elaborate designs were later introduced to needlefelts and tiles. Interior designers exploited combinations of plain and printed tiles to create spectacular designs, especially for large open areas.

Over the years, polypropylene became established as the most important fibre for needlefelts. Nylon is also widely used, and polyester is required for some flame-resistant products.

Needling has also been applied in the manufacture of primary carpet backing fabrics. A layer of fibre is needled on to the upper surface of a woven polypropylene primary backing fabric, the usual objective being to prevent the undyed polypropylene from showing ('grinning') through gaps in the pile. The needled fibre may be in a medium or dark neutral colour, as an unobtrusive background colour, or may be undyed nylon that will take up a similar colour to the pile yarn during piece coloration. Melt-bonding fibres have been needled into the surfaces of woven primary and secondary backing fabrics to enable laminating to be achieved without using adhesives.

1.5 INTO THE PRESENT ERA

1.5.1 Status of Tufting

The market for tufted carpets is now a mature one and there is little overall expansion. The country that mothered tufting, the USA, has a huge industry dwarfing that of any other country and produces about as much tufted carpet (about 1300 million m^2/year) as the rest of the world. Significant growth was still being experienced in the USA in the last decade of the twentieth century, mostly in the contract market and in rugs. In Europe, carpets were losing market share to hard and resilient floorcoverings, although there are indications that this trend has come to an end.

It is dangerous to generalise about the carpet industry because each country tends to have its own individual characteristics in terms of market situation and attitudes to production methods and product styles. For example, carpeted floors are the norm in the USA, whereas Sweden prefers hard floors; in German homes, oriental carpets are highly prized while British homes are usually carpeted wall-to-wall; New Zealand produces little but wool pile carpets; and the large Belgium industry (the world's second largest) exports over 90% of its production. Intercontuft [10] is a major source of detailed statistical information about the carpet industry and the following summary information is prepared from Intercontuft's statistics and estimates.

Table 1.1 points to the dominant position of tufting. Weaving, of course, produces much more valuable carpets than needling so these broad figures underestimate the financial importance of the carpet weaving industry in Europe. Textile floorcoverings are also made by bonding pile to backing fabrics, flocking and knitting, but these techniques are not featured in the statistical information in many countries.

Table 1.2 indicates the importance of nylon as a pile fibre for tufted carpets in Europe, but does not report the fact that polypropylene has been progressively taking market share from nylon as its quality and flexibility of production have improved. Wool is used only in the heavier pile weights, so its importance in kilograms is greater than indicated by the figures for areas of carpet tufted.

Figures for the USA are in pounds weight consumption and are not directly comparable. However, the situations for nylon and polypropylene are similar to those in Europe. Proportionally more polyester, but less acrylic fibre and wool are used than in Europe.

Table 1.1. The relative importance of different carpet manufacturing techniques in Western Europe and the USA in 1999 [10].

	Europe		USA	
	(million m²)	(%)	(million m²)	(%)
Tufting	681	61	1287	92
Needling	290	26	75	5
Weaving	137	13	41	3

Table 1.2. Production of tufted carpets in Western Europe in 1999 according to pile fibre type [10].

Fibre	(Million m²)	(%)
Nylon filament	283.3	42
Nylon staple	79.5	12
Polypropylene filament	226.0	34
Polypropylene staple	37.0	6
Wool	34.7	5
Polyester	6.6	1
Acrylic	2.4	0.5

Apart from the importance of cut pile, the significant point to be noted from Table 1.3 is that fine constructions (1/10 gauge and finer) were the most common ones produced towards the end of the century, indicating the level of sophistication of the European industry. In the early days of broadloom tufting, the most common gauges were 3/16 and 5/32.

Table 1.3. Production of tufted carpets in Western Europe in 1999 according to gauge of machine [10].

Gauge	(Million m²)	(%)
Cut pile		
1/4–5/32	30.7	5
9/64, 1/8	108.0	16
1/10	176.7	26
5/64	46.4	7
1/16	2.7	0.5
Loop pile		
5/8–5/16	3.7	0.5
1/4–1/8	66.3	10
1/10–1/16	238.1	35

In the USA at the end of the twentieth century, the major gauges for contract carpets were 1/10 and 1/8. For residential carpets, 1/8 gauge was the most popular, but 3/16 and 5/32 were also commonly used.

Despite the numerous advances in patterning techniques, most tufted carpets made in Europe in the 1990s were plain or semi-plain, having surface interest provided by yarns with melange, tweed or curled effects (see Table 1.4). Printing was the most important technique for introducing pattern, but was declining. Mechanical methods of patterning were less important, but the more advanced systems for producing multicoloured cut-pile carpets had been available to the industry for only a few years.

Table 1.4. Broad styles of tufted carpet produced in Western Europe in 1999 [10].

Style of carpet	(%)
Plain and semi-plain cut pile	37
Plain and semi-plain loop pile	32
Printed (cut pile and loop pile)	19
Patterned constructions	12

Again, the figures for the USA are classified on a different basis. The dominant style in the residential market was the saxony, with various patterned styles accounting for 23%. Printed carpets were much less important than in Europe.

In the contract market, patterned styles were much more important at 44%; level loop accounted for 24% and saxonies 27%. Overall, US carpets were heavier than European carpets.

There were tremendous variations between European countries in terms of preferences for the backing of tufted carpets but, overall, secondary fabrics took a market share of 50%. Foam backings were very important at 37% of the market, and felt backings were beginning to assume importance. Latex foam backings and jute were little used in the USA, although polyurethane foam was a small growth area. Special heavy backing materials, used for carpets destined to be cut into tiles, are not included in the figures for backings. The organic materials for tiles were principally PVC and bitumen, and glass fibre was often incorporated to enhance the dimensional stability of the tiles.

Tufting machinery

Machine developments have followed two distinct trends: towards high speeds for producing plain carpets such as print substrates; and towards greater patterning capability and flexibility. Improved precision of manufacture and interaction of tufting elements and lighter moving parts have enabled speeds up to 2000 rpm for loop pile and 1800 rpm for cut pile to be achieved. The two most common patterning principles (tension control and crossover of colours) have been combined in various ways to give more elaborate effects; and the individually controlled needle combined with sliding needlebar tufting has enabled multicolour cut-pile carpets, similar to Axminster styles, to be produced at modest speeds but still much faster than Axminster looms. Electronic drives and CAD systems have enabled larger pattern repeats and economical short runs to be achieved.

Sophisticated machine control systems have also contributed towards reliability and quality of production. Recall systems allow specifications to be selected from storage, setting machine conditions so as to enable carpets to be reproduced accurately; and machine monitoring systems aimed at minimising faults from tight or broken yarn ends are widely available. More comprehensive monitoring systems cover the whole carpet production sequence, with automatic (where possible) and manual input covering faults, waste, stoppages, etc, as an aid to mill management.

Trends in dyeing have been towards piece dyeing, including continuous processing, which offers opportunities for savings in water, energy and effluent. The former association of continuous dyeing and printing with long runs (and, therefore, high stock keeping and mass-market image) has been broken through the development of quick-change dye applicators and computer controlled jet printing. The other main trend in coloration has been towards polymer pigmentation – over 30% of carpets in the USA are now coloured in this way. At first, this trend was associated with polypropylene, but nylon is increasingly coloured in this way.

Performance in use

Attention to improved carpet performance in use has been focused on nylon pile. The fibre is inherently tough and resilient, but the original nylon carpets were easily soiled and stained and generated static electricity. Early approaches to hiding soil by changing the optical properties of the fibre have been supplemented by the application of finishes that resist stains and soil. Such finishes are often applied at the stage of piece coloration.

Static control is a common requirement for commercial carpets and has been usually achieved by blending a small proportion of conductive fibre in the pile and using conductive backing materials. A cheap and convenient post-dyeing treatment is available for wool (see Section 15.12.5). Static electricity does not seem to be a serious problem for polypropylene and polyester.

Safety from fire is a concern in many applications of carpets. Nylon and wool carpets can easily be constructed to pass tests involving ease of ignition and initial spread of flame simply by avoiding open structures that easily admit air. The behaviour of carpets in a conflagration is a serious issue in locations where escape is difficult (eg, transportation and high-rise buildings) and the carpet industry has to cope with a confusing variety of flammability specifications. Most specifications can be met using densely woven wool carpets, but this situation is not acceptable to the tufting industry. Simple solutions to flammability requirements are needed for the third millennium.

Retention of appearance in wear is now regarded as more important than resistance to abrasion, and important progress has been made with respect to retention of texture in cut-pile carpets. Melt-bonding fibres can contribute towards more cohesive tufts in all pile fibres, and their use spread rapidly in the last decade of the twentieth century.

1.5.2 Status of Weaving

A superficial look at the statistics would indicate that, in the late twentieth century, carpet weaving with an estimated market share of 6% worldwide had become almost insignificant. However, take out the USA which produces almost as much carpet as the rest of the world, most of it by tufting, and it is apparent that carpet weaving remains a significant industry in many geographic areas. In Europe, for example, the consumption of woven carpets was 16% by area of all pile carpets (tufted and woven) in 1999. Considering the much higher value per square metre of the largely wool woven carpets compared with the mainly synthetic tufted products, it may be estimated that weaving accounts for about a quarter of the financial turnover of the European textile floorcoverings industry. Moreover, as shown in Table 1.5, consumption of woven carpets was growing at a modest rate in a declining total European market (the expansion in European production of carpets towards the end of the millennium was based on exports). Although weaving is relatively much smaller in the USA (see Table 1.1), it should be noted that production of woven carpets there roughly doubled from a very low level over the five-year period to 1999.

Table 1.5. Consumption of carpets in Western Europe (million m^2) [10].

Year	All textile floorcoverings	Woven carpets
1993	915.8	84.7
1994	907.8	87.0
1995	896.4	87.3
1996	868.4	85.2
1997	863.3	90.9
1998	873.7	99.6
1999 estimates	875.4	105.0

Undoubtedly, tufting must be credited with the spectacular expansion of the market for carpets that occurred in the 1960s and 1970s. The advent of cheap tufted carpets encouraged domestic consumers to carpet their living and dining rooms wall-to-wall and to carpet areas such as bedrooms, bathrooms and kitchens; and architects and designers to expand the use of carpets in public, commercial and institutional buildings. At first, weaving participated to some extent in this overall growth but when the tufting industry moved up-market with its products, consumption of woven carpets began to suffer as a result of competition from good quality tufted carpets, and large areas of the weaving industry declined. Renewed interest in squares did, however, secure a strong face-to-face weaving industry in Western Europe, Eastern Europe and the Middle East.

Figure 1.7 provides an interesting window on a critical period for the carpet market. Worldwide economic recession in the late 1980s led to a decline in the demand for even the higher quality (wool) tufted carpets, but there was remarkably strong expansion in the hand-knotted sector, and India, Pakistan and Nepal increased their output of medium priced knotted rugs. Machine weaving of wool carpets was stable during the early part of this period: the late decline resulted from competition between knotted squares and woven squares. However, the weaving industry as a whole compensated by finding new markets for polypropylene squares.

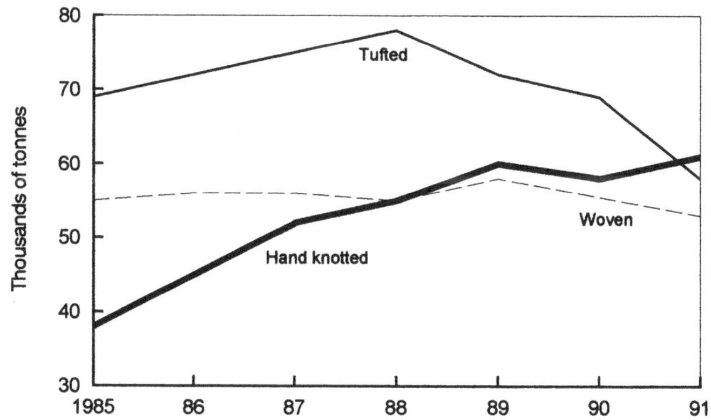

Figure 1.7. Wool consumption in the carpet market (Belgium, France, Germany, Netherlands, Italy, UK, USA) 1985–1991.

Quick response

Recent trends in the carpet weaving industry have been, to some extent, consequent on new technology. The widespread use of electronic jacquards and CAD systems has transformed weaving into an industry that can respond quickly to design trends and generate short runs of exclusive products economically.

Electronic jacquard-gripper looms in association with CAD systems do more than just enable traditional Axminster styles to be produced more quickly and in shorter runs. It is now practicable to generate very large design repeats (including designs to the shape of the room) and to use more colours per design. Creels with 12 frames of colour are being installed: they can be used to provide versatility in terms of rapid selection of colour (say 6 from a palette of 12 colours) when changing to a new design, or to weave carpets having 12 colours, once the province of spool looms or perhaps elaborate planting of colours in an 8-frame gripper loom. A further element of quick response and flexibility is provided by electronically controlled sampling systems that have been installed by many companies.

The new facilities to produce spectacular designs have proved attractive to some sectors of the market for commercial carpets, especially the hospitality sector, and weaving companies have geared themselves to further stimulate demand through their CAD facilities and associated quick response production systems. Undoubtedly, the new multicolour cut-pile tufting systems will have a significant impact on the market for patterned carpets, but in the late 1990s weaving appeared to be taking better advantage of this situation, with CAD/CAM being at an advanced stage of development.

Weaving machinery

The most soundly established systems for weaving carpets are face-to-face and gripper Axminster, with spool-gripper and wireloom weaving supplying important niche markets.

The spool-gripper system has a higher rate of output than the traditional jacquard-gripper system because of the shorter arc of travel of the grippers, and because no time is required in the cycle for colour selection. It is very much less flexible, however, so it is favoured for mass production of established designs, principally popular designs for the residential market. Because there is no limitation whatsoever on the number of colours, spool-gripper weaving is favoured for producing floral and chintz designs. Spool setting remains an expensive preparatory process, requiring each pile yarn to be threaded manually to the spool. Earlier attempts to automate the spool setting process achieved limited acceptance, but current research will benefit from modern computer technology and seems more likely to succeed.

Undoubtedly, existing wirelooms will continue to be used to produce high quality plain Wilton carpets and traditional Brussels squares. Body carpet in 2- and 3-frame designs for the contract market faces competition from more flexible and economical wide Axminster looms. Provided the fashion can be retained, wirelooms should prosper in the areas of textured loop-pile carpets, where they have no serious competition. A wire-less loom for producing loop-pile carpets and a special face-to-face loom for weaving sisal style carpets have been developed with this market in mind.

Face-to-face weaving has a firm position for weaving squares, both wool and synthetic, and is increasingly used for producing wall-to-wall carpets. Artistic modern designs are stimulating a squares market once focused primarily on classical designs. Electronic jacquards are now commonly used, and triple-rapier systems are enhancing output. Face-to-face weaving remains an excellent system for producing plain carpets having a beautifully level surface, and is also used for producing textured (friezé) plain carpets in competition with tufting.

Electronic jacquard-gripper Axminster weaving is clearly focused on the contract carpet market, taking advantage of its powerful design facilities and ability to produce short runs economically. Body carpet is no longer acceptable to some contract specifiers and seamless installations are a recent target. To this end, loom widths are being increased from 4 to 5 m; and two 2 m widths may be woven side-by-side on one loom. Although retrofit electronic jacquards improved output to a minor extent, new looms with electronic jacquards have greatly reduced gripper arcs, shortening the pile insertion time significantly. Handover rapier systems and projectile systems offer further increases in weft insertion rates, and it is possible that the output of future weaving looms could approach those of the more complex tufting systems [11].

Performance in use

An important element of the success of woven carpets in the contract market has been performance in use. The importance of the surface pile mass density factor (P^2/t) in determining wear life (resistance to abrasion) has been established for many years (see Section 20.4.1), and contract specifiers have progressively learned that heavy dense piles can most easily be created by weaving. Woven carpets are therefore favoured for locations subjected to heavy foot traffic, especially in the hospitality industry where design can be an equally important element. The integral nature of the pile/backing structure of woven carpets favours the different aspects of dimensional stability, in particular resistance to the 'pastry rolling' effect of traffic that may lead to expansion and rucking of installed carpet. Reliability of behaviour in the hands of skilled carpet fitters is a positive aspect of woven carpets.

Man-made fibres in woven carpets

Applications of artificial and synthetic fibres have had a significant impact on the carpet weaving industry, but new fibres have not played a decisive role in the development of the industry, as in the case of tufting.

Rayon was used as a diluent for wool in pile yarns in the years immediately following World War II. Now the principal diluent is the tougher and even lower priced polypropylene. Nylon was the leading blending fibre during the second half of the twentieth century, however. The 80/20 wool/nylon blend was first introduced in the UK to improve the wear life of the lower grades of Axminster carpets (the heavier grades in 100% wool lasted as long as needed in residential situations) and spread to some heavier cut-pile contract qualities (but not loop-pile constructions) and to other countries. More recently, 'performance' blending has moved towards 80/10/10 wool/nylon/melt-bonding fibre, with the objective of improving the retention of tuft definition and resilience and reducing fibre shedding, as well as enhancing resistance to abrasion.

Nylon in 100% form has been used only to a minor extent in woven carpets. Compared to wool, in pile masses appropriate to weaving, the fibre shows no overall superior performance features and is also more difficult to cut. Acrylic fibres were used quite widely in woven carpets in the UK and some other countries in the 1960s and 1970s, but interest declined owing to inferior performance compared to other carpet fibres. However, in Japan, where shoes are traditionally not worn in the home, woven acrylic carpets have remained popular because of the soft touch of their pile. In face-to-face squares, polypropylene has proved to be the most popular synthetic pile material. Markets for cheap prayer mats in Islamic countries were exploited by weavers using 100% polypropylene pile; and face-to-face squares in 100% polypropylene proved appropriate for the new consumer markets in Russia and some other countries formerly having centralised economies.

1.5.3 Carpets and the Environment

Carpets are generally considered to be particularly friendly to the indoor environment in terms of walking comfort, acoustic comfort, thermal comfort and safety from falls (see Section 20.3), but in the latter part of the twentieth century, strong attention was turned to their wider effects on the environment during manufacture and on disposal [12, 13, 14]. Germany and the USA led the way, and as Germany is a major importer of carpets, its legislation impacts on industries well outside its borders.

The policy of the European Community, expressed in the objectives of its voluntary ecolabelling scheme introduced in 1992, is:

- to promote the design, marketing and use of products which have a reduced environmental impact during their entire life cycle,
- to provide consumers with better information on the environmental impact of the products,

without, however, compromising product or workers' safety or significantly affecting the properties which make a product fit for use.

Anticipating this situation, two private labels, GuT and Oeko-Tex, were introduced in 1990. The Gemeinschaft umweltfreundlicher Teppichboden (GuT, the Association for Environmentally Friendly Carpets) quickly gained the support of the majority of European carpet manufacturers. For a carpet manufacturer to be able to use the GuT signet, the product must not contain specified contaminants, and must be tested for emissions and odours; and the manufacturing facility is audited for compliance with GuT standards and local legislation. GuT is also working to advance various issues through a comprehensive list of projects. As a result of the work of the dyeing group, members voluntarily banned the use of metal complex dyes on all nylon products for a period from January 1998, despite the difficulties this entailed, to achieve very minor reductions in metal contents in the effluent and in the carpets. Other changes in manufacturing procedures, or simply advice from the GuT on applications of carpet, may be expected in future.

Recycling is a major issue, stimulated by the German Recycling Management Act, which makes the manufacturer responsible for waste management, and the Technical Requirement on Municipal Refuse. GuT members are progressively abandoning the practice of disposal by landfill and are studying combustion with energy recovery, recycling and the logistics of waste management. Leading US fibre manufacturers are also promoting schemes for recycling used carpets [14]. Systems for reclaiming used carpet, mill waste and trade residues have been founded, and monomers, chemical feedstock, plastics, tiles and underlays are being produced from separated pile materials. Environmental issues are discussed further in Section 5.4.

REFERENCES

1. B Jacobs. *The Story of British Carpets*. British Continental Trade Press, London, 1968.
2. G Robinson. *Carpets*. 2nd edition, Textile Book Service, UK, 1972.
3. M L Eiland. *Oriental Rugs – A New Comprehensive Guide*. Little Brown, Boston, 1981.
4. *Heimtex*. Westdeutsche Verlagsanstalt GmbH, Herford, Germany.
5. T M Deaton. *Bedspreads to Broadloom*. Tapestry Press, Acton, MA, 1993.
6. Callaway Mills Co. *US Patent 3 342 152*, 4 October 1965.
7. G H Crawshaw and J Ince. *Textile Floorcoverings*. Textile Progress, 4, No.2, 1972.
8. Milliken and Co. *US Patent 3 969 779*,1974.
9. M V Locke. The Millitron machine: computer injection dyeing. In: *The Fabric Revolution*, The Textile Institute, Manchester, 1981.
10. Intercontuft, 3000 Tervuren, Brussels, Belgium.
11. K L Gandhi and G Stead. Floorings: trends in the 21st century. In: *Proceedings of World Textile Congress on Floorcoverings*, University of Huddersfield, 12–13 July 1995, 290–313.
12. B Burdett. Ecolabelling and the European GuT. In: *Proceedings of World Textile Congress on Floorcoverings*, University of Huddersfield, 12–13 July 1995, 151–161.
13. K Wise. The carpet industry – covering the future. *American Dyestuff Reporter*, 1997, 86(6), 15–21.
14. T Shaw. Environmental issues and the wool carpet industry. In: *Proceedings of Tifcon'90, Carpets – The Competitive Edge*, The Textile Institute, Manchester, 1990.

2 *Pile Fibres*

A wide variety of synthetic and natural fibres are applied for the use-surface of textile floorcoverings.

The most obvious difference between a carpet fibre and an apparel fibre is the higher fibre diameters generally used in carpet manufacture, but the many reasons for fibre choice are quite different in the two broad sectors of trade. The principal carpet pile fibres in current use are nylon, polypropylene and wool, whereas the leading apparel fibres, cotton and polyester, are used in the carpet industry mostly as backing materials. Polypropylene is unique in that it is widely used for both the pile and backing of carpets.

2.1 NYLON

Nylon (polyamide) was the first synthetic fibre to be used as carpet pile and has, for many years, been the leading fibre in the tufting sector, being used both as filament and spun yarns. It is also widely used as a blending fibre in the needlefelt and carpet weaving industries.

Two alternative polymers are used by different suppliers of carpet nylon. The original nylon 6.6, as developed in the 1930s by W H Carothers and his team at Du Pont, is produced by polycondensation of adipic acid and hexamethylene diamine; and nylon 6, developed in Germany only a little later, is made by self condensation of caprolactam. Details of the manufacturing processes for the polymers differ considerably [1] but in each case the product is polymer chips, which may be regarded as the starting material for fibre manufacture.

Figure 2.1 gives a broad picture of the steps required to produce nylon 6.6 in the principal alternative forms of staple fibre and filament. After the polymerisation, the procedure for producing nylon 6 fibre may be regarded as identical, except that details such as draw ratios and temperatures differ.

Both fibres have excellent elastic recovery, which translates into good pile resilience. The elasticity, coupled with high tenacity (over 4 g/dtex), means outstandingly high resistance to wear. Approximate melting points are: nylon 6.6 250°C, nylon 6 215°C. This difference is irrelevant to carpet product

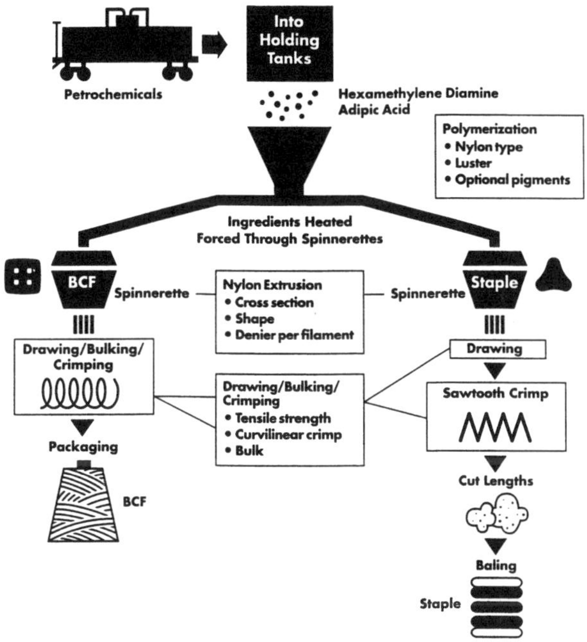

Figure 2.1. Overall flow diagram for the production of filament and staple nylon. (Courtesy of Du Pont (UK) Ltd)

performance, but points to a difference in molecular structure (crystallinity). Thus, while both fibres can be set by heat or steam to improve retention of tuft definition in cut-pile carpets, setting conditions are different (see Section 4.1). Also, both fibres are eminently dyeable with acid dyes (including metal complex dyes), but the more open structure of nylon 6 means that dye uptake is more rapid. The moisture regain of both fibres at 65% RH and 20°C is 4%.

Despite these remarkable advantages, the basic nylon is not the ideal carpet fibre. It is excessively lustrous for most product styles and is susceptible to soiling and staining in use. When people walk on nylon carpets, using commonly available footwear and in a dry atmosphere, charge separation occurs and when the electrical charge on the body is earthed, an unpleasant shock can be experienced. Nylon's resistance to sunlight is good, but may be considered inadequate under severe conditions of exposure (eg, in automobiles parked outdoors).

For these reasons, variants on the basic nylon fibre have been developed that have properties engineered to contend with the wide variety of circumstances that carpet may encounter in use [2]. An equally important reason for introducing fibre variants is to widen the variety of carpet products that can be made.

2.1.1 Filament v Staple Fibre

Fibre variants are determined at the stage of extrusion; and the most fundamental distinction is between continuous filament yarn and staple fibre.

Clearly, the production of filament yarn obviates the need for carding and spinning (see Section 3.2) and the filament route is attractive economically.

Because there is no opportunity for imperfections in filament yarns to be levelled by blending, great care has to be taken to achieve uniform dyeability [3]. For this reason, some companies prefer to use nylon yarns spun from carefully blended fibre when manufacturing plain carpets by the piece dyeing route. In the production of cut-pile tufted carpets, imperfect cutting of the tough fibre is likely to cause a more serious carpet fault with filament than with spun yarns. Another reason for choosing spun yarns is to simulate wool yarns more closely. However, loop-pile carpets from spun nylon and nylon blend yarns are often susceptible to pilling and cobwebbing in wear.

In practice, most loop-pile nylon carpets are made from filament yarns, while a substantial proportion of cut-pile carpet is made from spun yarns. Overall, filament nylon has the greater share of the tufting market.

The production of filament yarns is described in more detail in Section 3.1.

2.1.2 Production of Nylon Staple

Dried chips of polymer are fed into the main extruder (Figure 2.2), while chips of polymer plus additives, such as delustrants, UV stabilisers, antistats and pigments, are fed into a side-stream extruder which may be coupled to it. The concentrate in the side-stream extruder represents 3-10% of the throughput of the main extruder.

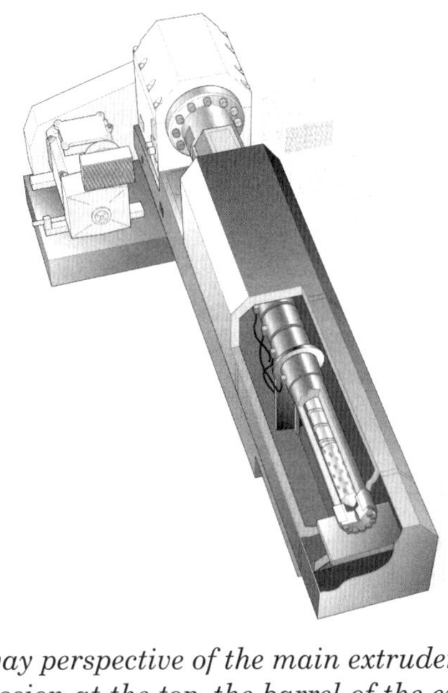

Figure 2.2. Cutaway perspective of the main extruder, showing the drive motor and transmission at the top, the barrel of the extruder in the centre, and the multiple spinnerets, fed via a manifold, at the bottom. The cooled hopper for the polymer chips is not shown. (Courtesy of Rieter Synthetic Fiber Machinery)

So that the polymer does not yellow, the overall design of the extruder is such that the polymer remains molten for only a short time, and an atmosphere of inert gas (dry nitrogen) is maintained to prevent oxidation. The chips melt in the heated barrel of the extruder and the screw mixes the melt with the side-stream material and develops pressure. Variants in screw geometry are available from the machinery manufacturer, the different versions being optimised for manufacturing particular fibres. A measuring head close to the spinnerets senses pressure and temperature and controls conditions in the barrel to provide the required values. As the molten polymer emerges from the tens of thousands of fine nozzles, it is cooled by a stream of air and steamed to condition it towards its moisture regain of 4%. The undrawn fibre is then subjected to cold drawing; it is stretched between rollers to four to five times its original length. Cold drawing orients the molecules towards the direction of the fibre axis, producing a remarkable increase in strength.

The heavy tow produced in this way is coated with spin finish, crimped using a stuffer box, chopped into staple and baled for delivery to the spinner. Cutting is traditionally done with a bladed roller opposed to a pressure roller coated with hard rubber. A newer system, devised by Barmag, continuously winds the tow around a cylinder. An inclined pressure ring rolls it off the lower part of the cylinder into the path of multiple blades that slice the tow at an angle (ie, the cutting speed is only about 5% of the tow inlet speed).

2.1.3 Key Fibre Properties

Different carpet products require their particular fibre specifications so when purchasing nylon fibre, it is vital to use the coding for the variant. Figure 2.3 is an example of a product description for Du Pont's nylon staple.

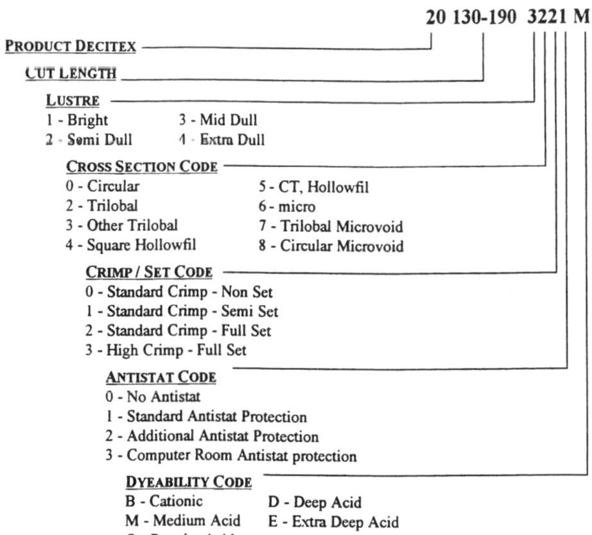

Figure 2.3. Example of Du Pont's product description for nylon staple. (Courtesy of Du Pont (UK) Ltd)

Staple length
The length to which the staple is cut depends on the spinning system for which it is intended. For example, fibre intended for blending with wool to be spun on the woollen system is typically 100 mm long, whereas for semi-worsted spinning lengths between 140 and 160 mm are more appropriate. Variable length fibre (eg, 130–190 mm) is preferred by some spinners.

Fibre titre
Nylon for mainstream carpet products is supplied in decitex ranges from 11 to 22, although higher diameters are used in some rugged contract carpets and needlefelts, and fibres as fine as 8 dtex are offered for soft velours and products such as bath mats, where wicking is a positive feature. Blending fibre is commonly 13 or 17 dtex, for compatibility with wool, while the fibre titre for 100% nylon yarns is determined largely by the softness or crispness of handle required in the carpet and, sometimes, by the fineness of the yarn to be spun: the semi-worsted system requires at least 70 fibres per yarn cross-section, so that fine yarns require fine fibres. Note that high diameter fibres offer a low specific area for soiling.

Speciality fibres may be produced to provide particular carpet characteristics. For example, Antron™ Microfibre, composed of three ultra-fine filaments aligned uniformly around a central core element which are fused together in the molten state, is designed to provide extra softness [4].

Lustre
All versions of carpet nylon contain white pigment, titanium dioxide, to modify the lustre of the fibre. The anatase version of titanium dioxide, rather than the rutile form used in plastics and paints, is the preferred fibre additive because it is less abrasive and is a bluer white. Inorganic coatings are applied to the anatase to aid dispersion in the polymer and, vitally important, to reduce UV degradation of the nylon. Manganese in the coating scavenges degrading radicals formed at the polymer/delustrant interface.

Delustrants have an important function in making soiling of the carpets less apparent. An understanding of this function may be obtained by considering dyeing theory: the depth of colour of dyed fibre varies with the Kubelka-Munk function, the absorption/scattering coefficient, K/S. To achieve a given colour, more absorption of light by dye is required if scattering of light by the fibre is high. Similarly, observable soiling is reduced if absorption of light by soil is counteracted by scattering of light by the fibre, notably by the delustrant in the fibre. Fibres are supplied with alternative levels of delustrant for manufacturing different styles of carpet.

As a supplement to titanium dioxide, microvoids may be introduced into the fibre. They provide a more subtle level of lustre and contribute to soil hiding. Microvoids of this kind are obtained by means of chemical additives in the polymer mix. Continuous voids ('artifical lumens') are a further development of microvoids and may be regarded as features of the fibre cross-section.

Fibre cross-section

Round cross-section is the principal one for nylon blending fibres, but many other configurations are additionally used in 100% nylon yarns. These include trilobal, triangular and shamrock cross-sections. Du Pont, in particular, has developed ingenious techniques for manufacturing fibres having continuous hollows, notably square cross-section with four hollows and convex triangular cross-section with three hollows. Figure 2.4 illustrates some alternative cross-sections. The group of round cross-sections shows how fibre diameter can vary.

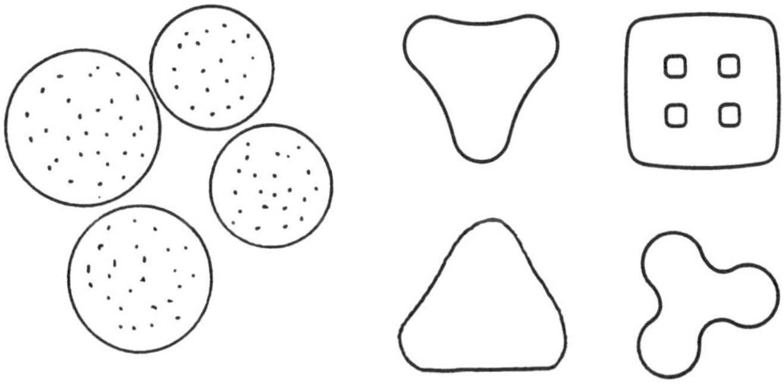

Figure 2.4. Some alternative fibre cross-sections used for nylon. Clockwise: round, trilobal, square with four hollows, high aspect ratio trilobal, and delta (triangular).

Reasons for moving away from round cross-sections are to increase the flexural rigidity of the fibre and to enhance lustre for particular styling purposes. The purpose of the hollows in fibres is to reduce apparent soiling by increasing the scattering of light, as shown in Figure 2.5. Hollow fibres are used most commonly in loop-pile filament constructions. Some cross-sections (eg, convex triangular) have been designed to reduce the adhesion of soil.

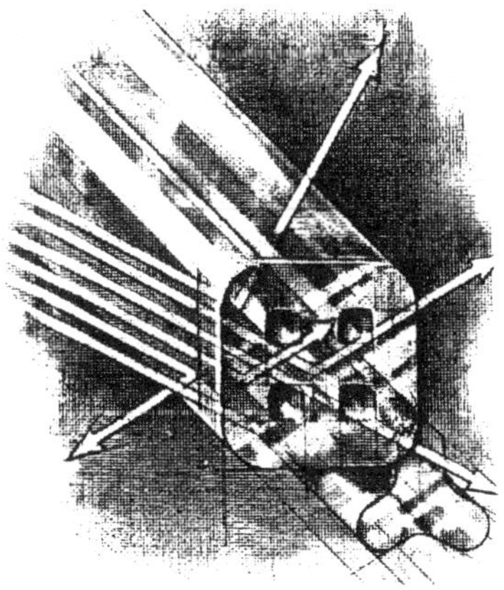

Figure 2.5. Artist's impression of light scattering by hollow fibres. (Courtesy of Du Pont (UK) Ltd)

A negative aspect of some of the alternative cross-sections is the need to use higher twist levels in yarns to compensate for lower fibre cohesion and higher flexural rigidity of the fibres.

Crimp
Crimp in staple fibres is usually introduced by stuffer box. Without set, the crimp simply serves to improve fibre cohesion during the carding and drawing stages of yarn production. If the crimp is set by heat or steam, it can contribute to yarn bulk.

Dyeability
In order to allow multicoloured effects to be produced at the stage of yarn or piece dyeing, suppliers of nylon fibre offer variants having modified dyeing properties. There are versions with a reduced number of available amide groups that have lower affinity for acid dyes; others with more available amide groups that have enhanced dye uptake; and some that have acid groups incorporated in the chains and are dyeable with cationic dyes but not acid dyes. Some manufacturers offer fibres having different levels of enhanced dye uptake.

Pre-coloured fibre
Pigmented nylon can be supplied to meet requirements for exceptionally high standards of colour fastness (see Section 15.1).

Static control
Versions of nylon that offer static control in the carpet, as distinct from during yarn production, incorporate small proportions of conductive fibre (see Section 2.6). Alternatively, spinners may elect to use ordinary nylon and blend in a conductive fibre themselves.

Stain resistance
Carpets that resist staining by food dyes are most often treated in carpet form with a resist agent to block dye sites and a topical fluorocarbon to inhibit penetration of the staining material (see Section 15.12.3). Some fibre suppliers offer variants of nylon that have already received such treatments. Special techniques are required to process them at the stage of dyeing or printing.

Soil resistance
A treatment for enhancing the soil resistance of carpet nylon is described in Section 15.12.4.

Flame resistance
It is not usual to modify nylon 6.6 or nylon 6 to improve resistance to fire: carpets made from the fibres are not easily ignited and their heat of combustion is lower than those of many other fibres. When exceptionally high standards of safety are required (eg, for aircraft carpets) the carpet pile can be Nomex™, an aramid fibre made from poly(m-phenylene isophthalimide). Nomex has outstanding resistance to fire, but is much more expensive than ordinary nylon. Coloration is by melt-pigmentation rather than by dyeing.

Fluorescence
An example of this speciality fibre is Du Pont's Footlights™ version of nylon 6.6, which contains an additive that glows under UV light. It is aimed particularly at

the market for carpets in the entertainment industry, where patterns visible under conditions of low illumination are sometimes required.

Resistance to bacteria
An example of a fibre designed for use in bacteria-resistant carpets is Du Pont's Lumena™ solution-dyed nylon, which has inherent antibacterial properties that are durable to wet treatments. However, to create a carpet free from bacteria that can be responsible for allergies, infections and odours, it is necessary to address the entire carpet system, including backing and latex.

2.2 POLYPROPYLENE

Polypropylene [5] entered the carpet industry as a constituent of the backing, but is now the second most important pile fibre in square metre terms, having totally displaced rayon as the principal pile material used in the lower priced grades of tufted carpet. It is the leading fibre in needlefelts, and entered the carpet weaving industry via the face-to-face system, where it is now more important than wool. Polypropylene is also valued as a material for outdoor carpets: melt pigmentation is the standard method of coloration and the required high standards of light fastness can only be achieved in this way. Polypropylene is commonly blended with other fibres, especially in the UK, where 50/50 wool/polypropylene blends are extremely popular.

Although polypropylene does not have the resilience of nylon, this may to some extent be compensated by increasing the pile density of the carpet and the fibre has many other merits in carpets [6]. The monomer is a by-product of the petroleum industry that can be isolated at low cost and less energy is required for conversion into fibre than for other synthetics. The handle of polypropylene places it in direct competition with wool, despite the difference in price level. Its low specific gravity means that it provides better cover in the carpet than other fibres (see Table 2.1), and the cover can be increased further by engineering high-bulk yarns with shrinkable fibres. Polypropylene does not absorb water so it is resistant to water-borne stains, although oily stains may be a problem. And the favourable position of polypropylene in the triboelectric series means that static generation is not a problem in carpet installations, despite its high resistivity. Extrusion conditions for carpet fibres are adjusted to engineer fibres

Table 2.1. Specific gravity and cover of different carpet fibres [6].

Fibre type	Specific gravity	Relative diameters for the same decitex
Polypropylene	0.91	1.0
Polyamide	1.14	0.89
Polyacrylic	1.17	0.88
Wool	1.32	0.83
Polyester	1.39	0.81

with tenacity of 3–4 g/dtex and extension at break of 80–100%, which translate into a tough fibre.

The basic polymer is rapidly degraded by light, and it was not until highly effective UV stabilisers were developed that polypropylene carpet fibre became practicable. Additionally, the progress of polypropylene as a carpet fibre was, for many years, retarded by the belief that dyeability was a necessity for the tufting industry, in particular, but once the industry concentrated on the development of flexible methods of pigmentation (see Section 15.1) a period of rapid expansion was initiated.

Progressive improvements in the fibre and its processing into carpets have resulted from research programmes conducted by a variety of organisations [5]. Three types of polypropylene pile yarn are used by the carpet industry: filament yarn, spun staple fibre and split film, the latter being used exclusively for textile sports surfaces.

2.2.1 Polymerisation

The production of polypropylene was pioneered in Italy, and was based around the patents of Natta and Ziegler filed in 1954, aimed at producing isotactic polypropylene for the plastics and fibre industries.

Professor Natta recognised that three types of polymer could result when propylene (readily available from the petrochemicals industry) was polymerised: isotactic polypropylene with all the methyl groups on one side of the main chain; syndiotactic polypropylene having the methyl groups on alternate sides of the chain; and atactic polypropylene with the methyl groups randomly distributed. Atactic polypropylene (APP) has an amorphous structure that is not amenable to fibre production: it is used as a material for backcoating carpet tiles. Isotactic polypropylene has a helical molecular structure and readily crystallises so that desirable fibre properties may be achieved. Syndiotactic polypropylene is not used in carpet fibres.

By heating propylene gas under pressure in the presence of a catalyst system, stereospecific polymerisation, leading to largely isotactic polymer, is achieved. The polypropylene produced in this way is extracted with organic solvent to remove low molecular weight material.

Molecular weight distribution, as well as molecular weight itself, can affect melt spinning behaviour [7]. Metallocene catalysts have been proposed for increasing the proportion of isotactic polymer, and therefore improving fibre properties [8].

Although most polypropylene is intended for pigment coloration, some copolymers are dyeable with disperse dyes, and nickel-modified versions are dyeable with special mordant dyes (see Section 15.5). A potentially important innovation is the use of double extruders for combining polyamide with polypropylene. In this way, a nylon sheath can be created to give a dyeable fibre; or polyamide fibrils can be introduced to improve physical properties.

2.2.2 Fibre Extrusion

Apart from a minority of dyeable fibre, polypropylene carpet fibre is pigmented to a wide range of colours, so flexible manufacturing plants suitable for relatively short runs are required. A typical staple fibre plant (Figure 2.6) has an output of 500 kg/h and is somewhat less sophisticated than a plant for producing nylon [9]. The polymer is hydrophobic and does not need to be dried; it is chemically inert and does not need to be melted under an atmosphere of nitrogen; there is no need for a side-stream extruder; and processing temperatures are not quite so critical as for nylon.

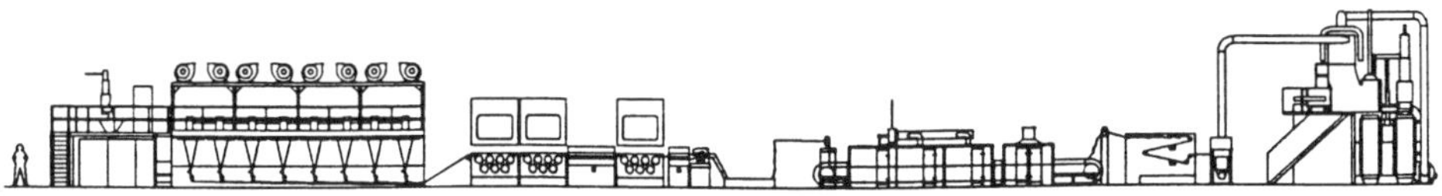

Figure 2.6. Horizontal configuration of staple fibre plant for the production of polypropylene. (Source: Plasticisers Ltd)

Bulk polymer is fed to the extruder throat at a constant rate and chips containing delustrant, UV stabiliser and other additives, plus up to four pigments (see Section 15.1), are individually metered into the material feed. The mixture is melted, pressurised, filtered twice and distributed to die plates. As it emerges from the multiplicity of fine holes, it is cooled by a stream of chilled air and drawn away from the die face at a high draw ratio by haul-off rollers. The tow is then drawn hot at a lower ratio to give the required physical properties. The material is then annealed, crimped as necessary, and cut into the required staple length.

Polypropylene fibre is mostly in round or delta cross-section, with a minor amount of trilobal fibre. For spinning, fibres in the titre range 10–20 dtex are generally used, with soft handling products down to 6 dtex and crisper products as coarse as 30 dtex. Fibres as coarse as 150 dtex are produced to give hairy effects in needlefelts. Coarse fibres are also used in durable products for automobiles and sports surfaces. Woollen spinning requires 100 mm staple and semi-worsted 150 mm.

2.2.3 Speciality Fibres

Shrinkable fibres can be produced by extruding copolymers of polypropylene and polyethylene under specified conditions on the same compact plant as used for conventional polypropylene staple [6]. By blending standard and shrinkable fibre, spinning the blend on the semi-worsted system, and relaxing the yarn at 120–130°C, a particularly bulky yarn is achieved. Moreover, the interlocking of fibres that takes place as the yarn is bulked contributes significantly to the stabilisation of twist.

Alternative high-shrink fibres are available to give softer, bulkier yarns, or yarns that are less bright. Melt-bonding fibres (see Section 2.5) also shrink on heating and are capable of generating bulk in a boiling dyebath. Some high-shrink polyolefin fibres generate bulk at temperatures around 110°C and can be used to create bulk in wool blend yarns that could be susceptible to yellowing at higher temperatures.

Among the fibre variants offered by Drake Extrusion is a product, Permafresh, that is odourless and has built-in antifungal and antibacterial properties. The objective is to prevent the development of odours that can be generated by fungi and microbes hosted by untreated pile that becomes wet.

2.3 ACRYLIC

Acrylic fibres became available to the carpet industry in 1956 and considerable volumes of acrylic carpet were being produced in the 1960s. The fibre peaked in popularity as carpet pile material in the 1970s and acrylic carpets have now virtually disappeared from the American market. They are, however, extensively used in some markets, notably Japan and the Middle East, and acrylic carpets are produced in some European countries. Acrylic products include Axminster carpets.

Although some remarkably good carpets have been made with acrylic pile, the fibre cannot claim leadership in any major respect. It does not have the toughness of nylon, the prestige of wool, the settability of polyester or the low price of polypropylene; on the other hand it has no serious defects as a pile fibre. Dyeings of excellent fastness to light and all other relevant fastness properties can be achieved in a wide range of colours, but the plasticity of the fibre at dyeing temperatures means that acrylic pile is highly susceptible to distortion during piece coloration, reducing its attractiveness to the tufting sector. Shrinkable acrylic fibre has found applications in bulked carpet yarns. Perhaps the strongest market attribute of acrylic carpet fibre is softness of handle, which accounts for its popularity in Japan, where shoes are traditionally not worn in the home, and in bath sets.

2.3.1 Fibre Production

Pure polyacrylonitrile fibres are virtually undyeable and their applications are in technical textiles. Dyeable fibres are made by copolymerising with monomers containing carboxylic or sulphonic acid groups. Fibres with up to 85% by weight of acrylonitrile units are defined as acrylic fibre; fibres having between 85 and 50% acrylonitrile by weight are defined as modacrylic fibres. Carpet fibres are dyeable acrylic fibres. Acrylic fibres may be blended with modacrylic fibres (containing chlorine) to meet a flammability specification, for example.

Acrylic polymer tends to decompose on melting, so fibres must be spun from solution. The most common solvent for this purpose is dimethyl formamide.

Wet spinning consists of pumping the polymer solution through spinnerets into a coagulating bath containing a liquid in which the solvent is soluble but the polymer insoluble. The filaments are merged together into a heavy tow which is rinsed and dried after emerging from the bath. The tow is heated, drawn, treated with fibre processing aid and mechanically crimped.

Dry spinning is carried out by extruding a hot solution of polymer through spinnerets into a stream of hot (eg, 400°C) gas which removes the solvent, leaving filaments of solid polymer. The tow is drawn hot, and may be relaxed to give stable fibre (the most common) or set to provide fibre having latent shrinkage of 20–25% for use in bulked yarns.

Both wet- and dry-spun fibre are used in the carpet industry. Conversion to staple fibre is by cutting, as distinct from the stretch breaking technique used for knitting yarns. Filament acrylic yarns are unusual in the carpet industry.

2.3.2 Availability and Applications

Fibre cross-section cannot be controlled in solution spinning as it is in melt spinning, so that cross-sections of acrylic fibres tend towards round (wet spun) or 'dog bone' (dry spun). Fibre titre is in the range 7–18 dtex. For semi-worsted spinning, staple length is typically 150 mm, but may be shorter for the finer fibres. Staple length for woollen spinning is 60–100 mm, depending on fibre fineness. Most acrylic fibre is supplied in ecru state, but some companies provide solution-dyed fibre.

Semi-worsted yarns are usually bulked yarns spun from a blend of shrinkable and non-shrink fibres. Examples of blends for bathroom carpets are:

Soft quality Nm 8/2
25% 9 dtex 150 mm bright
25% 9 dtex 150 mm matt
50% 6.7 dtex 120 mm bright (high shrink)

Super soft quality Nm 18/2 (or finer)
50% 5 dtex 150 mm bright
50% 6.7 dtex 120 mm bright (high shrink)

For broadloom tufted and woven carpets, the high-shrink fibre is more usually 8 or 10 dtex and the conventional fibre 12 or 18 dtex, depending on the count of yarn to be spun.

Bulking is normally accomplished by immersing the hanked yarn in boiling water for 10 minutes. The bath is cooled before the yarn is dyed. Solution dyed material is bulked by steaming in an autoclave. When the high-shrink fibre contracts, the non-shrink component is pushed outwards from the yarn core in a series of arches ('bee's wings') to increase the volume of the yarn. Fibre interlocking as the yarn bulks contributes significantly to stabilising the twist, improving tuft definition in cut-pile carpets and the retention of tuft definition in wear.

Monsanto first recognised the importance of this technique of setting [10] as a means of providing well-set yarns having better bulk and cover than conventional twist-set yarns which have torsional stress introduced via elevated levels of twist. Their "Traffic Control" yarns are spun from a blend of 9% high-shrink acrylic staple and 91% nylon staple.

Woollen spun acrylic yarns may incorporate a proportion of high-shrink fibre or may be spun entirely from stable (non-shrink) fibre.

2.4 POLYESTER

2.4.1 PET

Applications in the carpet industry for fibres composed of poly(ethylene terephthalate) (PET) have been limited because of poor resilience compared with nylon and the fact that the standard fibre is difficult to dye. Dyeing must either be carried out under pressure or with the aid of carriers, the latter procedure being regarded as environmentally unfriendly. However, the incorporation of co-monomers provides forms of polyester that can be dyed at atmospheric pressure without the aid of carriers, and it is these that are mostly applied in the carpet industry – in significant quantity in the USA. The poor resilience of polyester is often compensated partially by using high pile weights.

Positive attributes of polyester for carpets are its outstanding retention of set, resistance to aqueous stains and softness of handle. The cost of producing polyester fibre is lower than for nylon, but higher than for polypropylene.

Advantage has been taken of the settability of polyester to produce soft constructions of carpet via the piece coloration route. Nylon carpets that resist the mechanical action of the winch/beck are either loop-pile products or cut-pile products tufted from overtwisted yarns. The softness of handle, coupled with easy washability, has led to the widespread use of polyester fibre in bath sets. Softness and lustre have also been exploited in some broadloom tufted products.

Staple fibre rather than filament yarn is the principal form used by the carpet industry. Production is by direct spinning; melt from the polycondensation process is directly routed to the spinning machine [11]. Drawing, crimping, heat setting and cutting into staple are accomplished in one line. Multi-purpose plants are commonly used (ie, cotton types and carpet types can be made on the same line) but annealing of the tow must be carried out under carefully controlled conditions to achieve the specific settability required of carpet fibres. After crimping, the material is heat set at 170–180°C.

Polyester carpet fibre is produced in the range 11–22 dtex, typically in trilobal cross-section, bright or semi-dull. Staple lengths are appropriate for semi-worsted spinning.Polyester carpet fibre made from recycled bottles is used as the basis of an environmentally friendly route to carpet manufacture. Bottle grade PET resin is said to be superior to carpet grade PET because of the stringent requirements of bottle manufacturing [12]. In the USA, manufacture of carpets having pile constituted entirely from PET recyclate is well established.

2.4.2 PTT

Poly(trimethylene terephthalate) (PTT) was synthesised by Whinfield and Dickson as part of their research that led to the original commercialisation of PET fibre, but it was not economically viable at the time because of the high cost of the intermediate product, 1,3-propanediol (PDO). Relatively recently, Shell Chemical Company introduced a route for producing PDO by the hydroformylation of ethylene oxide with syngas (carbon monoxide and hydrogen) using a catalyst, and PDO became available at an attractive price [13].

Polymer is prepared by direct esterification of purified terephthalic acid and propanediol under pressure, followed by melt polymerisation at 250–275°C under vacuum. The molecular shape is said to account for enhanced fibre properties compared with PET.

Shell's fibre spun from PTT is known as Corterra. Du Pont's 3GT fibre is based on the same polymer but their PDO is made by a different chemical process (and biosynthesis is an alternative possibility). Conditions for texturing BCF Corterra have been published [14] and the fibre has been shown to be suitable for spunbonded fabric manufacture [15]. Corterra can be dyed satifactorily with disperse dyes at atmospheric pressure [16].

The results for BCF Corterra cut-pile carpets in floor trials [13] were virtually identical to those from comparable nylon 6 and nylon 6.6 carpets and were superior to those of heavier polyester (PET) carpets. Stain resistance was on a level overall with that of nylon that had been treated with stainblockers: the only problem was with oily items. PTT carpets generated very low electrostatic charges in both step and scuff tests.

Du Pont is a potential producer of PTT fibre [17].

2.5 MELT-BONDING FIBRES

Melt-bonding fibres, which had previously been used by the nonwovens industry, were first applied as carpet pile fibres in the mid-1990s, an early objective being to enhance the twist setting of wool-rich yarns [18]. By using a blend, typically 80/10/10 wool/nylon/bonding fibre, and heating the yarn to fuse the bonding fibres to each other, and possibly to the nylon, a network is created within the yarn, as illustrated in Figure 2.7, creating a more cohesive structure. The concept was applied around the same time to fully synthetic yarns. Later, melt bonding fibres were blended with cotton to improve cotton's potential as a pile fibre. In nylon and polypropylene carpet yarns, monocomponent bonding fibres have proved satisfactory because of the greater possibilities for bonding (melt-bonding fibres do not bond to wool). Polyester, polyamide and polypropylene melt-bonding fibres are all applied commercially as components of carpet pile yarns. Mostly, melt-bonding fibres have been applied in spun yarns, but they can also be used as components of BCF yarns.

Yarn processing conditions must be carefully selected in order to achieve the required results with the particular fibre combination (see Section 4.2).

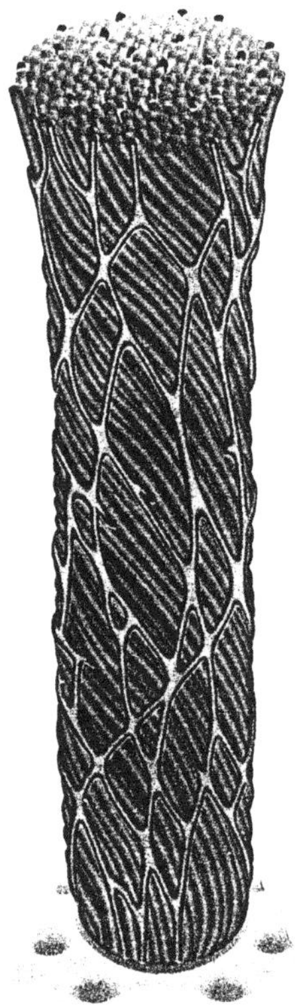

Figure 2.7. Artist's impression of a tuft of mainly wool fibres held together by a network of melt-bonding fibres. (Courtesy of Hoechst AG)

Advantages of incorporating melt-bonding fibres include:

- enhanced tuft definition in cut-pile carpets;
- improved retention of tuft definition after wear and/or cleaning;
- improved retention of tuft definition after piece coloration;
- reduced fibre shedding in the early stages of wear;
- the possibility of substituting singles yarns for plied yarns.

Bicomponent melt-bonding fibres are supplied by several companies. A typical sheath/core polyester fibre having a sheath with a melting point of 130°C is suitable for blending with nylon. For wool-blend yarns similar polyester sheath/core fibres are used, but the sheath melts at 110°C. Such fibres are relatively fine (4 and 6 dtex, respectively) and are supplied as 51 mm staple.

Commonly, single component melt-bonding fibres are composed of polypropylene copolymers. Bonding of 100% polyolefin yarn is achieved in steam at temperatures in the range 120–130°C, and lower temperatures (eg, boiling water) are effective in blends with wool. Some polypropylene copolymer bonding fibres shrink progressively at temperatures above 90°C, so they can also be used as bulking components. However, resistance to shrinkage of the yarn structure can modify the bulking effect, and a three component blend of conventional polypropylene, high-shrink polypropylene and fusible copolymer is probably more reliable for producing a high-bulk bonded yarn.

Polyethylene binder fibres for carpet applications have also been developed [19]. Fibres and continuous filament yarns can be produced from two alternative families of polyethylene, one melting at about 126°C and the other at about 106°C. Hercules Incorporated has commercialised a 280 dtex multifilament yarn from the low-melt temperature material, as well as staple fibre in a range of titres and cut lengths. Polyethylenes tend to cross-link during polymerisation to give gels that interfere with the spinning process. It is therefore important to use gel-free linear low density polyethylene (LLDPE).

Lomel T-780 multifilament yarn (Hercules) may be combined with BCF polypropylene yarn at ply twisting. During heat setting the low-melt filament shrinks to recover bulk that may have been lost in previous operations and fuses to lock the twist. It is claimed that low-melt filament or staple may be combined with wool or cotton to bulk the yarn and lock the twist; and that high- or low-melt types may be used with nylon or polyester staple.

Bonding fibres are usually offered in a limited range of colours, selected to render the fibre inconspicuous in a wide range of bulk colours.

2.6 CONDUCTIVE FIBRES

Antistatic versions of nylon actually contain a small proportion of conductive fibre or filament. An alternative proposition when manufacturing carpets from nylon is to purchase ordinary nylon and combine it with the required proportion of conductive fibre. When a conductive wool carpet is required, blending with conductive synthetic fibres is necessary, although there is a chemical route to antistatic wool carpets that are not conductive (see Section 15.12.5). It is not usually considered necessary to blend conductive fibres with polypropylene or polyester piles.

Note that in order to produce an effective antistatic nylon carpet, it is usually necessary to do more than provide a conductive pile: the backing must also be made conductive, either by incorporating conductive fibres or conductive latex, or both (see Section 21.2). All conductive fibres are dark in colour, but the proportions used in the pile are so small that they are not discernible, except by close search.

2.6.1 Stainless Steel

Carpet pile may be made conductive by incorporating as little as 0.2% of stainless steel fibre. Bekintex, Belgium, offers 9 dtex (12 μm) 90 mm steel fibre for this purpose. It is not easy to blend such a small proportion of fine fibre uniformly into a bulk lot of carpet fibre, so the material is marketed either as a pre-blend of Bekinox™ fibre or as a range of Bekitex™ yarns, rich in steel, that may be combined with pile or backing yarns in various ways.

Bekinox™ W 12/18 is a blend of 82% wool (30–35 μm, 100–150 mm staple) and 18% stainless steel fibre. It is recommended for combining with scoured wool in a blending plant at a minimum level of 1.1%, corresponding to a steel content of 0.2%.

Bekitex™ L80 is a fine (125 dtex) yarn composed of 95% nylon 6.6 and 5% Bekinox™. It can be incorporated in synthetic pile yarns by twisting or air entanglement; or introduced directly on the tufting machine, fed through separate tubes. Not all the yarns in the pile need contain Bekitex™ L80, but the detailed construction should be developed to meet the required specification for conductivity or body voltage generation.

2.6.2 Conductive Nylon

Nylon fibres may be made conductive by incorporating a core, sheath, stripe or multiple stripes of carbon. Some of the alternatives used commercially are illustrated in Figure 2.8. When making a conductive carpet, as distinct from reducing body voltages (see Section 21.2), it is important to use a fibre where the carbon can make surface contact, however minor the area of contact. An alternative nylon fibre is coated with silver. Thunderon™ nylon (Nihon Sanmo Dyeing Co) is flat or textured filament nylon that has been coated with copper sulphide after fibre manufacture.

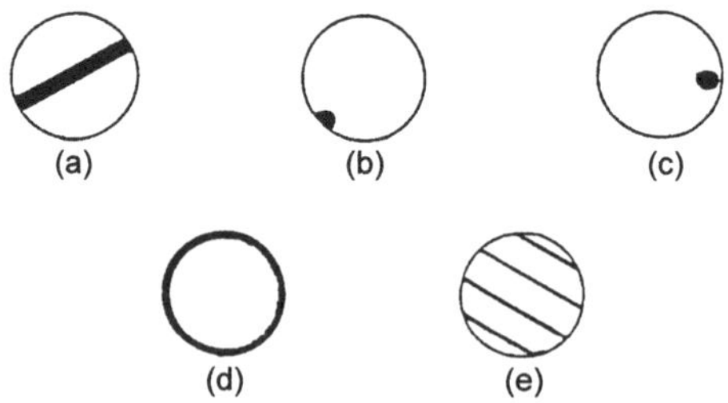

Figure 2.8. Some alternative configurations of incorporated carbon black in conductive nylon fibres: (a) stripe; (b) and (c) core; (d) sheath; (e) multi-stripe.

Conductive fibres of this kind may be incorporated in carpet pile at the stage of blending (pre-blending is essential if something approaching uniform distribution is to be achieved) or, in the case of filament products, at intermingling, cabling or twisting.

2.6.3 Other Conductive Fibres

Various fibres may be made conductive by treating them with copper compounds. Standard Thunderon™ is acrylic staple that has been coated with copper sulphide: filament acrylic versions are also available. Euro-static (Europa Corporation, Poland) is a generic name for polyester, polyamide and acrylic fibres that contain cuprous salts. Loose Euro-static fibre, sliver, tops and yarns are available. The copper compound is claimed to also provide antibacterial properties.

2.7 WOOL

Wool is the traditional pile fibre and its continued association with high quality products has made it the most prestigious. Its arguably best balanced range of properties for carpets provides technical justification for its position in the market. Outstanding aspects of wool are its good resistance to burning, and even if ignition takes place, the crisp intumescent char that forms can be removed with little trace from minor burns. Less widely appreciated is wool's excellent recovery from stress under the influence of moisture, so that wet cleaning revives the pile, as well as removing the soil. A negative aspect of wool is its poor resistance to abrasion when used in low pile weights: this problem can largely be overcome in cut-pile carpets by blending with 20% nylon.

The carpet weaving industry has largely remained loyal to wool, except for the cheaper grades of carpet. Blends of wool and synthetic fibres are, however, widely used. Wool is a minor, but important pile fibre in the tufting industry, being used in the upper and upper-middle quality levels. Good quality wool carpets are also made by fusion bonding (see Section 14.1) and warp knitting (see Section 14.3). Substantial quantities of wool are used in hand-knotted carpets.

2.7.1 Chemical and Physical Structure

Wool consists principally of spindle-shaped corticle cells (about 85% of the fibre) and cuticle cells that overlap like tiles on a roof. A non-keratinous cell membrane complex cements the structure together (see Figure 2.9). The ratchet-like surface structure accounts for the felting properties of wool that can be used to advantage in heavy count carpet yarns.

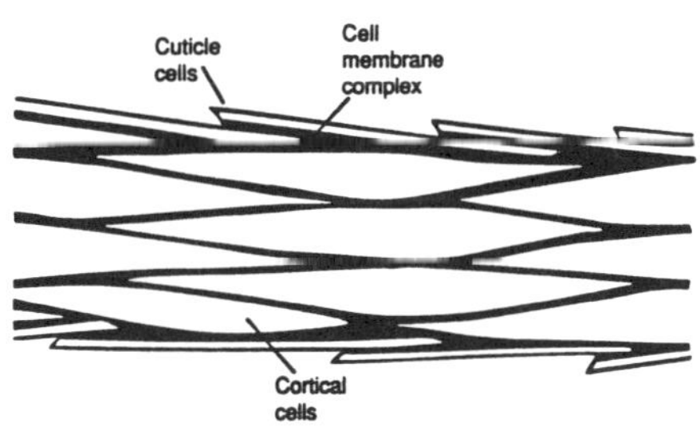

Figure 2.9. Simplified schematic diagram of the cuticle and cortex of the wool fibre [20].

The detailed structure of the fibre has been ascertained and is similar to that of hair, represented in Figure 2.10. Fibre crimp results from the asymmetric distribution of orthocortical and paracortical cells, which differ in structure. The cells themselves are composed of rod-like crystalline protein that has a low sulphur content embedded in a relatively amorphous matrix of proteins, some high in sulphur (from cystine residues) and some high in glycine and tyrosine. The combined protein material is known as keratin.

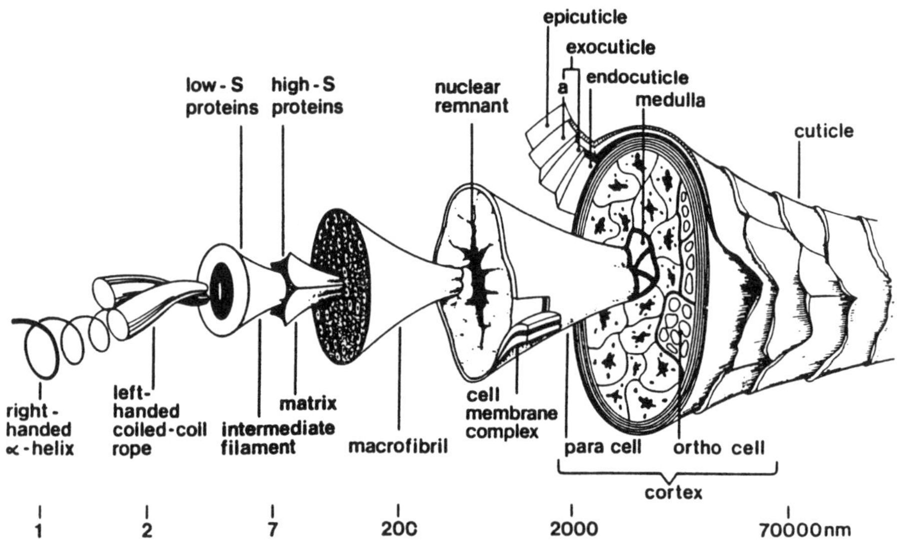

Figure 2.10. Artist's impression of the morphological components of wool. (Courtesy of Textile and Fibre Technology, CSIRO)

Dyes and chemicals diffuse into the fibre via the cell membrane complex and interact with the more open structure of the matrix protein rather than the crystalline regions. However, the matrix protein is reinforced by various chemical cross-links that must be broken if the fibre is to be made plastic for setting. Twist setting of carpet yarns requires molecular rearrrangement which involves 'unzipping' and 'rezipping' these cross-links (see Section 4.1).

2.7.2 Availability

Whereas wool for fine apparel is bred especially for that purpose, wools for carpets are selected from the coarser categories (33 μm diameter and over) which mostly come from sheep bred principally for their meat. Much carpet wool may, therefore, be regarded as a by-product. In New Zealand, which is the principal source of carpet wools that are traded internationally, attention is also paid to sheep breeding to benefit the quality of the coarser wools and most sheep are dual-purpose animals. World wool production statistics are not reliably categorised by fibre diameter, but it is known that substantial quantities of carpet grades are produced in other countries (Pakistan, India, UK and Mediterranean countries). Only small proportions are exported from the larger countries (Russia and China).

2.7.3 Wool Harvesting

Wool may be removed from the sheep at different points in the animal's life cycle to give different categories of wool:

- Full fleece wool is generated by annual shearing.

- Early shorn wool arises from the practice, common in some parts of New Zealand, of shearing three times in a two year period.
- Second shear wool is produced by shearing twice a year.
- Crutchings come from the hind quarters of the sheep and are trimmed off for reasons of animal health and management.
- Lambswool is shorn from animals around 6 months old.
- Hogget wool comes from animals 12–18 months old.
- Oddments (variously described) are wools not conforming to the characteristics of the main fleece.
- Slipe wools are removed from the skins of slaughtered lambs or sheep.

2.7.4 Wool Quality and its Significance

According to WRONZ [21], six measurable properties are necessary, but are quite sufficient, to allow the processing performance of a wool blend to be predicted. These are:

mean fibre diameter	(micron, µm, IWTO-12)
fibre length after carding	(mm barbe, NZS 8719)
bulk	(volume, cm^3/g, NZS 8716)
medullation	(% by number, IWTO-57)
base colour	(X,Y,Z, tristimulus values, IWTO-56)
vegetable matter content	(% by weight, IWTO-19).

The relative importance of these six properties depends on the type of carpet to be made.

Fibre diameter

Fibre diameter is of critical importance when spinning fine semi-worsted yarns as commonly used in face-to-face carpets, but in the production of heavy count woollen yarns is of minor importance for spinnability.

Higher fibre diameters give firmer handling carpets than finer wools. Preference for handle varies from country to country.

Fibre length

For the semi-worsted and worsted spinning systems, long fibres and fairly narrow variation in fibre length are necessary for smooth drafting and the production of regular yarns. The woollen spinning system can handle much greater variations in fibre length, but very long fibres are undesirable as they may disrupt splitting of the fibre web at the condenser.

In all spinning systems, longer fibres tend to enhance yarn properties, and excessive short fibre content can lead to undesirable fibre shedding during the early stages of carpet wear.

Fibre bulk

The bulk of loose wool is essentially a measure of fibre crimp. Although to a much lower extent than crimp, medullation also contributes positively towards fibre bulk because of the lower specific gravity of the fibres and because of the greater resistance to compression of medullated fibres. Figure 2.11 indicates that the bulk of different wool breeds varies widely.

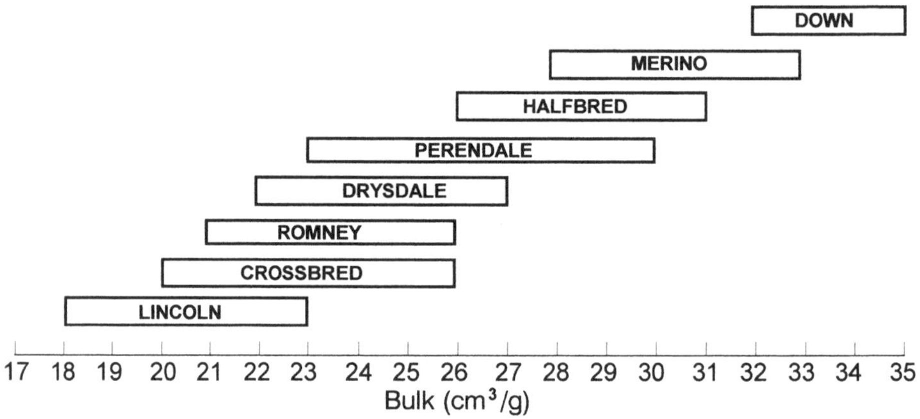

Figure 2.11. Indicative ranges of bulk values for different wool breeds.

Fibre bulk can also be taken as an inverse measure of lustre, since crimpy wools tend to be matt and straight wools lustrous. Bulky wools spin well on the woollen system and give high yields in carding because they provide a cohesive fibre web. Bulky wools tend to give bulky yarns, and this effect is particularly noticeable in semi-worsted yarns.

In carpets, bulky wools contribute towards good cover of the backing, enhanced tuft definition, a matt appearance and a resilient pile. When a dense lustrous carpet is to be made (eg, many rugs and carpet squares) less crimpy wool should be selected.

Medullation

The less refined breeds of sheep (eg, Chinese Yingtze, Indian Joria and many British moorland breeds) tend towards the bicomponent fleece of feral sheep (ie, having a fine undercoat with an outer coat of medullated fibre known as guardhair). The Drysdale breed of sheep was specially developed in New Zealand to provide wool with a high content of medullated fibre but free from undesirable kemp and dark fibres. The various types of medullation are shown in Figure 2.12.

According to Ince [22], medullation has an important effect on the physics of coloration. Because of high light scattering caused by multiple internal reflections, kemps appear virtually undyed and medullated wools appear to be dyed to a lighter colour than true wool dyed in the same bath. This effect can lead to streaks and a generally skittery appearance when producing plain carpets.

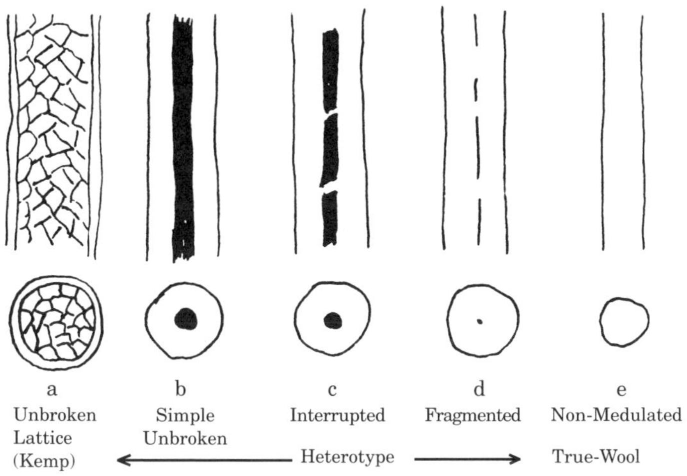

Figure 2.12. Types of medullation. (Courtesy of Dr J Ince)

However, medullation is more important than fibre diameter in providing a firm handle and resistance to compression in a carpet. Kemps are brittle and do not contribute positively to carpet properties. In some products a rugged appearance and a very crisp handle are considered highly desirable, for example in some grades of patterned Axminster carpets, where streakiness is not a problem.

Colour

Dark colours may be dyed entirely satisfactorily on discoloured or stained wool, but it is essential to have wools of good colour when dyeing pale bright colours. Even when dyeing the duller pastel colours, wool that is relatively free from yellow discoloration is needed, since the yellow components of wool colour are fugitive to light, so dyeings will fade off-tone.

Clearly, wool colour cannot be judged on greasy wool since the base colour is masked by impurities. Even scoured wool contains impurities that may be removed, at least partially, during carding, yarn scouring or in the dyebath, so that measurement of the base colour of the wool gives a better basis for forecasting the quality of a dyeing than visual examination.

Quite distinct from wool's overall yellow discoloration, the presence of dark fibres is an important aspect of wool quality. A plain carpet in a light colour can appear streaky if a few coloured hairs are distributed irregularly along the yarn.

Vegetable matter

Very high levels of vegetable matter in wool can, in the extreme, lead to a need for carbonising. Levels in most New Zealand carpet wools are such that they can be removed mechanically during carding. Some types of vegetable matter are, however, difficult to remove and may show as faults in finished carpet, especially in plain styles.

2.7.5 Marketing of Carpet Wools

In most countries, farmers' lots of greasy wool are sold at auction or privately to merchants or wool exporters. These organisations assemble and blend the farmers' lots into larger lots and have them scoured on commission (or in many cases in their own scouring plants). Such scoured blends, therefore, represent the most common raw material for the pile of wool carpets.

A few large carpet weaving plants have their own wool scouring and spinning plants, so that their raw material is greasy wool.

The reasons for preparing blends of carpet wool are:

- to provide sale lots of wool of sufficient size to be relevant to the needs of the industry;
- to provide continuity of supply by blending a raw material that may vary from farm to farm and from season to season;
- to level variations within a lot;
- to create a wool blend having the properties required for a particular style and quality of carpet.

Although some spinners may work closely with their wool supplier to create a specific blend for a specific carpet quality, the fourth purpose of blending is usually the responsibility of the spinner. Different suppliers' lots are blended to create particular blends for a whole variety of carpet styles. Price is an important consideration. For example, considerable savings may be made by preparing different blends for the different colours in a carpet: a white blend to be dyed to pale colours, a yellow-white blend for medium colours and a grey blend for dark colours.

2.7.6 Specification of Carpet Wools

Wools have traditionally been specified by sample. In New Zealand, objective specification is an important feature of wool trading. Greasy carpet wools are offered at auction on the basis of a sample accompanied by a test certificate indicating the yield of wool base, mean fibre diameter, colour and vegetable content.

Although pre-sale testing is now considered essential for the exporter buying greasy wool for carpets, because carpet wools are most commonly traded as scoured blends the system has only indirect value to the spinner. Deliveries of scoured New Zealand wool may, therefore, be accompanied by a Comprehensive Scoured Wool Package Certificate containing information useful to the spinner. The certificate includes: wool base; vegetable matter base; moisture content and regain; ethyl alcohol extractables; residual grease; ash; yield (seven alternative national and international calculations); fibre fineness; base colour; and optionally fibre length after carding and fibre bulk.

Wool exporters or their customers may set numerical specifications for scoured wool blends that can include limits based on a selection of measurable properties.

Provision of a sample to accompany the specification is considered essential.

It is advantageous for carpet yarn spinners to include at least a proportion of specified New Zealand wool in their blends as an aid to achieving continuity of processing performance and product properties.

2.8 SISAL

Sisal is extensively used in flat-woven carpets, usually in its natural colour. The style and colour of sisal carpets became so popular in the 1990s that their look was simulated in carpets made from other fibres.

Sisal is classified as a hard fibre and originates from the leaves of the plant *Agave sisalana*. Although the plant originated in Mexico and neighbouring countries, it is grown commercially in Indonesia and Africa. After harvesting, the epidermis and pulp are scraped away from the fibres in large machines: washing is carried out simultaneously. The decorticated fibre is then hung in the sun to dry and bleach for a few days, after which it is sorted and baled for marketing.

Sisal as supplied for spinning still has many of the characteristics of a leaf, but is composed of short (2.3 mm) fibre elements averaging 24 μm diameter. The fibre elements are flattish, polygonal in cross-section and have a lumen.

Sisal is composed of 72% cellulose and 14.5% lignin plus moisture, and inorganic and soluble organic material.

2.9 COIR

Coir is used in flat-woven styles of carpet similar to those made from sisal, but the products tend to fall into lower price brackets. Cut-pile entry mats are also commonly made from coir.

Coir is obtained from the shell of the coconut. Although coconuts are grown in many parts of the world, the fibre industry is centred around India [23] and Sri Lanka. To obtain the fibre, the husks are cut from the nuts by smashing them on inverted blades mounted in earth or sand. Optionally, the husks are crushed to facilitate retting, which may be accomplished by netting the husks into rafts, weighting the rafts with mud to sink them in shallow water and leaving them to soak for several days. The pith is partially dissolved and degraded during this time. After retting, the husks are beaten manually to break them up. Alternatively, the quartered husks are soaked for about 5 days in tanks of water while weighted with iron bars. They are then crushed between corrugated rollers and broken up further by holding them against rotating spiked drums. The torn fibres are washed, dried in the sun and hackled using fixed assemblies of steel spikes.

Coir is composed of fine (14–24 μm), short, hollow fibre elements but the fibres of commerce are coarse, roughly round fibres up to 150 mm long that are essentially assemblies of the fibre elements. Coir is a lignified fibre and is a yellow brown colour.

2.10 JUTE

Jute was at one time the most important backing fibre for both woven and tufted carpets. It is still important in some regions, but its leading position has been usurped by polypropylene. The largest outlet for jute – bags, sacks and bales – has also lost market share to polypropylene, and has been a victim of the move towards bulk handling. For these reasons, jute interests have been looking towards alternative market outlets.

An area where jute has achieved some success has been as pile material in woven rugs and carpets, principally squares woven on the face-to-face system [24]. A market for low cost bordered squares exists and is largely satisfied by products with polypropylene pile. The consumer's instinct, however, is to associate squares in traditional designs with natural fibres, and jute products have acceptance in this sense. Jute pile is soft and lustrous, provided the material is dyed as fibre and not in yarn form, and this is its principal attraction. When the pile becomes compressed, recovery can be promoted by shaking or suctioning.

Jute is quantitatively the second most important textile fibre after cotton. It is a bast fibre obtained from plants of the genus *Corchorus*, a herbaceous annual that flourishes in the hot, damp regions of Asia. The principal sources of supply are Bangladesh, India, China and other countries in SE Asia.

The plants grow to a height of about 5 m and the fibrous material in the stalks consists of individual cells of cellulosic material, 2–3 mm long, cemented together with lignin and various gums. Retting of the plants is carried out by soaking for 10–15 days in a sluggish stream of water or pits. Much of the cementing material is biochemically degraded and dissolved, leaving fibrous material up to 2 m in length that is dried, baled and marketed to spinning mills. The fibre length depends on the extent of the retting process, since full breakdown of the cement could result in unspinnable individual cells.

The basic cell of jute is irregularly polygonal (normally five to six sides) with thick walls and a lumen of oval cross-section. Many cells are cemented together to form a fibre (see Figure 2.13). In addition to cellulose, jute contains about 20% of lignin. The fibre is light brown and tends to darken further in sunlight.

Jute is a strong, stiff fibre and is virtually inextensible.

2.11 COTTON

Cotton was the first fibre to be used as pile by the tufting industry, but quickly lost its position when more suitable pile fibres were tried. Cotton then became used principally in bath mats, where its easy washability was important, and in squares. More recently, there has been a revival in the use of cotton pile in broadloom tufted carpets, based on blends with melt-bonding fibres and special yarn constructions. Cotton yarns are widely used for chain warps in woven carpets; and cotton primary backing fabric is used for some hand-tufted carpets.

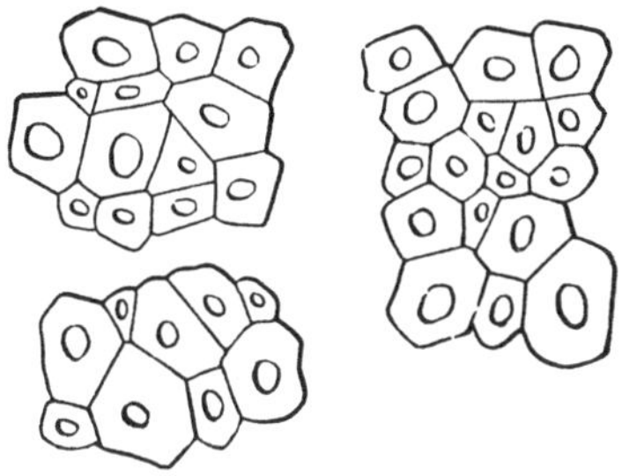

Figure 2.13. Cross-section of jute fibre.

Cotton fibre grows inside the seed pods of plant species in the *Gossypium* family. When the seed pod, or bolt, reaches maturity it bursts to reveal a wad of fine fibres attached to seed. After picking, the fibre is separated from the seeds by ginning, either in a saw gin (longer types) or roller gin (shorter types).

The cotton fibre is formed as a multi-layered tube. This collapses to a ribbon-like structure as the bolt opens and the fibres dry; the fibre twists into a convoluted form at this stage. A proportion of the fibre is immature. Immature fibres easily form neps and are difficult to dye. It is not easy to define the diameter of a fibre with such an unusual cross-section (Figure 2.14) so fibre titre is usually described in terms of micronaire values – equivalent fibre diameters measured on airflow apparatus.

Micronaire values of different cottons range from 10–22; and staple lengths are from 12–33 mm.

Figure 2.14. Cross-section of cotton fibre. (Courtesy of Cotton Incorporated)

REFERENCES

1. J G Cook. *Handbook of Textile Fibres, Vol.2 Man-made Fibres*. Merrow Publishing Company, Watford, UK, 1968.
2. G A Carnaby, J D Watt, W J Meade, D C Maddever, S J McNeil and E J Wood. 100 years of technology achievement by the global carpet industry. In: *Proceedings of The Fiber Society International Meeting*, Ashville, NC, July 1998.
3. R Pike. Carpet yarn processing from start to "finish". In: *Proceedings of Tifcon '95*, The Textile Institute, Manchester, 1995.
4. P Owen. Fibre and yarn developments at Domotex. *International Carpet Bulletin*, No.289, February, 1998, 8–11.
5. *Proceedings of World Textile Congress on Polypropylene in Textiles*, University of Huddersfield, 9–11 July 1996.
6. D G Ellis. Polypropylene in household textiles and floorcoverings. In: *Proceedings of World Textile Congress on Polypropylene in Textiles*, University of Huddersfield, 9–11 July 1996.
7. M Vanneste and F Godefroidt. Raw dealings. *Textile Technology International*, 1999, 31–34.
8. N D Scott. Metallocenes and polypropylene – new solutions to old problems in textile end uses. In: *Proceedings of World Textile Congress on Polypropylene in Textiles*, University of Huddersfield, 9–11 July 1996.
9. D G Ellis. Plant supplier offers integrated service. *Textile Technology International*, 1993, 42–48.
10. J Southern, J-P Yu, W Baggett and R Miller. Fundamental physics of carpet performance. *Journal of Applied Polymer Science, Applied Polymer Symposium* 47, 1991, 355–371.
11. A Heintze and T Gries. Status and prospects of polyester carpet fibres. *Chemical Fibers International*, 1997, 47(4), 292–294.
12. H Hansler. Recycled PET. *Textile Horizons*, March 1999, 24–25.
13. H Chuah. Corterra poly(trimethylene terephthalate) – a new polymeric fibre for carpets. In: *Proceedings of Tifcon '96, Carpets – The Current Perspective*, The Textile Institute, Manchester, 1996. Also in *Chemical Fibers International*, 1996, 46(6), 424–428.
14. F Werny and H Chuah. Texturing of Corterra™ BCF performance in carpets, *International Fiber Journal*, 1996, 11(4), 628–664.
15. H S Brown, P K Casey and J M Donahue. Poly(trimethylene terephthalate) for fibers. *Corterra™ Polymers*, Shell Chemical Company, 1997.
16. H L Traub, P Hirt and H Herlinger. Dyeing properties of poly(trimethylene terephthalate) fibres. *Melliand Textilberichte*, 1995, 76(9), E179-180.
17. J Herlihy. '97 fiber and yarn report. *Carpet and Rug Industry*, 1997, 25(6), 10–15, 57–58.

18. J D Watt. Stabilising wool carpet yarns by melt bonding. In: *Proceedings of Tifcon '93 – Covering Europe*, The Textile Institute, Manchester, 1993.
19. R J Legare, R K Gupta, R E Kozulla and G I Rehberg. Polyethylene binder fibres for carpets and related applications. *Carpet and Rug Industry*, 1996, 24(6), 46–48.
20. J Rippon. In: *Wool Dyeing*, D M Lewis (ed), Society of Dyers and Colourists, Bradford, 1992.
21. G A Carnaby *et al*. *Computer Blends I–V*, WRONZ Communication Nos. C81, C82, C83, C95, C110, 1983–1989.
22. J Ince. The search is on for the ideal carpet wool. *Wool Record*, 1976, 130, No. 3386, 75–77.
23. Video. *Coir – the Golden Fibre*, Indian Coir Manufacturers Association.
24. M M M Rahman. Face-to-face jute carpets – problems and prospects. In: *Proceedings of World Textile Congress on Floorcoverings*, University of Huddersfield, 12–13 July, 1995.

3 Pile Yarns

The carpet industry utilises both filament and spun yarns as pile materials. In the tufted sector, filament yarns are the more popular in terms of square metres of carpet, although spun yarns tend to be used in the heavier weights of carpet. All of the filament yarn is bulked (BCF). About a third of the volume of BCF yarn is heat set for cut-pile carpets and the majority of the spun yarn goes into cut-pile carpets. The situation for woven carpets is the inverse: most pile yarns are spun,the main exception being the widespread use of BCF polypropylene in face-to-face carpets. Most of the woven constructions are cut pile, requiring the yarn to be set if the individual tufts are to be defined and if this tuft definition is to be retained during wear (see Section 4.1).

3.1 FILAMENT YARNS

BCF yarns are textured essentially by the hot-fluid-jet technique, although some older stuffer-box machines may remain in use. There are three basic production routes:

- Two-step process: first spinning and winding, followed by drawing and texturing.
- Single-step process: spinning, drawing and texturing in one machine.
- Direct spinning: polymerisation, spinning, drawing and texturing.

The two-step process is the most flexible, allowing the production of various yarn types (eg, differential dye types, mixed decitex or different colours of melt-dyed polymer simply by combining on the creel). Spin-draw-texturing is particularly suitable for the production of standard yarn types in larger quantities, although the flexibility of this single-step system has been improved over the years. The trend is towards increasing use of the single-step system [1,2]. Machinery is mostly suitable for all carpet fibres (nylon 6.6, nylon 6, polypropylene or polyester) but some slightly simpler machinery is dedicated to the production of BCF polypropylene. Direct spinning lacks flexibility and is not used in the carpet industry.

3.1.1 BCF Single-Step Process

The single-step process starts with polymer chips and incorporates the sequence shown in Figure 3.1. Figure 3.2 is a schematic of a typical vertical plant, consisting of extruder (upper gantry) feeding the spin beam and spin pack, quench duct and spin finish applicator leading vertically downwards, drawing rollers to orient the polymer, texturing unit with cooling drum, intermingling system and take-up unit. In the bottom left of the illustration is the control/monitoring unit, usually with mimic panel and digital read-out of processing conditions, to ensure consistent product quality. Machines having a horizontal configuration are also available. Design of the overall machine must take into account the need for a practicable string-up procedure.

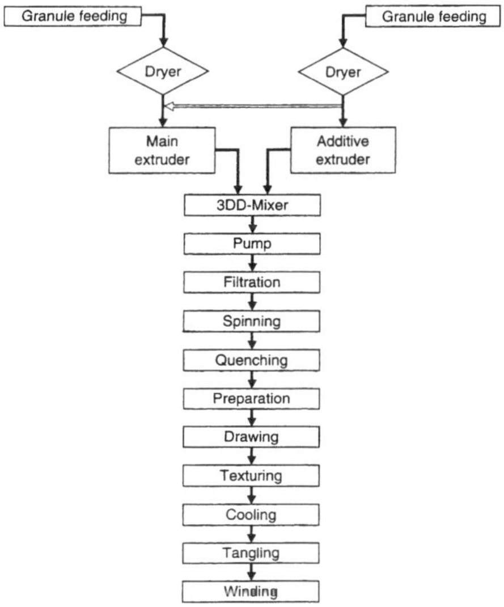

Figure 3.1. Processing steps included in a single-step machine for producing BCF yarns [2].

A production unit of this kind typically produces at 3500 m/min and can produce multicoloured yarn as well as solid colours and ecru yarns for subsequent dyeing.

As there is no blending following the production of BCF yarns, every aspect of the process must be tightly controlled to avoid irregularities in the final carpet. The most critical application of BCF yarns, one that is avoided by some companies, is for carpets that are to be dyed to a plain colour, where any variation in polymer, heat history, moisture relations, frictional effects, etc, may affect dye uptake or dye strike. Other styles of carpet may suffer from texture streaks if processing conditions at all stages are not precisely controlled.

It is equally important that subsequent yarn processing needs to be well controlled if irregularities in dyeability and texture are not to be introduced.

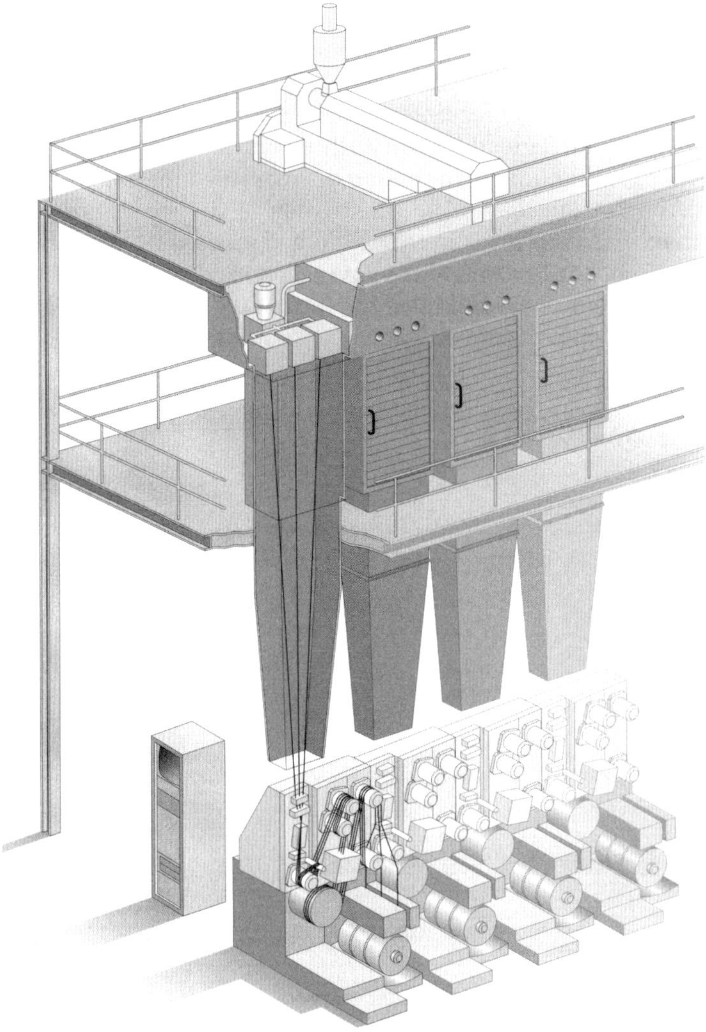

Figure 3.2. Schematic drawing of a plant for producing three-end BCF yarn by a single-step process. (Courtesy of Rieter Synthetic Fibre Machinery)

Yarns must be from the same merge and should be processed in sequence of delivery to eliminate ageing effects, cable twisting should be controlled and conditions in heat setting must be very precisely maintained.

Extrusion

In the extruder, polymer chips are melted, compressed and forwarded by a screw system and are concurrently homogenised. Nylon must be dry and has to be melted under nitrogen or other inert gas to avoid oxidation. Polyester chips must also be thoroughly dried before extrusion. At this stage, additives are introduced, which may include delustrant (usually titanium dioxide), UV stabilisers, antistats, flame retardants and pigments for colouring the yarn (see Section 15.1), using a subsidiary extruder.

Spin beam and spin pack

Each head of the machine may produce two, three, four or six ends of yarn that are merged after texturing. Figure 3.2 shows a three-end machine. The ends may be identical, may differ physically in terms of filament decitex or cross-section, or may be in different colours. Pressurised melt from three extruders is distributed along the spin beam to the corresponding three spin packs on each head. To ensure physical uniformity of the polymer at each head, the path lengths must be identical. The design of the spinnerets in the spin packs determines the filament decitex, cross-sectional shape and possible hollow channels.

Quench duct

As the polymer streams emerge from the spin packs they are cooled by a controlled stream of air. Moisture may be introduced at this stage for static control and to move the moisture content of nylon towards equilibrium at an early stage. Airflow must be sufficiently high to ensure rapid cooling, but not so high as to disturb the filaments as they solidify.

Spin finish applicator

At the bottom of the quench duct the filaments are bunched together and pass over a lick roller or metering device where they are coated with a spin finish. The functions of the finish are to lubricate the yarn to control static and to provide yarn cohesion. The composition of the spin finish (eg, blends of phosphate esters and ethoxylated products) is critical since it must not only fulfil its primary functions but also should not be susceptible to processing problems (eg, spray-off, gelling, volatilisation) and should easily scour off before dyeing. The trend is towards reducing levels of application from around 1 to 0.25% of carefully engineered products, the principal objective being to reduce the levels of pollutants from subsequent aqueous processing.

Drawing

The molecular orientation of the polymer towards the yarn axis is improved by drawing at a ratio of 3.0–4.5, depending on the polymer type, thereby increasing yarn strength and decreasing the extensibility. Drawing is accomplished by two duos of godet rollers, tapered in opposite directions and geared so that yarns repeatedly wound around them (6–10 times) experience increasing surface speeds. The yarns arrive from the quenching duct and are drawn by the first godet duo, which applies light tension and heats the yarn for drawing. The second duo accelerates the yarn, controlling the draw ratio, and preheats the yarn to plasticise it for the texturing process.

For the texturing process to proceed uniformly and to ensure that variations in dyeability are not introduced, temperature control must be extremely accurate (±1°C), including in the axial direction. For example, Rieter's system for achieving temperature control is based on the 'heat pipe' principle. Fluid is distilled within the Rievap hollow roller in such a way that it condenses on the inside of the shell, maintaining the shell at the boiling point of the fluid. Barmag uses a high-frequency heating system.

Hot fluid texturing
Immediately following drawing, the yarn passes into the texturing unit, the principle of which is shown in Figure 3.3. A jet of hot air (or in some cases, superheated steam) disturbs the column of yarn and creates three-dimensional crimp and simultaneously forms the yarn into a plug downstream from the jet. In practice, the texturing chamber is tapered like a cornet (inset). Pressure falls as gas is released through slots in the tapered section and the plug continuously falls on to a cooling drum, quenching the yarn and setting the crimp.

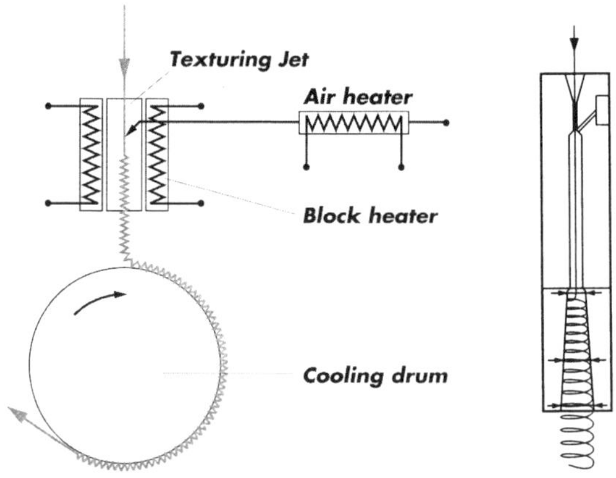

Figure 3.3. Principle of hot fluid texturing [3]. Inset: configuration of the texturing chamber (Courtesy of Rieter Synthetic Fibre Machinery)

Intermingling
In the previous steps, the individual ends passed in parallel through the process but retained their separate identities. If an intermingled yarn is required, the three (or two, four or six) ends are passed together through a tangle chamber where they are subjected to jets of cool air. Intermingling to an extent of 30–40 tangle knots per metre gives yarn with suitable cohesion for subsequent processing. If ends of different colours are intermingled, a 'marbled' effect is produced. Different types of jet can produce different visual effects. Ends may by-pass the intermingling unit, for example if it is intended to cable them to achieve a clearer distinction between the colours.

Take-up unit
The final yarn is taken up by high-speed winders on to cross-wound packages. Doffing is automatic or manual with package lifter. The requirements are for good physical stability during handling and transport, good unwinding performance, an accessible thread tail, and a correctly placed thread end. Package weights are 4–5 kg.

3.1.2 BCF Two-Step Process

The individual items of equipment for the two-step process are the same, in principle, as those used in spin-draw-texturing, except that undrawn yarn spun to a given specification is wound up after spinning. Packages of undrawn yarn are then creeled up on the draw-texturing machine.

At the draw texturing stage, yarns differing in colour, dyeability and/or decitex are creeled up on the one machine. Intermingling is a key step in the sequence and sophisticated systems allow special yarn effects to be produced. The objective of the process is to provide maximum versatility.

Two-step filament yarn production is less productive but more flexible than the single-step process. It is used by carpet companies looking to create their own individual products, rather than by large manufacturers of branded synthetic fibres.

3.1.3 Dedicated Systems for Polypropylene

BCF polypropylene yarns are produced by spin-draw-texturing on machinery that has the same essential components as the machine that is suitable for all the principal synthetic carpet fibres. However, polypropylene is not so susceptible to oxidation as is nylon and has virtually zero moisture uptake, so that it is not necessary to melt the chips under nitrogen, nor to control moisture in the system. Godet rollers may be electrically heated. Plants suitable for polypropylene only may, therefore, be made a little less expensively than universally applicable plants. Polypropylene yarn is usually produced in pigmented form, since the fibre is essentially undyeable apart from a few modified forms.

Polypropylene split film yarns were once fairly widely used as carpet pile, mainly to overcome patent situations. Their applications are now confined to textile sports surfaces and other forms of artificial grass, where their flat cross-section, coupled with green coloration, simulates the appearance of grass.

Polypropylene film is stretched to increase molecular orientation to the extent that the film has some tendency to fibrillate. Usually, the stretched film is cut into ribbons and is then split to the required width in a second process (scribing, cutting, abrading or air blowing). Linear densities are in the range 2500–4000 dtex.

The cut ends of split film pile tend to be abrasive and can cause friction burns when sports players fall. To alleviate this problem, split film may be textured by the knit-de-knit setting process, so that the pile curls and a high proportion of sharp ends are buried.

3.1.4 Dedicated System for Polyester

Most polyester for carpet applications is produced as staple fibre. BCF polyester carpet yarns were introduced at a relatively more recent date: in particular, PTT (Corterra) is produced as BCF yarn (see Section 2.4.2). Single-step draw-texturing equipment, as used for nylon, is suitable. Hot air texturing is carried out at 170–210°C.

Polyester is more commonly available as partially oriented yarn (POY) than are nylon or polypropylene, so POY is the starting material for the process.

The draw-texturing process is based on the Fibre M process of Heathcoat [4]. The yarn path is vertically upwards from the drawing rollers through the texturing unit. Two (or possibly four) ends are processed together, depending on the yarn linear density to be produced. They are subjected to a jet of superheated steam and take on a three-dimensional crimp when they pass into a wider section of tubing. The first part of this tube is heated to promote setting, and a longer period of cooling follows. Control of bulk is achieved by sensing the height of the column of crimped yarn with feedback to the steam jet heater. Raising the steam temperature increases the bulk and raises the height of the column: too high a column causes the supplementary heat in the steam jet to be reduced.

The process is a two-stage process and consequently has the flexibility to combine different yarn types and colours. Production speeds are up to 1000 m/min.

3.1.5 Cabling and Twisting

Direct cabling has largely displaced ring twisting and two-for-one twisting for the production of multi-ply BCF carpet yarns.

Two-for-one twisting requires, first, assembly winding of the individual components followed by insertion of twist on a separate machine (Figure 3.4). Two turns of twist are inserted for each stage of the twisting element: one turn between a stationary feed package and a rotating disc, and a further turn between the disc and a balloon guide.

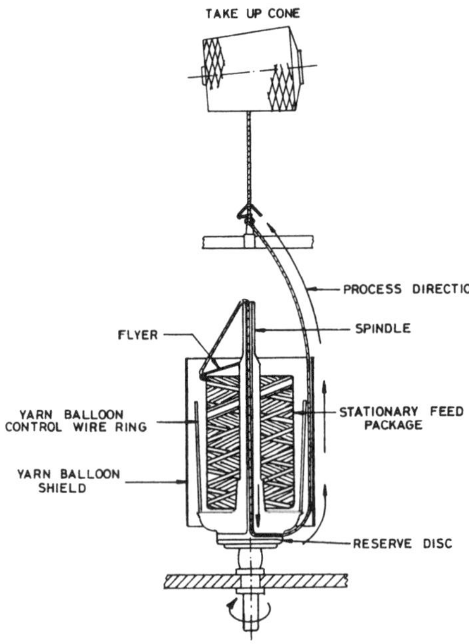

Figure 3.4. Principle of two-for-one twisting. (Courtesy of The Textile Institute)

Strictly, cabling means the twisting together of two or more plied yarns. In the case of BCF yarns, the term direct cabling is applied to the production of folded yarn from twistless singles yarns. Direct cabling is performed on a machine superficially similar to a two-for-one twisting machine (Figure 3.5). One package of supply yarn is laid on the spindle rotor and yarn is drawn axially from it without inserting significant twist. The other package is stationary and yarn is withdrawn axially from it into a hollow spindle shaft, and from there in a balloon around the outside of the first package. The two ends combine at an eye and are twisted together, one turn of cabling for each revolution of the spindle.

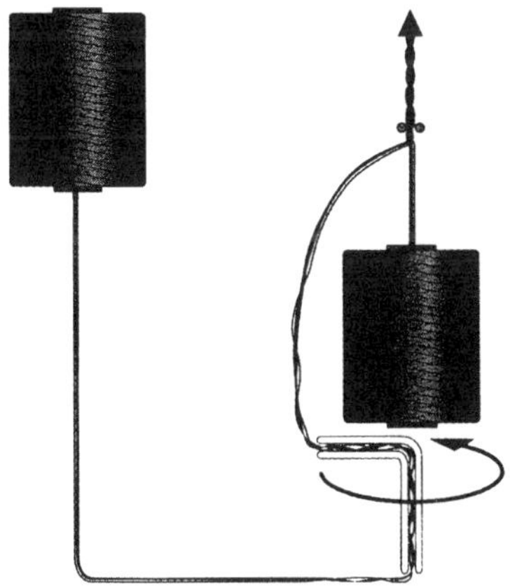

Figure 3.5. Principle of direct cabling. (Courtesy of Volkmann GmbH & Co)

High-speed cablers operate at 6000–7000 rpm, inserting twist levels between 160 and 320 tpm. A potential disadvantage is that yarn tensions at 0.4–0.5 g/dtex are much higher than in twisting, so that problems can arise in terms of deposits of spin finish, polymer dust, wear on machine parts, hairy packages and end breakages. These problems can be overcome by using high quality feed yarns, suitably engineered spin finishes and controlled atmospheric conditions (20°C and 50–60% RH). Current trends are towards yarns with very low levels (0.3%) of non-fuming spin finishes, used in conjunction with porcelain 'undulators' that restrict yarn ballooning, without the abrasion associated with conventional balloon restrictors [5].

3.1.6 BCF Yarn Products

BCF nylon carpet yarns are produced in linear densities ranging from 650 to 5400 dtex, with filament decitex in the region 11–30. The finest ranges tend to be used in cut-pile products (fine velours) and the very coarsest ones in loop-pile carpets, but the principal distinction between cut- and loop-pile yarns is in the twist direction of the cabled yarns. Most loop-pile tufting machines require S twist and most cut-pile machines Z twist direction.

Product variants can be introduced at several of the stages of yarn production and in subsequent cabling and heat setting processes.

An indication of the wide range of possibilities is given in Figure 3.6, which also explains Du Pont's description system for BCF nylon products. In addition to finished yarns that are ready for tufting, most fibre manufacturers will supply feedstock for draw-texturing by the carpet manufacturer. Such yarns may be as fine as 650 dtex.

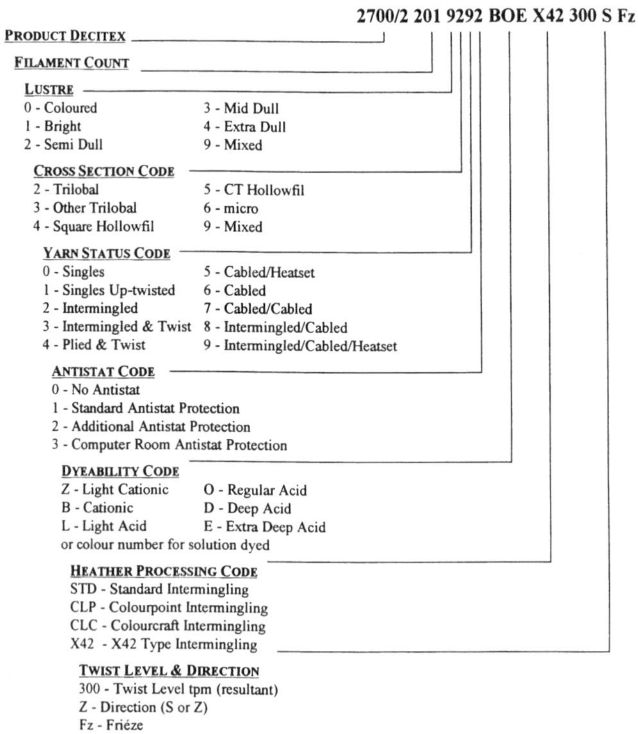

Figure 3.6. System of Du Pont's BCF nylon product description, indicating the principal available options. (Courtesy of Du Pont (UK) Ltd)

BCF polypropylene yarns are produced in the range 1100–4000 dtex, typically in round, trilobal or delta cross-sections. Hollow filaments have been proposed. A popular yarn is 2900 dtex f120 in three colours for 1/8 gauge tufted carpet, cut pile, loop pile or cut/loop. A similar singles yarn, and some lighter ones, but in solid colours, are used in face-to-face woven carpets.

Polyester filament yarns have found niche markets, especially in Japan and the USA, but with the price of PET chips intermediate between nylon and polypropylene, the market position is not clearly defined. Speciality products, such as FR polyester, have clearer market positions. Polyester has inherently better resistance to burning than polypropylene, and is a little better than nylon. Its properties can be enhanced further by copolymerising with other monomers. Corterra (PTT), which is claimed to have the resilience and handle of nylon without its problems of staining, is being offered to the carpet industry in the form of BCF yarns.

3.2 SPUN YARNS

Spun yarns are produced principally on the semi-worsted or woollen systems, although the worsted system is used for some fine weaving yarns (eg, R140 tex/3 (Nm 21/3)) simply because it is difficult to spin such fine yarns from uncombed sliver. Generally, synthetic carpet yarns are spun on the semi-worsted system and most wool yarns are spun on the woollen system. However, synthetic fibres can be spun on the woollen system and wool is also spun on the semi-worsted system. Some tufting yarns are produced by unconventional spinning systems.

Although the woollen system has fewer processing steps, semi-worsted machinery is more productive in medium and fine counts of carpet yarns. Relative conversion costs move in favour of semi-worsted as counts become finer. The semi-worsted system requires raw material having a high mean fibre length and minimum short fibre content, however. The woollen system is more flexible in its requirements for raw material, so that the carpet manufacturer is free to blend wools that provide the pile character required in the carpet, and can incorporate a proportion of shorter, cheaper wools (eg, crutchings, slipes). Woollen yarns are usually bulkier than semi-worsted yarns (although highly crimped wool and synthetic fibres provide similarly bulky yarns on either system) and their more random fibre orientation contributes towards settability of twist, providing tuft definition in cut-pile carpets. The woollen system can easily be used to introduce nep, slub and flame effect material to provide fancy yarns. Such considerations lead to woollen spun wool yarns being the principal ones used in Axminster, Brussels and wire Wilton carpets made from counts such as R620 tex/2 (Nm 3.2/2) and R780 tex/3 (Nm 3.9/3) and for tufting yarns for 1/8 gauge and coarser. Wool tufting yarns for 1/10 gauge carpets may be woollen or semi-worsted.

Semi-worsted wool yarns are most commonly used in the range R200 tex/2 (Nm 10/2) to R170 tex/3 (Nm 18/3) for fine face-to-face carpet squares, dense loop-pile tufted carpets and some fine gauge cut-pile carpets. Multi-ply worsted yarns (eg, R433 tex/2/2, Nm 9/2/2) are applied in a few lustrous styles of Axminster and Wilton carpets, and woollen yarns are used in some face-to-face squares.

3.2.1 Woollen Spinning

The woollen system [6] has three essential steps: blending, carding and spinning. Because slubbings (rovings) for spinning are formed by splitting the carded web at the condenser of the card, it is critically important for the fibre web to be uniform. The blending and carding steps must, therefore, be very carefully controlled. Drafting waves are a minor problem because draft ratios at the spinning frame are low: regular slubbings are likely to give regular yarns.

Blending

The principal objective of blending [7] is to ensure that each 10 kg portion of the material fed to the card is identical in composition – the card and its feed hopper can homogenise this quantity of fibre, but no more. This requirement applies to synthetic fibres as well as to wool (staple as supplied may vary in dyeability and there may be deliberate variations in fibre characteristics). Even single component batches require at least one levelling process. Reliably uniform dyeability is, in fact, one advantage of spun synthetic yarns over BCF yarns. Some other functions of blending apply only to wool.

Fibre for blending is often in its natural colour. Sometimes the fibre is dyed as loose stock, in which case an important objective of blending is to combine a number of different dye lots and merge them into a uniform colour. If a tweed yarn is to be produced, several different dyed colours plus coloured effect material are blended.

The blending process consists of a number of essential steps: assembling blend components in the required proportions, opening, binning, oiling.

Blend components

Carpet wool is normally provided in scoured form and will have been pre-blended at the scouring stage. If a component has not been blended in this way, it should be pre-blended. Individual blend components should not constitute less than 20% of the total blend. Components smaller than this should be pre-blended with one of the main components.

After any necessary pre-blending, the blend components are assembled in the required proportions and fed into the system. A simple way to accomplish this is to place all the bales of fibre required to make up the blend alongside a conveyor and manually feed the components on to it. (Calibrated hoppers, as used in the semi-worsted system (see Section 3.2.2), are sometimes used in the woollen industry.)

Synthetic fibres or densely dumped wool bales may be difficult to separate manually and a bale digester is advantageous for feeding the fibre automatically. Bands and covers are removed from the bales and the packed fibre is loaded on to the bale conveyor. The press-packed structure is broken up by the spiked sheet of the bale digester, a levelling roller and a stripping roller. The resulting small lumps of fibre pass directly to the opening machine.

Opening

Depending on the character of the material to be processed, the opening machine may be a teazer, picker, step blender or fearnaught (cockspur). These machines have cylinders bearing coarse teeth that open the lumps of fibre, allowing more intimate contact between the various components. The configuration of the various machines varies widely, but the fearnaught has a configuration similar to a single-cylinder card (refer to Figure 3.10), having workers, strippers and a doffer, but with very coarse teeth instead of card clothing.

A secondary function of the opener in the case of wool is to remove dust, dirt and some vegetable matter, which may be extremely important for some blends. After opening, the blend is normally conveyed pneumatically in ducts.

Binning

The principle of the blending bin is illustrated by the traditional system of preparing a blend by layering the individual components on a floor, slicing the assembly vertically and mixing the vertical slices (Figure 3.7).

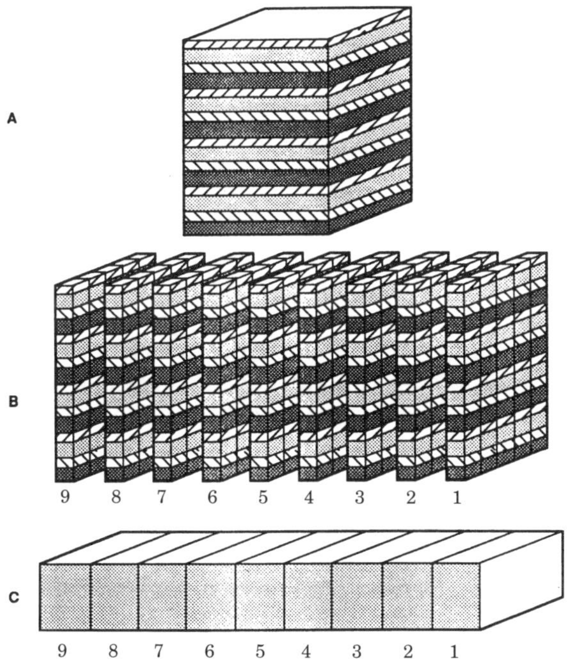

Figure 3.7. Principle of layer blending. (A) shows ideal layering of the components in the bin; (B) shows the vertical sections into which the blend is divided; and (C) indicates the uniformity of supply provided if each of the sections 1–9 is thoroughly mixed [6].

In practice, a modern blending bin is usually rectangular (cylindrical or octagonal bins are sometimes used) and is filled by the mixture supplied from the opener by means of a travelling spreader. A bin can accommodate a blend up to 5000 kg. When the blend has been layered in this way, it is removed by a bin emptier, bearing a spiked lattice that slices vertically upwards, and is conveyed to another similar bin.

Binning twice is normal. For plain carpets, where streaks are a serious potential defect, it may be wise to blend the material three times.

When the blend is required for carding it is removed from the final bin (by slicing vertically again) and conveyed to the hopper of a further opener. This 'willeying' process gives further intermingling of the fibres.

Oiling
A fibre processing aid (traditionally oil but now one of a number of synthetic substances) is applied during blending in order to facilitate carding. Fibre processing aids have several functions:

- to prevent the generation of electrostatic charges,
- to reduce fibre-to-metal friction,
- to reduce fibre-to-fibre friction,
- to enhance fibre cohesion.

The yield of yarn is significantly improved by the optimum choice of fibre processing aid.

Application is with water, which is itself an excellent antistat. The level of water depends on the fibre being processed. For wool, sufficient water is applied to raise the regain to 15–18%, which improves the flexibility of the fibre for carding. The level and type of the fibre processing aid is recommended by its manufacturer and depends on the overall processing routine to be followed, the nature of the wool blend, its residual grease content and the type of synthetic fibre in the blend. Levels of 2% or more on the weight of fibre are usual when the yarn is to be scoured or dyed, processes that remove residues. When stock-dyed yarn is to be processed dry and the fibre processing aid remains on the yarn, special products that do not attract soil and are effective at very low levels are recommended.

Application of the solution of fibre processing aid is typically by an in-line mist lubricator associated with the pneumatic conveyor immediately following the final opener. The level of application is governed by control equipment. An example of an alternative technique is Garnett Controls' Rollaflow system which continuously monitors, via sensitive weigh rollers, the mass of fibre fed to the fearnaught and controls the flow of lubricant, water and additives applied at the oiler mounted in-line with the fearnaught.

After oiling and conditioning (applying moisture), the blend is fed to storage bins located near the cards, or is baled for transport to the carding room.

Figure 3.8 shows a comprehensive blending plant, representing one of a number of alternative configurations.

Carding
The functions of the woollen card are to:

- reduce the tuft structure of the willeyed blend to individual fibres;
- mix the fibres;
- remove impurities;
- form the fibres into a uniform web;
- divide the web into slubbings (rovings) of the required weight per unit length.

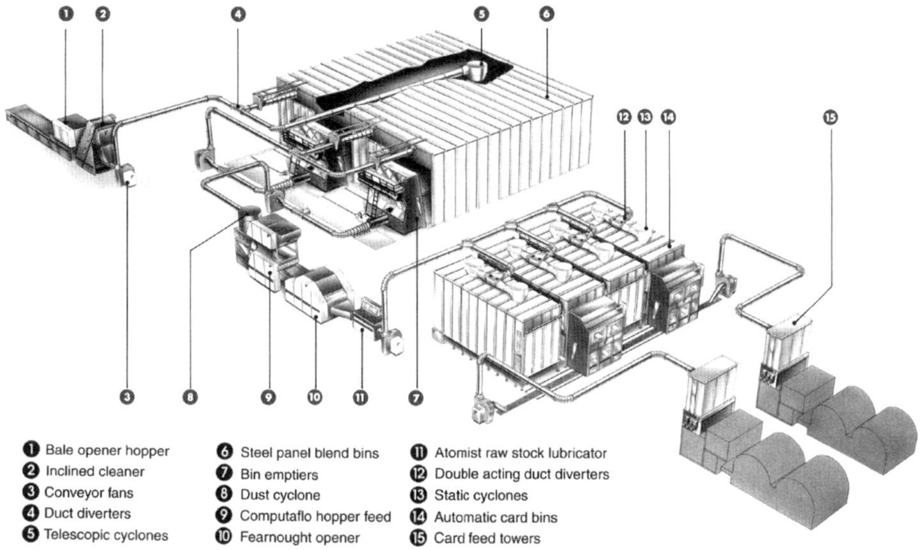

Figure 3.8. Example of a blending plant suitable for woollen carpet yarn production (Courtesy of Pneumatic Conveyors (Huddersfield) Ltd)

Essential components of the card are: a feed unit, a scribbler section (usually composed of two swifts), an intermediate feed, a carder section (one or two swifts) and a condensing machine. For wool, there is also a burr-crushing machine usually located at the end of the scribbler. A layout of a typical card as used in the carpet industry is shown in Figure 3.9.

Carding action

The basic components of a card are shown in Figure 3.10, which also indicates the relative speeds of the various components. In industrial carding machines there are up to four swifts, and there are typically four pairs of workers and strippers on each swift.

The essential carding action results from the relative motion (one slow and one fast) of two card-clothed surfaces with the points of the clothing configured towards each other. Point-to-point action results in the tufts of the willeyed blend being separated, one part being retained on one roller and the other on another (eg, swift and doffer).

On an industrial card there are up to 30 places where point-to-point action can take place, but fibres are retained on the clothing for a time, so that a typical fibre passes carding points 100 to 300 times. The delay factor also results in carded fibres being combined with other fibres that have more recently arrived on the swift, accounting for the blending action of the card.

Point-to-back action of the card wire results in the fibres being stripped from one surface and transferred to the other (eg, stripper and swift). The 'fancy'

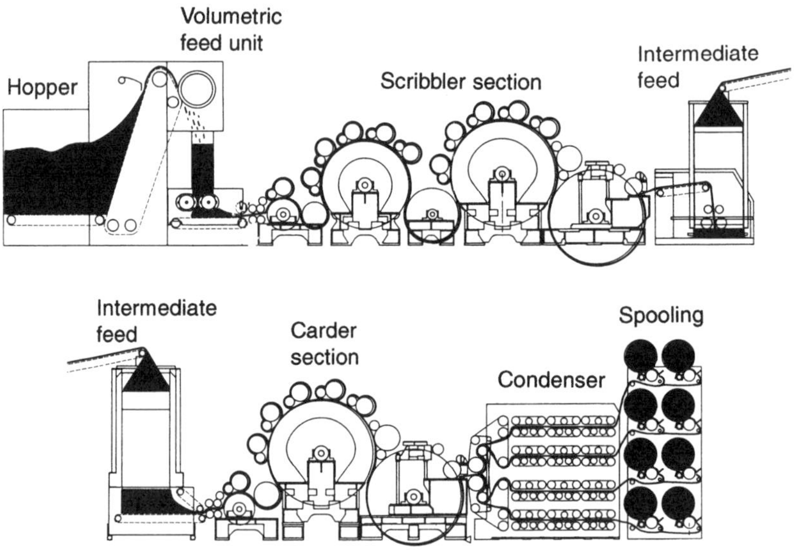

Figure 3.9. Layout of a typical woollen card suitable for processing carpet yarn fibre (Courtesy of Houget Duesberg Bosson SA)

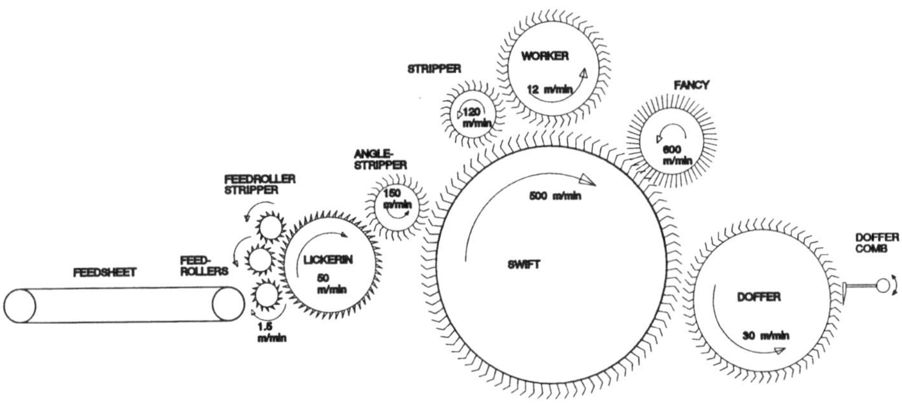

Figure 3.10. The key components of a woollen card, with indication of typical speeds.

roller has a unique back-to-back action which simply raises the carded web higher in the clothing of the swift, so that fibres are more readily caught by the doffer wire and removed.

Card clothing

The rollers of cards may be clothed either with flexible clothing (fillet wire) or 'metallic' wire (sawtooth wire). For spinning wool carpet yarns, the early parts of woollen cards are usually clothed with robust sawtooth wire and the later parts with flexible wire. Sawtooth wire, which is rigid, is more widely used for processing strong synthetic fibres.

Coarse gauge wire is used in the first part of the scribbler section and progressively finer wire is used towards the second part of the carder section. 'Settings', the spacings between rollers determined by feeler gauges, also become progressively closer moving along the card. The objective is to apply a gentle action in the early stages of carding and increase the intensity down the card, so spreading the opening and blending actions over the entire card.

Feed units

Fibre must be fed to the card at a specified rate in order to obtain the required slubbing count. There are two ways of achieving such control: by a combination of hopper and weigh pan or hopper and volumetric feed.

The hopper consists of a bin and an angled spiked lattice that lifts tufts of fibre out of the bin and forward to the weigh pan or volumetric feed unit. The functions of the hopper are to reduce the average size of, and further mix fibre tufts and feed the blend towards the feed sheet.

To take advantage of the blending action of a hopper, two hoppers are often used in tandem.

Using a weigh pan system, the hopper feeds fibre to the pan, which is mounted on a load cell and which tips on to the feed sheet when the required mass has been loaded. To achieve accuracy, there is a fast feed rate for the bulk of a pan load, followed by a trickle feed to reduce the effect of 'in-flight' fibre. Modern control systems are available that compensate for the build-up of contaminants in the weigh pan and deviations from the intended mass of fibre in the pan. For example, Garnett Controls' Microweigh XL2 system uses a multiple compensation system that includes spiked lattice speed control, in-flight fibre control and drop compensation. The latter allocates a specific area of feedsheet to a specific weight of fibre by automatically adjusting the drop frequency. Feed roller control is also available. The Microweigh XLM has the additional feature of an inbuilt moisture measurement system that automatically makes adjustments to the feed in response to variations in moisture content.

Volumetric feed systems have the advantage of providing a constant supply of material to the feed sheet, but there is an obvious potential for error from variations in the density of the fibre blend. HDB's Servolap principle is to scan the material on the feed sheet with gamma rays, measuring its mass in this way, and a microprocessor adjusts the drive to the feed rollers to compensate for variations.

Garnett Controls' Rollaweigh system (Figure 3.11) incorporates a series of lightweight sensor rollers mounted within a load cell weighing assembly. Within the weigh zone, an independently driven short-term control system takes into account fluctuations within the zone. This system is also commonly used to control woollen-type cards in the needlefelt industry (Section 13.4) and there is feedback from the output of the card, so that the operator simply enters the

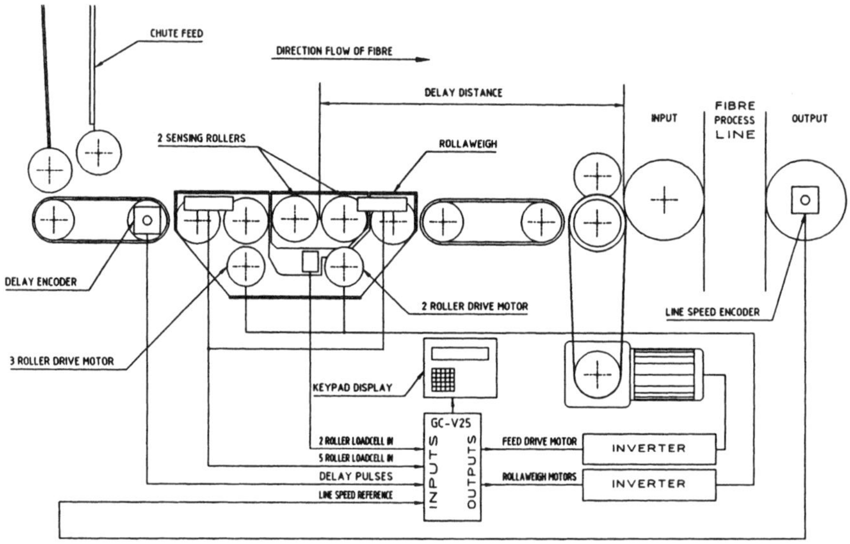

Figure 3.11. Principle of the Rollaweigh card feed control system. (Courtesy of Garnett Controls Ltd)

required production rate and the system automatically adjusts processing parameters, including feed rate, to achieve the result.

Removal of impurities (wool)

Some impurities in wool fall out as the fibre tufts are opened in the early stages of carding. Others become embedded in the web but can be removed if they are crushed into smaller particles. The web must not cushion this crushing action, and a convenient point where the web is adequately thin is after the scribbler section of the card.

A pair of hardened pressure rollers at this point (Peralta or web purifier) crush vegetable matter and depilatory residues (on slipe wool), facilitating their removal later in the carding process.

Web purifier rollers may be omitted if only synthetic fibres are to be processed.

Intermediate feed

After the scribbler section of the card and the optional web purifier, the fibre web is doffed and conveyed laterally and then vertically away. This intermediate web is then layered laterally on to the feed sheet of the carder section, so blending the material across the card, an action that would be otherwise lacking.

There are two general types of intermediate feed: the Scotch feed, in which the scribbled sliver is gathered into a rope; and the broad-band feed, in which the sliver is layered before being conveyed away laterally in the form of a broad band.

Layering of the web on to the feed sheet of the carder section must be uniform and reproducible (eg, unaffected by air currents) if the final web is to be uniform.

Tape condenser
The fibre web from the last part of the card is held against a parallel array of rubber tapes mounted so that alternate tapes move upwards or downwards, so splitting the web into delicate ribbons, or slubbing. The width of condenser tapes is selected to cover a particular range of slubbing counts.

Tapes may be series tapes, where one tape produces one end of slubbing, or figure-of-eight tapes, where one tape produces two ends.

Immediately after splitting the web, the fragile ribbons are consolidated by rubbing between aprons. Rubbing aprons comprise pairs of wide belts that convey the ribbons forward, but reciprocate transversely to give a rolling action. High production cards often have two or three sets of rubbing aprons in tandem to ensure that the slubbings are adequately consolidated. Multi-level banks of aprons are required to accommodate the large number (over 100) of ends of slubbings generated.

Note that the slubbings produced have no twist and are still delicate, despite the rubbing.

Spooling
The slubbings are wound on to banks of spools by surface drive.

Many carpet cards are fitted with the WRONZ/IWS High Density Spooling system (HDS). The HDS attachment consists essentially of a grooved plate mounted over each drive drum. Each groove carries an end of slubbing, providing friction and a consolidating effect, allowing the slubbing to be wound under greater tension than normal. The spools of slubbing are, therefore, denser and heavier. The HDS attachment is claimed to provide a number of advantages:

- 50% increase in doff length at carding.
- 3.5% increase in carding efficiency.
- 5% increase in carding throughput (as a result of higher draft).
- Fewer end breaks at the condenser.
- Firmer spools, less prone to damage.
- Fewer end breaks at spinning.
- Improved yarn properties.

Control of carding
Successful woollen yarn manufacture depends critically on control of the carding step. Specialised knowledge is required and full consideration of quality control may be obtained from a publication *Woollen Yarn Manufacture* in The Textile Institute's Manual of Textile Technology series of publications [6]. The following aspects must be given careful attention.

Raw materials: The fibre blend must be appropriate to the type of yarn to be spun and detailed attention must be given to factors affecting the uniformity of blending.

Fettling: When spinning coloured yarns, the card and associated equipment must be thoroughly cleaned between lots. Ducting to the card must also be cleaned. For undyed lots, a programme of fettling appropriate to local circumstances should be established to ensure that build-up of fibre does not interfere with the carding action.

Slubbing count: Rate of feed, coupled with delivery speed, should govern the average slubbing count. Edge slubbings will be off-count and are automatically rejected. At equilibrium running conditions the count should be checked by weighing. Automatic in-line devices are available that monitor and adjust the count. Periodic variations in count can also be measured in-line. Calculations based around the periodic length of the variation can trace the fault to a particular component of the card (eg, eccentric roller or tight condenser tape).

Reproducible settings: Some modern cards have a central panel that controls the speeds of the principal components. The settings for each batch can be recorded on floppy disk so that they can be reproduced at a later date.

Ring spinning

Condenser spools are mounted on top of the spinning frame on positively driven let-off drums and the slubbings are directed to the drafting unit. Figure 3.12 illustrates the layout of a drafting unit. Because the slubbings are twistless, a false-twist unit (F) is required to provide strength and to control drafting. Drafts in the range of 20 to 30% are governed by the ratio of speeds between the front rollers (FR) and back rollers (BR).

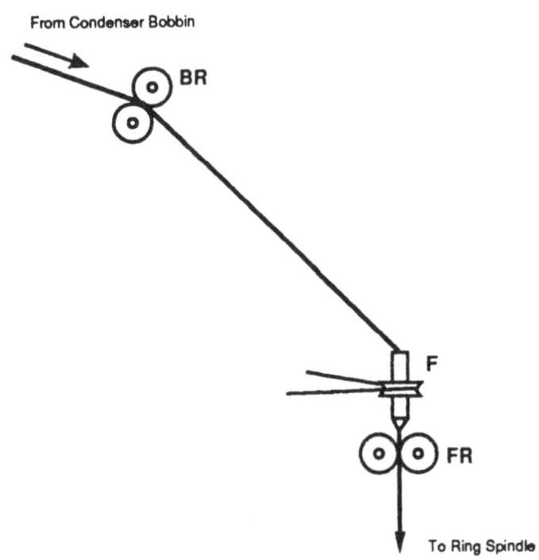

Figure 3.12. Layout of drafting system, with false twisting unit (F), for a woollen ring frame [6].

Levelling of the yarn is achieved by the principle of drafting against twist. False twist tends to run to the thinner places, gripping the fibres and inhibiting drafting, while the thicker, less twisted places are drafted more. The level of false twist must be appropriate to the fibre length of the blend if drafting is to proceed smoothly.

Real twist is inserted into the yarn by means of a collapsed-balloon ring system. Optimum dimensions of the system appropriate to the count to be spun are recommended by machinery manufacturers: finer counts are best spun on smaller packages, whereas counts as heavy as 1000 tex are spun on 140 mm rings in package weights around 1.4 kg.

The trend in modern machinery is for a central microprocessor to control all parameters (speeds, draft, twist, etc) via DC motors. Pneumatic threading and automatic cop lifting are common. Fully automated machines incorporating supply of condenser spools, automatic piecening and mechanical doffing are available.

Twisting

Ring frames designed to accommodate large packages are most commonly used for plying heavy count carpet yarns. End break detectors stop a spindle if any of the singles components or the plied yarn breaks. Ring diameters are typically 180 mm and tube heights 600 mm. Alternatively, two-for-one twisting (see Figure 3.4) or the Hamel two-stage twisting system may be used.

Reeling and winding

Wet processing of carpet yarns is usually carried out in the form of hanks (skeins), since processing in the form of cross-wound packages tends to result in loss of bulk. To limit the number of joints introduced at backwinding, cross-reeled 'jumbo' hanks, typically 380 mm wide and weighing 3 kg, are used. Hank weights up to 6 kg may be produced. Reeling is direct from twist packages.

Backwinding is into cheeses or cones, as required by the carpet tufting or weaving machine. Jumbo hanks must be handled in a controlled manner for smooth backwinding. Figure 3.13 shows a backwinder with duo-expander for holding the hank, one of the common alternatives. Yarn jointing is by means of air splicers or fusion splicers (for synthetic fibres) for the carpet tufting industry. Weaver's or fishermen's knots are often used for weaving yarns.

It is more important to provide the tufter or weaver with the number of yarn packages ordered than to supply the exact weight.

Lubrication

It is not usually considered necessary to lubricate tufting yarns in the way that knitting and sewing yarns are lubricated with wax: the lubricant is provided from the primary backing fabric.

3.2.2 Semi-Worsted Spinning

The semi-worsted system consists of the following steps: blending, carding, multiple gilling, roving (for the finer counts only) and ring spinning. Semi-worsted

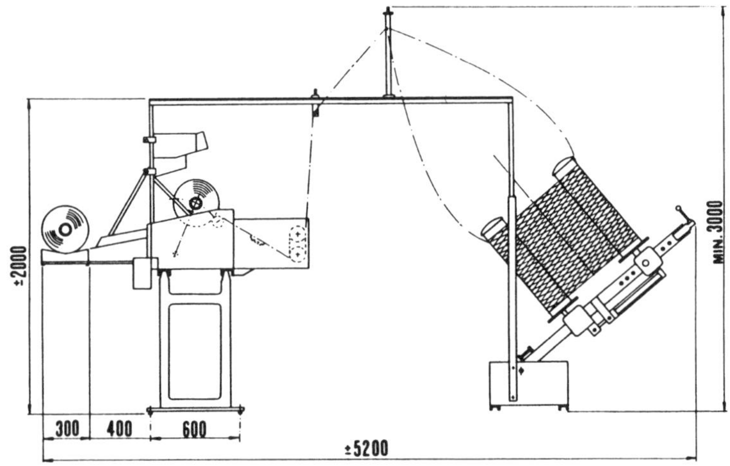

Figure 3.13. Backwinding system IDS-HK with duo expander. (Courtesy of Gilbos NV)

yarns for face-to-face woven carpets are usually dyed in hank form. For other applications, wool yarns are usually stock-dyed and there is no wet processing after spinning. The blending step, therefore, is commonly used to blend stock-dyed lots of nominally the same, but slightly different colour to achieve an accurate and uniform colour.

Blending

Blend components

In the case of wool, blend components should be selected for performance in the drafting systems that are an essential feature of the semi-worsted system. For medium-count yarns this means a mean fibre length (hauteur) not less than 75 mm and a short fibre content (<30 mm) below 25%. When spinning close to the limits (often the case with face-to-face yarns) fibre diameter should be selected so as to provide more than 60–70 fibres per yarn cross-section (depending on the details of the equipment and the skill of the spinner).

Blending equipment

Continuous blending, as distinct from the batch blending system used in woollen yarn manufacture (Section 3.2.1), is most commonly used in the semi-worsted industry. Calibrated feed hoppers, one for each blend component, are arranged to feed the various components at specified rates on to a common conveyor. Because the conveyor moves slowly, the blend components are layered over each other.

Because semi-worsted wool blends are usually more refined than woollen blends, less emphasis is placed on dust removal. A picker opener, capable of opening densely packed synthetic staple fibre, is usually used to mix the assembled blend arriving from the conveyor (Figure 3.14). The mixed blend is then fed pneumatically and layered in a blending bin of the type used in the woollen system. The bin is emptied from the end in vertical slices and the blend is layered in a second bin.

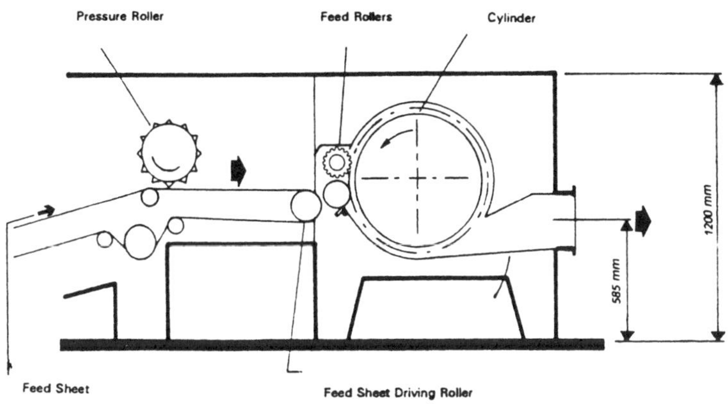

Figure 3.14. Outline of a picker opener, commonly associated with blending systems for semi-worsted processing [6].

Moisture to condition the blend and a fibre processing aid (0.3–0.5%) are applied at the stage of blending.

When blending stock-dyed lots into a solid colour, the blending procedure is virtually identical to that for undyed fibre. Mixture colours may be blended three times if an intimate effect is desired. Less intimate effects may be achieved by combining solid colours at the gill.

Carding

The semi-worsted card does not have the function of preparing rovings so can be shorter than a woollen card. Its principal functions are to separate the fibres, blend them together and, in the case of wool, remove vegetable matter. Note that additional blending takes place at the gilling stage.

A typical semi-worsted card for carpet yarn manufacture (Figure 3.15) consists of a feed hopper, forepart, single swift, drafting unit and coiler, although alternative configurations are used. Note the reduced complexity compared to the woollen card illustrated in Figure 3.9.

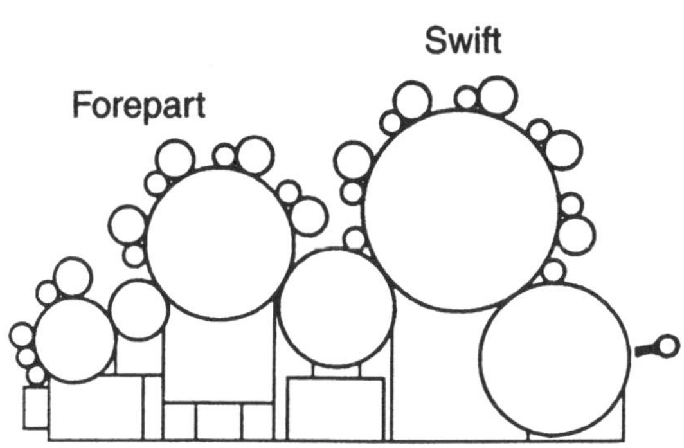

Figure 3.15. General configuration of a semi-worsted card. (Courtesy of Thibeau et Cie)

Feed hoppers are generally of the simpler volumetric type, rather than weigh pan. Adjustment of yarn count is made later in the processing sequence, so that sliver weight is not critical. Gamma-ray feed control units are often applied, however, to eliminate short and medium term variations in sliver weight.

Note that Morel units, rather than burr crushing units, are used to remove vegetable matter. The design of the Morel roller exposes vegetable matter on its surface and burr beaters flick the foreign material out of the fibre web.

The final product of the semi-worsted card is a single heavy sliver (sometimes two slivers).

Drawing

For fine count face-to-face weaving yarns, drawing consists of three (sometimes four) sequences of gilling into cans followed by a roving operation. Heavier count yarns are spun direct from cans of sliver: there are only the three gilling operations.

A schematic diagram showing the principle of a gill box is shown in Figure 3.16. Up to 10 cans of sliver (from the card for the first operation) are assembled at the creel and fed into the back rollers of the machine. The front rollers are driven perhaps eight times faster than the back rollers so that the sliver is drafted. Drafting is controlled by pins mounted on two chains, to give a quiet operation. Any irregularity in an individual sliver is diluted by the process of combining with other slivers. Machines may have draft autolevelling, automatic piecening and pneumatic threading.

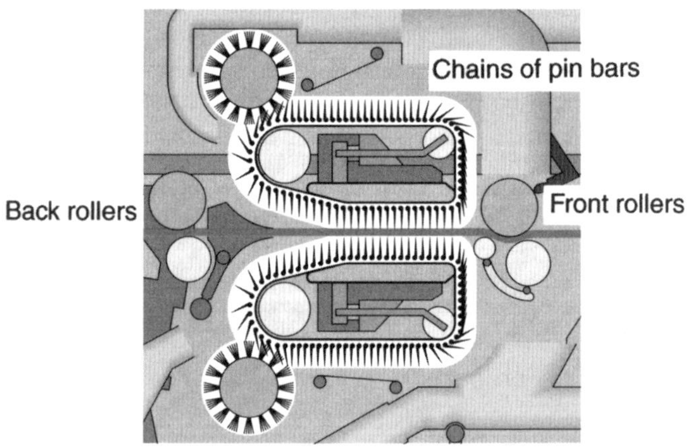

Figure 3.16. Principle of the intersecting gill box. (Courtesy of N Schlumberger and Co (UK)Ltd)

Cans of sliver from the first gilling operation are used to feed a second, similar, machine. Usually the first machine delivers one sliver and the second machine delivers two slivers. The third operation is similar, but the sliver delivered may be split into four components.

Roving frames used to draft material into lighter weights for spinning to fine counts may apply twist or simply consolidate the slivers by rubbing. Roving machines have multiple heads. One type drafts the slivers on aprons and then applies twist by means of a flyer. An alternative type has pin drafting (on a smaller scale than a gill box) and rubs the slivers to consolidate them into rovings, in much the same way as on the condenser of a woollen card.

Spinning

The essential difference between a woollen ring frame and one for spinning semi-worsted carpet yarns is the drafting system. A single-zone drafting system, commonly used for the heavier count range, applies drafts in the range 18–28; a double-zone system can apply drafts up to 200. The principle of operating a semi-worsted line is, therefore, to operate the card and drawing machines on standard settings and to achieve the required yarn count by adjusting the draft.

Ratch settings should be shorter than the longest fibre (otherwise faults will be caused as fibres are stretched and catapult forward on release) but should be close to maximum fibre lengths. Short fibres floating between the back and front of the drafting zone are liable to draft irregularly to give irregularities in the yarn. Aprons or soft upper rollers (or both) in the drafting zone lightly grip the short fibres, controlling their movement and contributing towards smooth drafting. Figure 3.17 shows a system having multiple soft rollers in the drafting zone.

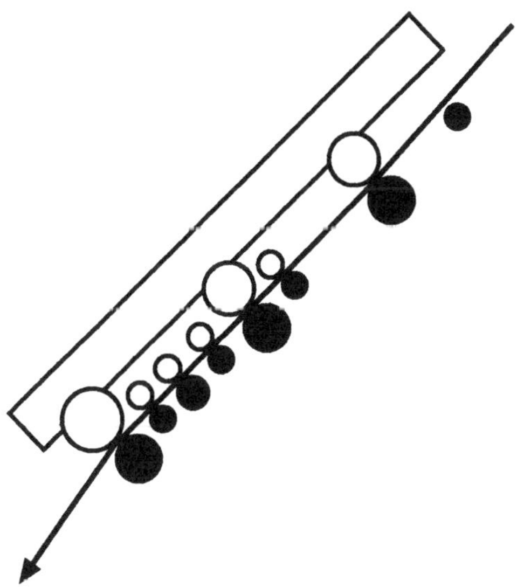

Figure 3.17. Example of a 7-cylinder drafting system. (Courtesy of Houget Duesberg Bosson SA)

As with woollen spinning, machinery should be selected to give optimum productivity for the counts of yarn most commonly spun. This means that the finer counts are spun on smaller packages, where higher spindle speeds are attainable. Heavy count yarns fill the spinning tubes quickly so that doffing is more frequent: larger packages mean less time spent in doffing and this more than compensates for the lower spindle speeds.

The trend is for all variables in the spinning process to be adjustable by DC motors controlled by a central panel. Changing of gear wheels to adjust yarn parameters has been rendered obsolete. Settings for particular yarns are recorded digitally and may be reproduced accurately. Various levels of automation are available to suit the requirements of the individual mill and the labour costs in the country.

Twisting, reeling and winding
Similar machinery to that used in the woollen yarn industry is applied for the further processing of semi-worsted yarns.

3.2.3 Unconventional Spinning Methods

Wrap spinning and friction spinning have been proposed for the production of pile yarns, but have not achieved significant penetration into the industry.

The PLYfiL process of Suessen [8] has been applied specifically for the production of 'fine saxony' yarns. American saxony carpets are heavy (up to 2 kg/m^2) carpets having strong tuft definition, tufted from heat-set yarns of about R400 tex/2 (sometimes 3-ply); but the European fine saxony is a lighter carpet (0.7–1.2 kg/m^2), usually tufted from heat-set yarns in the range R168 tex/2 to R60 tex/2. Such yarns are expensive to spin in the conventional way, and are generally considered to require singles clearing.

Suessen's system involves a rapid spinning technique and one fewer step in the processing sequence compared to the conventional semi-worsted route (Figure 3.18) and consists of:

1. Sliver preparation to 4–6 g/m, as for conventional semi-worsted spinning.
2. PLYfiL spinning/assembly winding.
3. Two-for-one twisting.
4. Rewinding with clearing of faults.
5. Heat setting and winding on to tufting packages.

On the PLYfiL machine, a 5-roller system drafts long-staple nylon sliver to the required yarn linear density and the fibre assembly is consolidated by an air vortex, rather than by twist. Accessible fibre ends are wrapped round other fibres that form the yarn core in an S or Z direction, as required. Delivery speeds are around 300 m/min. Spindles operate in pairs, and two ends are wound on one package (assembly wound) ready for twisting.

The singles yarns do not have high strength: strength is introduced when the yarns are ply-twisted. Because the singles yarns are virtually twistless, lower levels of folding twist than normal are required to create the overtwist situation that is necessary to give strong tuft definition. A typical PLYfiL fine saxony yarn of R92 tex/2 has 320 tpm folding twist. Heat setting is required to set the twist, preserving tuft definition during carpet wear.

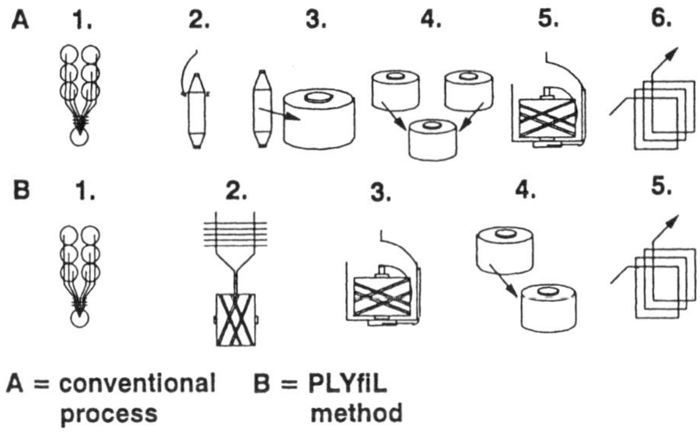

Figure 3.18. Processing steps for conventional and PLYfiL heat-set carpet yarns. (Courtesy of Spindelfabrik Suessen)

3.2.4 Core Spinning

Exceptionally heavy counts of yarn for hand-crafted rugs and some textured loop-pile wireloom carpets are often produced by spinning wool around a core of jute, cotton or wool yarn. Equipment is often improvised by the spinning companies themselves. For example, a flyer roving frame may have a modified creel that carries the core material as well as one or more ends of sliver; no draft is applied and the sliver simply wraps around the core. Alternatively, an end of sliver may be wrapped around the core using a can-coiler mechanism.

3.2.5 Spinning of Hard Fibres

Fibres such as sisal and coir are used in large volumes for flat-woven carpets. The processing system for sisal consists of multiple opening, drawing, spinning, shearing, possible twisting, and package winding. There is no carding. Coir is too short for the hard fibre system and is mostly spun by hand or on simple hollow-spindle machines.

Opening

Opening is accomplished on a heavy-duty machine, such as the Mackhigh Goods Machine. Essentially, the fibre is worked and drafted on a massive open gill bed having robust pins. The material is fed by hand, longitudinally aligned, on to a feed sheet and it is taken by feed rollers, optionally with a spreading regulator, to a slow section gill bed where it is pressed into the pins by bars mounted on cylindrical frames. The slow section has a reach of 1.5 m and moves at about 1 rpm. A fast section, with a reach of 1.8 m, has bars pushed along continuous tracks, applying draft in the region 10–12. The sliver that is created is withdrawn and condensed by delivery rollers at up to 90 m/min and coiled into piles.

A full set of machinery consists of four goods machines, but it is common to use the same machine two, three or four times, depending on the required total output of the set. Perhaps six piles of sliver are fed to the goods machine for the second, third and fourth machines (passages), so that the material is progressively cleaned, regularised and drawn into finer slivers.

Drawing
Drawing frames are fed by four or six piles of sliver (depending on the yarn to be spun) to each head, and a frame has a small number of heads (typically three). The slivers are laid parallel and fed to a long, inclined open gill bed and drafted in the range 8–16 by drafting rollers. The output sliver is condensed into tubular conductors and coiled into cans. Drawing frames are fitted with choke, knot and lump detectors that stop the machine by operating microswitches.

There are two such drawing steps on all sisal lines, and for carpet yarns there is a third step on a light drawing frame to produce the appropriate sliver weight and to further regularise the sliver. Note that the speed of a drawing frame must be reduced for processing stainless sisal for carpet yarns, as distinct from oiled sisal or oiled manilla for ropes or twines.

Spinning
Flyer frames with open gill drafting are used for spinning. For carpet yarns, bobbin dimensions are 9×5 in. (229×127 mm) or a little smaller, depending on the counts to be spun. The reach of the machine depends on the material to be processed, but is long (eg, 1 m) by the standards of synthetic pile fibres or wool. Flyer speeds are up to 3000 rpm. Counts spun are typically around 1000 tex for singles weft, or a little finer for two-ply weft or three-ply carpet warp yarns.

Twisting and winding
Sisal carpet yarns require shearing to reduce hairiness. This is accomplished with helical blades (as for shearing carpet pile), and may take place on the singles yarn, on the folded yarn at twisting or on both singles and folded yarns. At a minimum, the twisted yarn should be sheared if good weaving efficiencies are to be achieved.

Mackie's Twinemark machine combines the twisting, shearing and winding operations, and was designed specifically for carpet yarns in two-, three- or four-ply. It is a hollow spindle machine that imparts two turns of twist for one revolution of the discs. With a disc speed of 3000 rpm, 6000 turns of twist per minute are inserted. The machine is designed to produce finished yarns in the region of 1300–3000 tex.

3.2.6 Jute Spinning

Jute is used in the backings of both woven and tufted carpets, and is also used as a pile material in some of the lower grades of face-to-face woven rugs. Examples of pile yarn linear densities are 210 tex×2 and 330 tex×2.

The jute processing system consists of spreading, carding, drawing, spinning, twisting and winding. The following description is based on machinery supplied by Mackie.

Spreading
A jute spreading machine opens, cleans and drafts the fibre in the same way as the goods machine for sisal, but the feeding and delivery arrangements are somewhat different and only one passage is required.

A tared truck holding the raw fibre is placed on a platform scale near the feed sheet and its weight is registered on a dial. A leader hand is set to the weight of the batch and is geared to the machine speed. The operative feeds fibre from the truck at a rate sufficient to keep the two hands in line, ensuring a uniform rate of feed. Emulsion is applied under pressure to the sliver at the delivery. The sliver is taken up in the form of large diameter broad rolls. To ensure that the rolls have firm selvedges, narrow leader rolls of prepared jute are introduced at each side of the feed sheet.

The rolls of sliver are conditioned for 24–48 hours: heat is generated spontaneously, assisting breakdown of the gums that contaminate the fibre.

Carding
Carding involves three machines: a breaker card, an intermediate card and a finisher card. Each is a single-swift card with workers, strippers and doffer. Six rolls of spread jute are fed to the breaker card and the uniformity of the web is improved, with drafting, down the set.

Drawing
Drawing consists of three passages through chain gills. Doublings are 4 to 1 on the first two passages and 2 to 1 on the third passage. Drafts are in the region of 8.

Spinning
Spinning is usually on apron draft frames fitted with Baxter flyers, but Mackie's range of jute spinning equipment includes ring spinning frames operating at speeds up to 8000 rpm and counts as fine as 140 tex.

3.2.7 Cotton Spinning

Cotton fibre is very short and is spun on a dedicated system not normally found within the carpet industry. Detailed information about cotton spinning is available from The Textile Institute's *Manual of Textile Technology* series [9].

3.2.8 Hand Spinning

Although most hand-knotted carpets are made from machine-spun yarns, a substantial minority utilises hand-spun wool yarns for the pile; coir yarns for flat-woven carpets and entry mats are mostly spun by hand.

The European and Indian styles of spinning wheel have been introduced in some of the newer areas of hand-knotting, but the much older spinning whorl (Figure 3.19) has survived the test of time in the traditional regions.

Whereas the hand-carded wool is drafted between two hands for the treadle/wheel-driven spindle to give yarn with the characteristics of worsted yarn, yarn from the whorl is drafted against twist after the fibre is drafted in the hands, so that it is somewhat like woollen spun yarn. Carded fibre is attached to a notch in the shaft of the whorl; the heavy whorl is then given a spin and released. Fibre is drafted and fed from the hands and is twisted and further drafted by the whorl as it falls. When the whorl reaches the ground, the spinner winds the

Figure 3.19. Spinning by hand using a whorl in Iran. (Photograph: Warren Meade)

newly created length of yarn on the shaft, fixes the end in the notch and repeats the whole cycle. (Mule spinning, to some extent, simulates this technique.)

A technique for producing woollen style yarn, which is of more recent origin, is used for spinning extremely heavy yarns for hand-woven rugs. The spinning operative takes an armful of scoured and opened (but not carded) wool and loops a staple over a hook that is rotated by an electric motor. She (the operatives are usually female) walks slowly away from the rotating hook, feeding the fibre into the point of yarn formation as she goes, and some levelling occurs as thick, loose places are drafted. When several metres of yarn have been produced, the material is detached and a further cycle is commenced. The relatively short lengths of yarn are adequate for producing weft-faced flat carpets on handlooms.

Coir yarns, as used in flat-woven carpets, are most commonly spun by hand by a similar method, except that the spindles are usually turned by hand by another operative. The fibre is first cleaned mechanically by passing over a small wooden swift, and is then spun by the armful into lengths of up to 10 m. Two- or three-ply yarns are produced by rotating the individual yarns and transferring the twist to the plied yarn in a controlled manner by means of a hand held tool that moves the point of convergence progressively towards the rotating ends. The plied yarn is spliced with longer lengths for use as warp.

REFERENCES

1. H Dietrichs. Interesting new application areas for BCF spin-draw-texturing technology. *Chemiefasern / Textilindustrie*, 1993, 43/95, 172–175 (E17–E19).
2. F Bäcker and K Schäfer. *Technologies for the Economical Production of BCF Yarn Today and Tomorrow*. Barmag AG: Developments, Trends, Technologies, 10/1995.
3. W Nabulon. *The Most Ideal Machine Features to Ensure High Quality BCF-Yarn*. Rieter Synthetic Fibre Machinery, Germany.
4. D K Wilson. Use of texturing jets to produce carpet face and backing yarns. In: *Proceedings of World Textile Congress on Floorcoverings*, University of Huddersfield, 12–13 July 1995, 215–219.
5. J M Kauffmann and G Münker. Direct cabling of low-finish BCF: environmental and productivity benefits. *International Carpet Yearbook*, 1998, 29–33.
6. R T D Richards and A B Sykes. *Woollen Yarn Manufacture*. Manual of Textile Technology, The Textile Institute, Manchester, 1994.
7. G W Pfeiffer. Up-to-date fibre preparation systems for the manufacture of floorcoverings. In: *Proceedings of World Textile Congress on Floorcoverings*, University of Huddersfield, 12–13 July 1995, 62–81.
8. J Fischer. The fast route to fine saxonies. In: *Proceedings of World Textile Congress on Floorcoverings*, University of Huddersfield, 12–13 July 1995, 54–61.
9. W Klein. *Short-staple Spinning*. Manual of Textile Technology, Vols 1–5. The Textile Institute, Manchester, 1987–1993.

4 Yarn Processing

Under the heading of yarn processing fall the steps that are essential for producing some carpet constructions or for particular processing routes. Chemical treatments that are designed to introduce or enhance particular carpet properties are usually carried out in conjunction with dyeing or printing, and are considered in Chapter 15.

The principal process that is applied to carpet yarns is setting of the twist of yarns that are to be used in cut-pile carpets. Many millions of kilograms of nylon yarns are set annually, and polypropylene yarns for the better grades of cut-pile carpets are also set. Wool yarns for cut-pile carpets also need to be set, but the mechanism of setting is somewhat different than that for synthetic fibres, and a special process is sometimes used.

Concurrently with setting, melt-bonding can be activated, creating a fibrous network that enhances the level of set, minimises fibre shedding in wear and improves the resistance of the carpet to crushing.

Bulking of polypropylene and acrylic yarns that are composed of shrinkable and unshrinkable fibre is also carried out on the same equipment as heat setting, although the process should be regarded as set-release, rather than setting.

Other common yarn processes include yarn scouring, which may also be regarded as a bulking process, and felting, a process unique to wool.

4.1 TWIST SETTING

If tuft definition is required in the finished carpet, the twist of the yarn must be set before the tufts are cut so as to prevent the cut ends from bursting excessively during subsequent processing (eg, piece dyeing or printing). Equally important, setting improves the retention of carpet appearance during wear and protects the texture against the effects of accidental spillage of liquids.

As fibre setting is essentially a stress relaxation process, best results are obtained if the fibres are deliberately stressed by applying unbalanced twist. Setting of highly twisted yarns to produce friezé carpet textures represents a particular favourable case.

Synthetic fibres can be set by means of heat alone, although steam is usually used as the heating medium. Wool requires heat and moisture to allow the rearrangement of disulphide and ionic cross-links in the protein chains.

4.1.1 Water Setting

Wool and wool/nylon yarns

The action of a boiling dyebath is adequate to partially hydrolyse disulphide cross-links in the matrix component of the wool fibre (see Section 2.7.1) and to weaken the ionic bonds, rendering the fibre plastic. Twist setting is achieved as a useful consequence of hank dyeing.

Stock-dyed wool yarns may be set in a similar way by boiling in a blank dyebath for at least 20 minutes. The bath is first set to pH 5 with acetic acid to inhibit bleeding of dyes and minimise fibre damage. To prevent yellowing of white or pale coloured wool in a mixture colour, hydroxylamine sulphate or other proprietary dyebath anti-yellowing agent may be used in the liquor.

For a very high level of set, 0.25 g/l sodium metabisulphite (to increase the proportion of free thiol groups in the wool) may be used as an alternative to acetic acid. After boiling with bisulphite, the yarn is rinsed and treated with hydrogen peroxide to neutralise residues of the reducing agent.

Acrylic yarns

The most convenient way of setting/bulking acrylic yarn containing high-shrink fibre is to immerse it in boiling water for 10 minutes in a Hussong-type dyeing machine. After bulking the yarn in this way, the bath is cooled and set for dyeing.

4.1.2 Batch Autoclave Setting

Autoclave setting of carpet yarn is most commonly carried out in the form of hanks (skeins). In this form, optimum bulk is developed. There are, however, a few cases where yarn is preferably set on cone or cheese, for example to prevent kinking of singles yarn or where a particularly compact yarn is required.

The processing sequence is shown in Figure 4.1. Yarn is reeled into hanks and is tumbled to develop bulk. Wool yarns may have been previously scoured so that bulk has already been developed and the tumbling step is omitted. Wool friezé yarns benefit from tumbling, which optimises the required kinking of the yarns.

An alternative to processing in hank form is processing in the form of tubular knitted fabric. The objective is to set the yarn in the curled configuration of the knitted loop. Such yarn gives a textured carpet pile, but one that is not as tightly curled as pile made from hard twist (friezé) yarn. After setting, the tubular fabric is de-knitted on an ordinary backwinding machine, but with the knitted fabric plaited in boxes alongside the machine.

The autoclaving conditions are critical, since variations in moisture content of either nylon or wool can result in variations in the level of set and dyeability. In

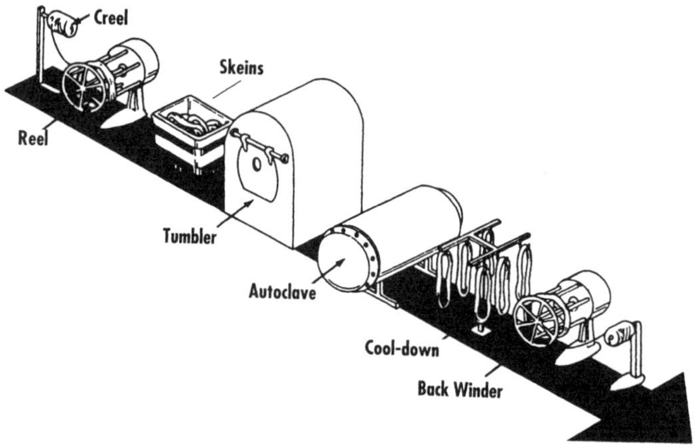

Figure 4.1. Schematic diagram of the processing sequence for batchwise autoclave setting. (Courtesy of Du Pont (UK) Ltd)

the extreme case, wet spots on wool yarn can result in hydrolysis and weakening of the fibre. Yarns should be conditioned to a uniform moisture content before processing, and precautions should be taken to avoid water coming into contact with the yarn in the autoclave. These include traps and baffles in the steam supply; a warm-up cycle to minimise condensation; a sloping false roof which will direct any condensation away from the yarn; and lightweight frames, which condense only minor amounts of moisture, to support the fabric trays that hold the yarn.

To ensure that steam penetrates the goods thoroughly, ensuring uniform treatment, air must be removed from the yarn. Alternate vacuum and brief steam cycles are used to remove air before the setting process itself. For some applications, double pre-vacuum/steaming is used to take into account the fact that steam condensing on the yarn entrains residual air during the first cycle, to some extent impeding the penetration of further steam. Steaming for the setting process itself has a duration of 5-10 minutes. Typical steaming temperatures are:

nylon 6.6	130°C
nylon 6, polypropylene	120°C (depending on the brand)
wool	105°C
wool/polypropylene	110°C
polyester	140°C
acrylic	105°C

A final vacuum cycle removes most of the condensed steam and partially cools the yarn. After the pressure has been equilibrated, the yarn is removed and allowed to cool completely before backwinding on to cones for tufting.

For polypropylene and acrylic spun yarns, the autoclaving treatment is more a bulking process than one of setting the yarn twist. Such yarns are commonly composed of blends of shrinkable and non-shrinkable fibres, and autoclaving releases the set of the shrinkable fibres, which concertinas the non-shrinkable component into a crimped form, bulking the yarn and locking the twist. Different brands of shrinkable polypropylene differ in the shrinkage forces they can exert at particular temperatures, so the manufacturer's recommendations for steaming should be followed carefully.

4.1.3 Continuous Autoclave Setting

The Superba TVP continuous pressure steamer is widely used for twist setting of carpet yarns. Figure 4.2 outlines the basic components of the machine. (In reality, the TVP machine is a twin-channel machine with sophisticated componentry that requires a scale drawing, too large for inclusion on the pages of a book, to describe it fully.)

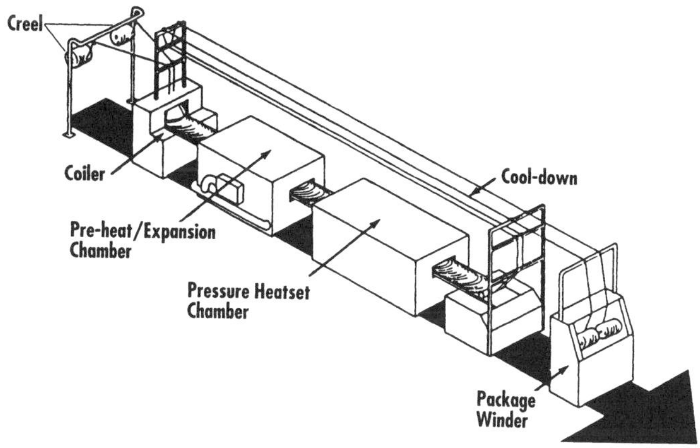

Figure 4.2. Schematic diagram of a continuous yarn autoclave. (Courtesy of Du Pont (UK) Ltd)

Multiple ends of yarn from a creel are coiled on to a perforated stainless steel belt which conveys the coiled yarn through the various treatment steps. The yarn is first bulked by steaming in saturated steam at atmospheric pressure before it moves to the autoclave. Specially engineered seals at the entry and exit of the pressurised chamber make the continuous autoclave a practical possibility. A useful optional accessory is the circulating blower system (CBS) which is designed to maximise heat transfer through the mass of yarn on the conveyor. The heating medium is saturated steam at a temperature (for nylon 6.6) of 132°C. After setting, the yarn passes through a dryer to remove condensed moisture. The coils are then transferred to an accumulator (not shown in the diagram) which enables setting to continue in a uniform manner when it is necessary to stop the winding unit for any reason. Uncoiling follows, and the warps of yarn are traversed over the length of the machine to condition them to a uniform moisture content before take-up on a multi-head winder.

Conditions for controlling setting on the Superba TVP machine, including the critical case of yarns destined for piece dyeing to solid colours, have been published in considerable detail [1]. Although the pre-bulker should, in effect, condition the yarn uniformly for the setting process, it is possible for the yarn to become locally wet. For example, condensed water can come into contact with the yarn during stoppages and create changes in dyeability. Water splashes from the steam saturator can also cause this defect. For full development of bulk, the steam must be saturated, however, since dry steam does not raise the temperature to the required 98–99°C in the time available.

Time is also required for the optimum temperature to be reached in the autoclave, but this can be shortened considerably with the CBS option which circulates the steam. As in the pre-bulker, steam should be saturated and air excluded, and steam pressure must not be allowed to fluctuate, since a drop in pressure can cause the coiled yarn to become saturated with moisture. It is also important that the interior of the autoclave and the transport system be well maintained in order to avoid snagging of filaments and consequent yarn irregularities.

At take-up, yarn packages should be wound to medium density. Too high a density leads to loss of bulk, while low density packages may become deformed in handling or transit.

For critical applications, it may be necessary to regard the yarns processed through the twin channels of the machine as separate lots.

The Superba TVP machine, because of its use of saturated steam, is suitable for setting wool yarns as well as synthetics.

4.1.4 Continuous Setting in Superheated Steam

A widely used machine for continuous setting in dry steam is the Hörauf-Suessen GVA yarn heat setting plant. For full setting, it operates at higher temperatures than are required in saturated steam, but can be used at lower temperatures for bulking high-bulk acrylic or polypropylene yarns, and dry-spun wool yarns. The dry conditions in the unpressurised steam chamber make it unsuitable for twist setting wool yarns. An important feature of the machine is that it allows complete freedom for the yarn to relax and bulk.

The general layout of the machine is shown in Figure 4.3. Four (or six) ends of yarn are reeled together on four moving ropes to form a continuous hank, and the yarn is transported through the steaming chamber in this form. The path of the tapes allows the yarns to shrink in the early stages of steaming. Figure 4.4 shows the cantilevered 'mast' with its transport system: the initial cross-section of the hank (A) is changed to (B) to allow relaxation to configuration (C) as the yarn passes through the steam chamber (GKK). Figure 4.3 does not show the fact that in the actual machine six masts are located in parallel through the setting chamber, so that a machine processes 4×6 = 24 or 6×6 = 36 ends of yarn simultaneously. The masts can be stopped independently, but it is not

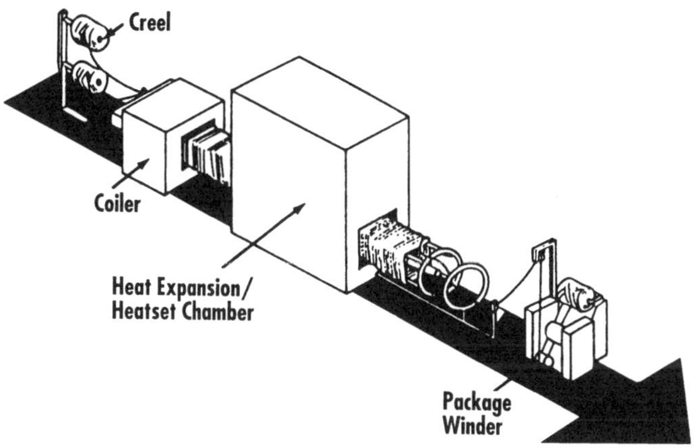

Figure 4.3. Schematic diagram of the Hörauf-Suessen GVA heat setting machine. (Courtesy of Du Pont (UK) Ltd)

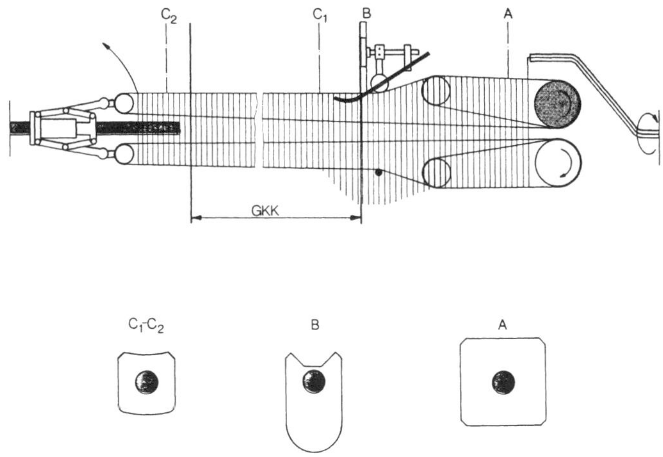

Figure 4.4. A 'mast' of a Hörauf-Suessen GVA machine, showing how the continuous hank is formed and how the configuration allows for yarn shrinkage. (Courtesy of Hörauf-Suessen)

necessary to stop a mast for an end break at the winder. Yarn simply accumulates at the end of the mast, and the winder accelerates on restart to take up the accumulated yarn.

Two general purpose machines are available: the GVA 2500 works with four or six ends of yarn per mast and the GVA 5000 is a high production machine taking six or eight ends per mast. The GVA 2500 PP is dedicated to bulking and setting polypropylene, BCF or spun, and has a different configuration for the ropes on the mast. The GVA Texturing System is for setting friezé yarns. Yarn from the creel passes into a texturing box that develops kinks, and transport through the setting chamber is via a conveyor belt. A mast-type yarn accumulator is located between the setting chamber and the take-up winders.

4.1.5 Continuous Bisulphite Setting of Wool

The WRONZ Twistset™ (Chemset™) process[2] is carried out continuously on a machine available from ADM Group (formerly Annett and Darling), New Zealand. Yarn from a creel is coiled on to twin mats (18 or 24 ends of yarn in each coil, depending on the model of machine) on a single perforated conveyor that transports them through four wet processing bowls, each with a squeeze head. The coiled yarn is then dried, passes to twin accumulators (to enable the machine to operate continuously during doffing, etc), is uncoiled and wound by multi-head auto-doffing winders. The layout of the machine is illustrated in Figure 4.5.

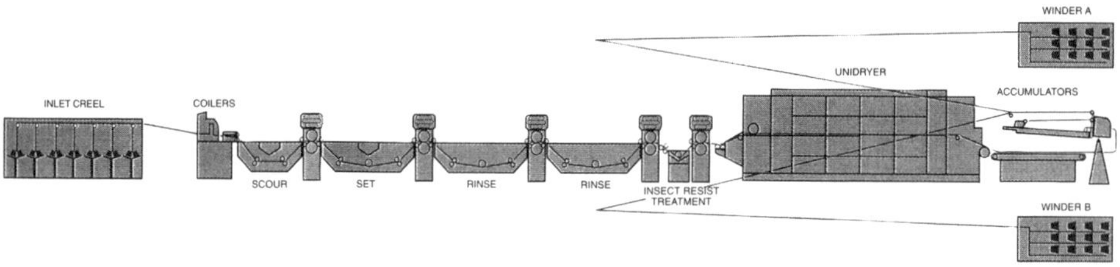

Figure 4.5. Components of the ANDAR / WRONZ Twistset™ machine.

The first bowl of the machine is used for scouring; the second contains typically 10 g/l sodium metabisulphite adjusted to pH 7; and the third and fourth bowls are used for rinsing away residual bisulphite. Treatment with bisulphite is for 2 minutes at 85°C.

The function of the bisulphite is to increase the level of ionised thiol groups in the protein of the wool matrix, facilitating thiol-disulphide interchange between the protein chains. Molecular rearrangement takes place through an 'unzipping/ rezipping' action, allowing twist stresses to relax.

An important technical advantage of the Twistset™ process is that a fairly high level of unbalanced twist (fibre stress) can be applied without yarn kinking occurring, because the yarns are lightly constrained on the conveyor during the process. This enables a higher level of twist set to be achieved than when processing in hank form, where the yarn is not constrained in this way. Another technical advantage is that bisulphite setting provides a mild bleaching action, brightening colours, whereas steam setting tends to dull colours. Claimed processing advantages of the Twistset™ process include avoidance of multiple handling of yarn, fewer yarn joints and excellent uniformity of treatment.

4.1.6 Setting Wool Yarns During Scouring

By modifying existing tape, brattice or suction-drum scouring sets (see Section 4.3.1), processing conditions analogous to those pertaining in the WRONZ Twistset™ process can be created. To achieve an adequate residence time, scouring and bisulphite treatment are carried out simultaneously, using the first two bowls.

Setting that is adequately uniform for stock-dyed yarn can readily be achieved. Mixture colours are advantageously brightened.

4.2 FUSING OF MELT-BONDING FIBRES

Yarns from all the common pile fibres can be engineered for enhanced performance of cut-pile carpets by making use of heat-fusible fibres to create a fibre network within each tuft (see Section 2.5). The benefits claimed include:

- enhanced twist set (durable tuft definition in cut-pile products);
- improved resilience;
- increased yarn bulk (better cover of the carpet backing);
- reduced fibre shedding.

Selection of fibres and processing conditions are usually such that the bonding step can be carried out in conjunction with a normal twist setting operation.

The concept of applying melt-bonding fibres, which were originally developed for other purposes, to stabilise carpet yarns emerged around the same time in the nylon and wool industries [3].

Wool
A typical wool-rich yarn for bonding is composed of 80% wool, 10% conventional nylon and 10% bicomponent polyester fibre. The bicomponent polyester (eg, Unitika Melty Type 4080 or Hoechst Type K54) has a sheath with a fusing temperature of 110°C (dry) and a sticking temperature of 80°C (wet), selected to minimise yellowing of wool during fusing. Recommended fusing conditions include batch autoclave for 3 minutes at 110°C; continuous autoclave for 1 minute at 116°C; WRONZ Twistset™ with bowls 3 and 4 at 90°C; or hank dyeing at 98°C. Fusing can be accomplished in dry steam, but this is not recommended because it does not set the wool.

Nylon
Sheath/core fibres can also be used for bonding nylon yarns, and versions with melt temperatures of 110 and 130°C are equally suitable. Alternatively, monocomponent polyamide bonding fibres may be used. Whereas only the minor nylon component of a wool-rich yarn is available for bonding with the fusible fibre, opportunities for bonding a 100% nylon yarn are greater, and the monocomponent fibre gives effective results. Fusing may be carried out in saturated steam in batch or continuous autoclave, or in superheated steam. Fusing cannot be accomplished in the dyebath.

Polypropylene
Polyolefin melt-bonding fibres may be 100% polyethylene [4] or copolymers of polypropylene and polyethylene [5]. For example, Drake Extrusion's (formerly Plasticisers Ltd) Fixset™ fibres soften and bond to themselves during steam setting in the temperature range 120–130°C. As they fuse, the fibres shrink up to 30%, so that they can also be used to create bulked yarns. The actual level of shrinkage achieved depends on the resistance to contraction, as well as on the

temperature, so that yarn construction has a significant influence on the development of bulk. As well as 100% polypropylene yarns, polyolefin bonding fibres can be used to bond yarns of other fibres.

4.3 OTHER YARN PROCESSES

4.3.1 Yarn Scouring

Carpet yarns can often benefit from a scouring process even though spinning can be carried out with a very low level of a fibre processing aid that has been chemically engineered so as not to promote back-soiling in the carpet, and good housekeeping in the spinning mill can minimise accidental soiling. In the case of wool, levels of woolgrease (the principal residue that detracts from the good soil resistance of the clean fibre) are usually reduced, and the hot wet treatment relaxes fibre strains introduced at spinning, allowing fibre crimp and yarn bulk to develop. An additional reason for scouring in the case of stock-dyed yarns is to remove loose dyestuff, particularly from the darker colours. If yarn scouring is to be part of the processing routine, carding of wool can be carried out with higher levels of fibre processing aid, enhancing yields of sliver or slubbing.

Scouring in the dyeing machine

Provided they are not heavily soiled, synthetic and wool yarns that have been spun using a water-soluble fibre processing aid can simply be given a light scour in the hank dyeing machine prior to dyeing.

A typical procedure is to set the batch at 60°C with 0.5 g/l biodegradable (aliphatic) nonionic detergent, enter the yarn and circulate the liquor for 20 minutes. The yarn is drained thoroughly and rinsed before dyeing.

Tape scour

A tape scouring set is designed to continuously process wool yarns in hank form at rates up to 1000 kg/h. The machine consists of four (sometimes five) steel bowls, each with its own squeeze head, mounted in line. Two sets of tapes travel through the machine, one set returning over the machine and one under it. The upper and lower tapes are located alternately and not in pairs, so as not to restrict access of liquor. The lower tapes form a table at the entry end: hanks are fed on to this section at a uniform rate and the upper tapes move down to grip them lightly and transport them through the bowls and squeeze heads. Much of the scouring action takes place in the nips of the squeeze heads.

Typical processing conditions are:

Bowl 1 Detergent 0.25 g/l; sodium bicarbonate 2.5 g/l; temperature 55°C.

Bowl 2 Detergent 0.25 g/l; sodium bicarbonate 1.5 g/l; temperature 50°C.

Bowl 3 Water at 45°C.

Bowl 4 Water at 40°C.

For longer runs and greater uniformity in scouring action, a flowback system from bowl 4 to bowl 3 to bowl 2 to bowl 1 may be used. A heat exchanger recovers heat from bowl 1 liquor and applies it to fresh water fed to bowl 4.

The above procedure is capable of removing residual grease, spinning lubricant, mineral soil and protein debris from a wool carpet yarn having a moderate level of residual grease (eg, containing some unscoured slipe wool). If the original yarn is low in grease and soil, the bicarbonate may be omitted. A densely constructed yarn with a high level of residuals may require sodium carbonate, rather than bicarbonate.

Scouring should be thorough and uniform, otherwise streaks may be apparent in plain carpets. The scouring procedure should, therefore, err on the side of severity.

Brattice scouring
An alternative to a tape scouring machine is one in which the hanks are transported on a metal brattice. Liquor is jetted on to the hanks so the mechanical action is more severe than in a tape scour. A brattice machine may be regarded as a heavy duty scouring machine.

Package-to-package scouring
The ANDAR/WRONZ package-to-package yarn processing machine (see Section 4.1.5), designed initially for twist setting of wool carpet yarns, is also highly suitable for just scouring. It provides all the technical advantages of scouring without the yarn handling costs and other negative features associated with processing in hank form.

4.3.2 Felting of Wool Yarns

Felted wool yarns have been produced for many years, usually by spinners using equipment they devised themselves, to allow carpet styling possibilities to be extended to include the use of exceptionally heavy counts of yarn. Core-spun (see Section 3.2.4) singles yarns are among the constructions that are normally felted. Conventional multi-ply yarns up to R3000 tex intended for felting are usually spun and folded using reduced levels of twist. Particularly bulky yarns may be produced by twisting slubbings or rovings together and felting them. Woollen spun yarns often respond to felting better than semi-worsted yarns, depending on construction.

Yarn felting provides a more cohesive tuft end in cut-pile carpets, reduces fibre shedding and often improves yarn bulk. In some respects, it is analogous in function to twist setting, but for a heavier count range.

Felted yarns around 10 mm diameter, usually core-spun, are used in hand-crafted flat-woven carpets in a number of countries. In Europe, felted yarns are used in coarse loop-pile wireloom carpets and in components of the design in Axminster carpets having patterns of texture, rather than patterns of colour.

Felting of wool materials depends basically on the ratchet-like scales on the surface of the fibre (see Section 2.7.1). Key factors in accomplishing felting are moisture, to facilitate extension and recovery of the fibres, and mechanical action, which causes fibres to move towards their root ends, consolidating the structure. Additional factors favouring felting are heat, alkaline pH (or very acid pH) to plasticise the fibre, and a lubricant (eg, soap).

Figure 4.6 illustrates fibre migration (a to b or c to d, depending on fibre orientation) during felting. A wool fibre readily slips past other fibres when moving in the direction of its root because the scale edges do not catch. On the other hand, a fibre forced to move in the direction of its tip will have its scale edges caught by other fibres oriented in the reverse direction and slippage will be impeded. The result of these differences in the amount of the slippage will be either a moving together of fibre contact points (Figure 4.6 (a) and (b)) or fibre buckling (Figure 4.6 (c) and (d)), leading to increased fibre entanglement and compaction, and shrinkage of the structure

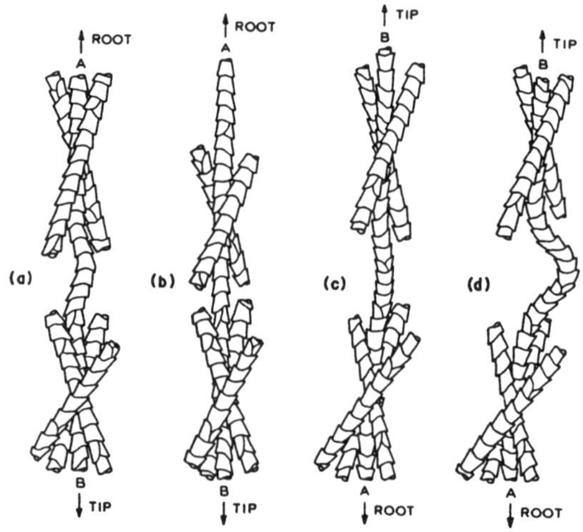

Figure 4.6. Shorter's mechanism of felting [6].

Formerly, hanks of yarn were milled in warm soapy water, loose in milling stocks (mechanical action provided by wooden hammers) or in strings in wooden dollys (mechanical action provided by wooden rollers). Such processes tended to give rapid and uncontrolled felting so, in the 1970s when demand for felted yarns was strong, a number of processes more amenable to control were developed.

Solvent batch
The use of a horizontal-drum solvent processing machine was patented by Böhler and Weber in 1965. Hanks of yarn with loose tie-bands are first scoured in pure perchloroethylene; fresh solvent is introduced and a measured amount of milling solution, consisting of water and lubricant/emulsifier, added; the machine is run for a specified but short time (eg, 5 minutes); the liquor is dumped and the yarn is hydroextracted, dried at 70°C and deodorised by circulating cold air. The machine incorporates environmentally friendly features.

The mechanical action is provided by raising hanks in turn on the baffles of the drum and dropping them, loaded with heavy perchloroethylene, on the rest of the batch. Control is provided principally by the measured amount of moisture in the milling solution, with time as a secondary factor. A microprocessor is programmed to control the cycle.

Aqueous batch

A horizontal laundry washer is used in a process developed jointly by Thos Broadbent and Son and Stephenson Thompson Textile Chemicals. The felting step in the cycle is with 2 g/l Sodalkon 215 K at 40°C, for 3–10 minutes. Rinsing and drying follow. Once a satisfactory cycle for a particular yarn quality has been established, the cycle is programmed so that it can be reproduced.

Continuous process

The continuous Periloc™ machine was developed jointly by the Vezelinstitut TNO and IWS in the 1980s. Yarn is impregnated with felting medium (eg, 0.5 g/l Sodalkon 215 K) and passes through a flexible tube where it is subjected to mechanical action by a rotor. The yarn contracts lengthwise and bulks to fill the tube. Untwisted sliver, rovings or slubbings can also be processed on the Periloc™ machine, with the aid of a false-twist unit.

The Periloc™ machine was originally manufactured in the Netherlands, but a decline in the fashion for felted yarns made its future uncertain.

Another continuous machine, the Tatham Woollen Yarn Processor, manufactured in limited quantity in the early 1980s, was originally developed by WRONZ. Wet slubbings are passed through a cascade of horizontal rubbing aprons where they are consolidated into a firm structure before drying in-line.

REFERENCES

1. R Pike. Carpet yarn processing from start to “finish”. In: *Proceedings Tifcon ’95 – 25 Years of Carpet Technology*, The Textile Institute, Manchester, 1995.
2. *The WRONZ Chemset Yarn Processing System*. WRONZ Techline No T1, 1988.
3. J D Watt. Stabilising wool carpet yarns by melt bonding. In: *Proceedings Tifcon ’93 – Covering Europe*, The Textile Institute, Manchester, 1993.
4. R J Legare, R K Gupta, R E Kozulla and G I Rehberg. Polyethylene binder fibres for carpets and related applications. *Carpet and Rug Industry*, 1996, 24(6), 46–48.
5. D G Ellis. Polypropylene in household textiles and floorcoverings. In: *Proceedings World Textile Congress on Polypropylene in Textiles*, University of Huddersfield, 9–11 July 1996.
6. S A Shorter. *Journal of the Society of Dyers and Colourists*, 1923, 39, 274–275.

5 Backing Materials

The backing of a carpet is an essential component and must be correctly constructed if the carpet is to perform satisfactorily in use. In carpet weaving the backing fabric and the pile are formed concurrently, whereas tufting utilises pre-constructed backing fabrics.

5.1 WOVEN CARPETS

The backing structure of a woven carpet consists of chain warp, weft and stuffer warp. Face-to-face carpets do not usually have stuffer yarns in the warp: if they do, yarns similar to chain warp are used. The chain warp and the weft interlaced together form the basic backing fabric; the weft also holds the pile tufts in place. Useful supplementary contributions to the dimensional stability of the carpet and weftwise positioning of the chain warp yarns and pile are provided by the stuffer warp yarns that lie straight in the backing. Typical weave structures are illustrated in the chapters dealing with particular weaving processes.

5.1.1 Chain Warp

Each pair of warp ends is constantly interchanged during the weaving process and is also subject to contacts with the stuffer and pile yarns. The key properties required of a chain warp are:

- Flexibility and high tensile strength to permit reliable working in the loom.
- High energy absorption, enabling it to accept peak forces during weaving.
- Low hairiness, to prevent excessive interaction during the shedding action.
- Fairly high inter-yarn friction, contributing to the binding of the overall structure.

It is usually necessary to provide two separate supply beams of chain warps to the loom because, except in most wireloom constructions, the upper and lower warps do not follow symmetrical paths and sometimes are let off at different tensions so as to achieve one relatively straight warp and one that is sinusoidal. Some face-to-face carpet structures have trios of chain warps, rather than the usual pairs (one is straight and alternatively may be regarded as a stuffer warp). Beams are wound having the respective lengths of yarn required for the upper and lower (and possibly middle) warps.

The most commonly used materials are cotton, cotton/polyester and polypropylene. Polyester filament is occasionally used. Jute yarns, commonly used for weft and stuffer warps, are only rarely used for chain warps because they are excessively hairy, virtually inextensible and provide lower tensile strength than other readily available materials.

Cotton
Cotton in 100% form has been the traditional chain warp for all types of woven carpet for many years. Linear densities range from R205 tex/3 (Nm 12/3) to 170 tex/4 (Nm 24/4), depending on the type of carpet to be woven.

At beaming, cotton yarns are sometimes sized to reduce hairiness, thereby facilitating weaving. Sizing, if any, should be light so as to avoid creating too sleek a yarn that does not bind the structure firmly.

Cotton/polyester
Polyester is often blended with cotton[1] for warp yarns in order to improve spinning efficiencies, especially in the finer counts, and to increase yarn strength, thereby reducing the risk of breaks at weaving. Cotton/polyester chains may be treated by the weaver in a similar way to 100% cotton yarns.

Polypropylene
Polypropylene chain warps may be fibrillated tape or roller-embossed film (see Section 5.2.1), lightly twisted (60 tpm) to give a low cost round yarn.

Yarn linear densities are usually in the range 90–133 tex. Round fibre is not used because it does not grip the wefts as well as tape. To obtain increased dimensional stability, heat-set yarn may be used, although in Axminster and Wilton carpets the stuffer warp is the prime determinant of dimensional stability in the warp direction. Polypropylene chain warps are not normally used in face-to-face carpets, where there is often no stuffer warp.

Polyester
Multi-filament polyester is favoured as chain warp in some carpet constructions, notably to meet flammability specifications where polypropylene is unacceptable. Yarn linear densities around R92 tex are used, twisted to about 100 tpm. A particular advantage of polyester is that it packs densely, so that longer warps may be beamed, providing longer runs between beam changes. Polyester is strong, resists abrasion in weaving and does not catapult if a warp yarn breaks.

5.1.2 Weft

Clearly, the weft of a carpet should have adequate strength to hold the carpet together, but since yarns usually have more than adequate strength, weavability is a more important criteria. Dimensional stability is the key carpet property provided by the weft.

Jute yarns have proved highly satisfactory as weft because of their high modulus and low extensibility. Improved spinning techniques have enabled the traditional 500 tex/2 yarns for Axminster and Wilton carpets to be replaced by singles yarns in the range 344–550 tex. For the finer face-to-face yarns two-ply jute yarns in the range 140 tex×2 to 560 tex×2 are preferred. Linen is sometimes used in high quality traditional Wilton carpets.

Polypropylene tape yarns in the range 133–444 tex are used as weft in Axminster carpets; some qualities are twisted, some not. To ensure adequate dimensional stability in the carpet it is necessary for the polypropylene weft to be heat set. Setting may be achieved by autoclaving the yarn on spools, processing under tension at 120°C, or by a sequence of hot air, hot (200°C) godet rollers and cold godet rollers at the stage of tape manufacture. Polypropylene is increasingly used in Axminster carpets as weft, but only to a small extent in face-to-face carpets.

5.1.3 Stuffer Warp

The stuffer warp is held straight during weaving with a view to providing dimensional stability in the warp direction.

Jute yarns, similar to those used for weft (perhaps identical), are highly satisfactory as stuffer yarns. Singles or plied yarns are used.

Polypropylene tape or split film yarns are used in the range 133–500 tex, and must be twisted to minimise abrasion of the chain warps during weaving. Heat setting is essential for good dimensional stability.

Cotton yarns, identical to those used in the chain warp, are sometimes used as stuffer yarns. Glass yarns are used in aircraft carpets to satisfy the stringent requirements for dimensional stability and flammability.

5.1.4 Selvedge Yarns

Melt-bonding fibres may usefully be incorporated in the selvedges of the carpet. They fuse in the dryer during back-coating to give a strong selvedge that does not fray.

5.2 PRIMARY BACKING FOR TUFTED CARPETS

Primary backings have the essential function of constituting the sheet material on which a tufted carpet is based. The backing must support the tufts as they are formed, facilitate the formation of tufts in a uniform way and grip them adequately so that they do not become disturbed during any processing or handling taking place before the application of adhesive. A primary backing necessarily suffers loss of physical properties from the action of the tufting needles but should retain adequate strength, and the structure after needling should contribute positively towards good dimensional stability of the finished carpet. Advanced tufting technologies, especially crossover tufting (see Section 7.2), place greater demands on the primary backing fabric than plain tufting.

Lubrication for the tufting process is provided from the primary backing fabric, whereas in other stitching processes (knitting, sewing) the lubricant is applied to the yarn. In addition to these basic requirements, special backing fabrics are engineered to meet the needs of particular carpet structures or processing routines. The backing fabric must be free from irregularities that could distort the structure of the pile. Price is an important consideration: the industry is more inclined to cost a carpet on the basis of the visible pile than the hidden backing.

Backings may be woven or nonwoven (eg, thermobonded, spunbonded). Woven primary backing fabrics are the more popular, with a market share of more than 75%. Nonwovens tend to be selected for specific purposes.

5.2.1 Woven Primary Backing Fabrics

The cotton canvas primary backings first used by the tufting industry have disappeared from general use, and survive only in the gun-tufting industry. Jute primary fabrics of 360–450 g/m^2 having a broken twill weave may also be regarded as part of the industry's history, although they are still used in the jute producing areas of the world and elsewhere where they may fill a niche for tufted carpets composed entirely of natural fibres. The structure of woven jute primaries is liable to deflect needles, a problem which was not serious when relatively coarse carpets were the most common, but which leads to major defects in the finer gauge tufting that is now the mainstream of the industry. Primary fabrics woven from polypropylene tapes overcame many of the technical problems associated with jute and were rapidly accepted once they had been engineered to meet the requirements of the carpet industry.

Production of tapes

Polypropylene tapes are produced by slitting film or by roller embossing. The first (and most widely used) process consists of extruding film, cooling it and slitting it into tapes by drawing it over standing blades using a multiple roller system [2]. Molecular orientation, which results in an increase in strength, is achieved by drawing the tapes through an oven using a high-speed roller system; stabilising (annealing) the material often follows in the same line. The tapes are taken up on cross-wound packages weighing up to 10–12 kg.

The second method (known as REF – roller embossed film) has two variants. In the hot method, the extruded film,while still hot, is embossed into a parallel rib structure by passing it between two pressure rollers heated to a temperature above the crystalline melting point of the polymer. The embossed film is then stretched in an oven, where the thin membranes between the ribs rupture: molecular orientation is achieved concurrently. In the cold process, the extruded film is chilled and then embossed by hot rollers but at a temperature below the crystalline melting point of the polymer. Higher speeds can be used in the cold process.

Polymer from both the slitting and embossing processes is so well oriented that the tapes tend to fibrillate, which facilitates penetration by the tufting needles.

Tapes from the REF process may be profiled with micro-ribs, which is also believed to benefit the tufting process.

Weaving of polypropylene tapes
Backing fabrics are woven [3] in plain structures, either on projectile machines or handover rapier looms having widths of 4300, 4600 or 5450 mm. Dobbies are not required and tappet motions are sufficient. Weaving from beam, rather than creel, is recommended and weft feed units are required on projectile looms to supply the long weft shots reliably.

Lubrication
A low level of spin finish is applied to the tape to prevent electrostatic problems during its production, but this does not necessarily provide adequate lubrication. Low tape/metal friction is essential for the tufting step [4]. Fabric that has not been lubricated is seriously damaged by the action of the tufting needles and loses about 60% of its strength. Application of about 1% of a suitable lubricant reduces the strength loss to 10–20%. Although the lubricant may be applied to the fabric, it is more conveniently applied during tape manufacture, and weaving efficiencies benefit too.

Dimensional stability
Polypropylene tapes as conventionally produced can shrink up to 8% under the influence of heat (ie, during carpet finishing or conversion of carpet into tiles), so precautions must be taken to produce backing fabrics that are dimensionally stable.

Stability to heat may be achieved by heat setting the fabric or the tapes before weaving. 'Zero shrinkage' tapes can be produced by modifying the drawing procedure so that the material is annealed in line or by a separate heat setting process. Strains, however, may be introduced at weaving so very stringent demands for dimensional stability may require both weaving from zero shrinkage tapes and heat setting of the fabric on a stenter.

Selection of backing fabric
A wide variety of constructions (all in plain weave) are offered by suppliers of woven polypropylene backing fabrics, and it is important to select one appropriate to the carpet being manufactured, since the surface appearance of the carpet may be affected by the backing material [5]. Irregular orientation of tufts resulting from the incorrect choice of backing material is particularly apparent in plain carpets.

Patterning or moiré effects, in the form of waves in the pile surface, may result from the superimposition of two grids, the two grids being the structure of the backing fabric and the tufted structure formed by the rows of needles and the stitch rate. Such interference patterns are well-known phenomena in optics. Generally, moiré effects can be avoided by selecting a backing fabric having a pick density that is greater or less than 10% of the stitch rate (eg, a fabric construction having 51 picks/dm could exhibit moiré effects when tufted in the range 45–56 stitches/dm).

'Needle deflection' by the backing fabric can result in uneven pile height, which may be exhibited as holes or rills in the pile, and perhaps even 'grinning' of the backing fabric through the pile. Such problems are more often caused by the deflection of the tape by the tufting needle than by the deflection of the needle by the tape. Ideally, a needle should penetrate the tape at the point of contact. However, it is possible that the needle may be deflected so that it penetrates between the tapes, to give a tuft that lies at a different angle from its neighbours. More likely is the possibility that the needle shifts the tape, usually the narrower and more easily moveable warp tapes. When the tufting needle withdraws, the tape tends to return to its original position, moving the tuft and creating the same effect as needle deflection. Because such effects are most likely to occur in the warp direction, surface defects in the form of long rills are usually parallel to the warp.

Bowing and skewing of woven backing fabrics during handling after tufting can be a problem when manufacturing or printing carpets in geometric designs. The selection of a soundly constructed fabric, coupled with appropriate carpet guidance mechanisms, can usually obviate such problems.

Recommendations for the selection of woven backing fabrics for different carpet constructions are given in Figure 5.1.

Speciality backing fabrics

Speciality backing fabrics are available that are designed to overcome the problems outlined above, as well as for other reasons.

Fibrillated tapes. Needle penetration may be facilitated and needle deflection problems obviated by fibrillating the tape. Fibrillating may be accomplished mechanically by means of a rotating needle roller, although other methods, such as the use of air jets, blowing agents and pre-embossing, have been proposed. Fibrillated tapes are widely used, for example Don and Low's Donbac™ range of primary backings has weft tapes fibrillated as standard, and constructions with fibrillated tapes in both warp and weft can be supplied on request.

Profiled tapes. Tapes embossed with micro-ribs (eg, REF) are more likely to be penetrated by tufting needles at the point of contact rather than deflected. They are also susceptible to fibrillation during processing, which is desirable.

Pre-needled fabrics. Standard fabrics may be subjected to the action of felting needles on a conventional needling loom (see Section 13.6) at an intensity of 60–90 penetrations/cm^2. Tufting needles can penetrate pre-needled fabrics more easily and the incidence of tape deflection is decreased. However, the fabric is a little more susceptible to skewing and bowing because of the deliberate damage created by the action of the felting needles.

Fibre-capped fabrics. Fibre fleece of 30–50 g/m^2 is needled on to the surface of the woven polypropylene fabric. The surface fibre is often offered in a limited range of subdued colours, and is designed to eliminate the risk of grinning of the backing through low density pile. Alternatively, the fleece may be dyeable

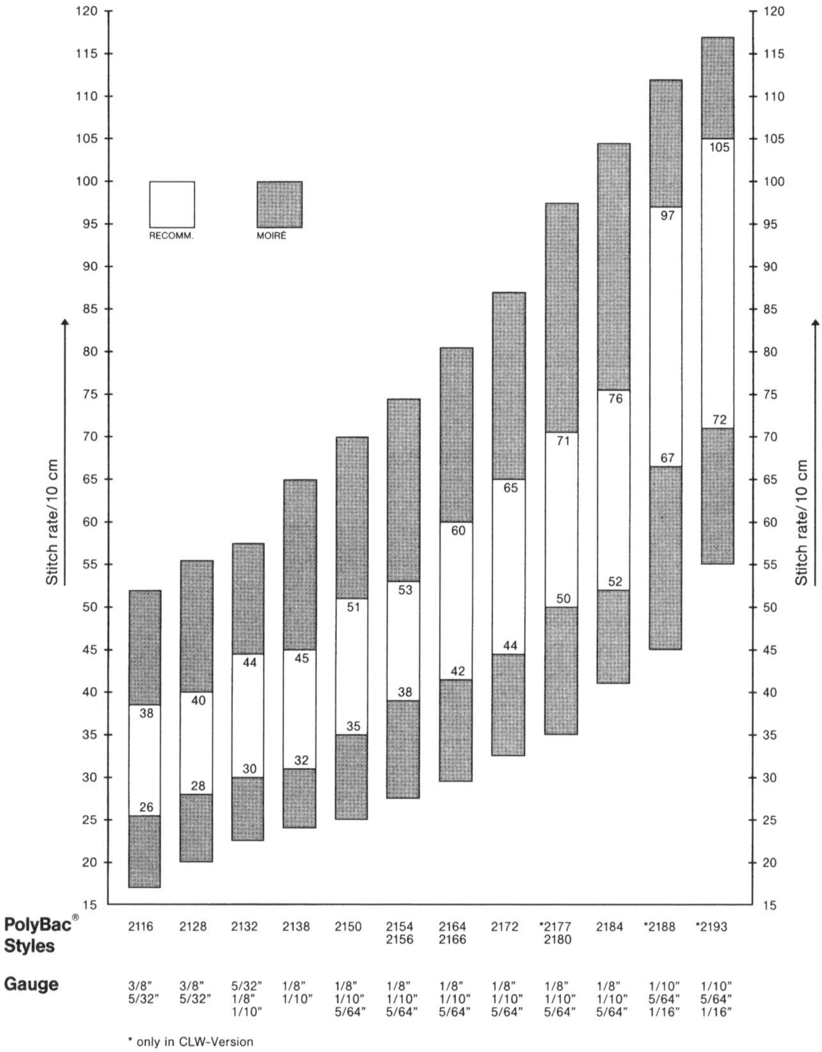

Figure 5.1. Recommended stitch rates for avoiding moiré effects in various gauges. (Courtesy of Amoco Fabrics)

fibre that takes on the colour of the pile in piece dyeing. Fibre-capped fabrics facilitate uniform tuft placement, and the presence of the surface fibre may make a positive contribution in this respect. The surface fibre also helps grip the tuft in an upright configuration.

A relatively recent development in fibre-capped fabrics is to use about 150 g/m^2 of melt-bonding fibres as the fleece material [6]. The objective is to be able to create carpets that are free from latex and filler and are, therefore, more acceptable for recycling (see Section 5.4). The primary fabric is used inverted relative to the older established fibre-capped fabrics (ie, the back stitch is located in the fleece). Such fabrics are intended to be used in conjunction with secondary fabrics also bearing a layer of fusible fleece. At finishing, the materials are brought together with the fusible layers facing each other; the material is heated from the back to 140°C to fuse the binder fibre; and the assembly is laminated by pressure rollers, concurrently binding the tufts and pile fibres.

Calendered fabrics. A different approach to producing a more isotropic fabric that does not give rise to problems of tape deflection is to calender woven polypropylene fabric at 120–140°C, flattening the surface and closing the gaps between tapes. A further level of sophistication is to bind a layer of coloured or dyeable fleece to the surface of the woven fabric by calendering, rather than needling.

Calendered fabrics have enhanced geometrical stability, making them less susceptible to skewing and bowing, and edge fraying. In cases where pile yarns have low frictional properties, calendered fleece products may be used reversed so that the fleece surface can better grip the tufts.

A problem of calendering material in the 4–5 m plus widths associated with tufting is to achieve uniform pressure across the fabric. The Hot-S Roll of Küsters is advantageous in this respect: it utilises the well-established principle of Küsters' mangles in using internal hydraulic pressure uniformly across the nip (see Section 13.9).

Substrate for artificial turf. Textile sports surfaces and other forms of artificial grass (see Section 21.7) must have exceptionally high resistance to UV degradation, and this requirement extends to the backing. Special woven backing fabrics containing elevated levels of UV stabiliser are offered for this purpose: they are usually black (so that they do not grin through the green pile) or may have a brown fleece surface.

Antistatic backings. Conductive yarns (see Section 2.6) may be incorporated in the warp, spaced to give the required electrical properties, as an alternative to using conductive latex in the production of antistatic carpets.

5.2.2 Nonwoven Primary Backing Fabrics

A number of alternative processes for manufacturing nonwoven primary backing fabrics are available: needling followed by adhesive bonding, needling followed by thermobonding, thermobonding by the use of bicomponent fibres, and spunbonding. All have the advantage of giving an isotropic product.

The principal merits of nonwoven primaries are that a single fabric is suitable for a very wide range of fabric constructions; there are no problems of tape deflection so the materials are particularly suitable for fine gauges; the products are dimensionally stable and are not susceptible to skewing or bowing; and the fabrics do not fray at the edges. A negative aspect is that the needle holes do not close up well after tufting so that tuft grip is not as good as with woven primaries.

Polypropylene and polyester are the principal fibres used in nonwoven backings.

Polypropylene nonwoven primaries are generally used for residential tufted carpets and some contract products, for example, Du Pont's Typar™.

Polyester nonwoven primaries provide all the advantages of polypropylene nonwoven primaries and have the additional merits of resistance to temperatures up to 180°C, better flammability classifications and exceptionally good dimensional stability. Advantage is taken of these characteristics in the manufacture of automobile carpets, tiles and contract carpets [7].

Automobile carpets need to be moulded to the shape of the floor-well and other three-dimensional components of the interior. A common procedure is to use carpet backcoated with polyethylene or other hot-melt material. Panels are cut to shape; the backcoating is softened using infrared heating; and the panel is pressed into a mould having the shape of the relevant automobile interior component, where it cools. Carpets tufted on a polyester primary backing are easily moulded and are not damaged by the thermal process involved.

The manufacture of carpet tiles involves the application of a heavy polymer coating under hot conditions, so that the thermal stability of polyester primary fabrics is advantageous. Equally important, carpet tiles need to have high dimensional stability if they are to lie flat, and polyester nonwoven primary fabrics contribute positively to this requirement. Non-fray properties are important, considering that each 500 mm tile in an installation has 2 m of seam length.

Although polypropylene primary fabrics are entirely adequate for the residential broadloom market, nonwoven polyester primaries are increasingly used for contract markets. In large installations, small changes in dimensions and distortion of designs are very apparent, so the outstanding dimensional and geometric stability of nonwoven polyester is valued. Seams need to be stronger in contract installations, too, and the non-fraying properties of nonwovens provide a good basis for sound seams. Of decisive importance, some contract installations specify flammability requirements based on radiant panel tests that subject the backing to intense heat. For example, when a range of nylon pile carpets was subjected to DIN 4102 Part 14 Radiant Panel Test, one tufted on a polypropylene primary showed a critical radiant flux of 0.35 W/cm^2, whereas a carpet tufted on a polyester primary showed a value of 0.64 W/cm^2, placing it comfortably above the limit of 0.45 W/cm^2 for building material Class B1 [8]. In the residential market, heavy nonwoven polyester is favoured for bath mats, which are required to be washable and to lie flat.

The Lutradur (Freudenberg) range of polyester backing fabrics is spunbonded (spunlaid), that is, the material is made direct from polymer chips. Filaments from the spinneret are laid on a conveyor belt in the form of a web and are bonded by heat. The technique of laying gives a random distribution of filaments and, therefore, an isotropic backing material.

5.3 SECONDARY BACKING FOR TUFTED CARPETS

Most tufted carpets made in the USA and a high proportion of European carpets have a secondary fabric layer. Foam backings are extremely popular for residential carpets in Europe, and some contract carpets that are intended for direct-stick installations have a 'unitary' backing (a generous layer of unfoamed latex).

The technical reasons for laminating a secondary backing fabric to tufted carpet are to enhance its dimensional stability, improve its strength and contribute to its weight. Knitted carpets without a secondary fabric are quite unstable. Needled products sometimes have a secondary fabric laminated to them, but this is not essential, except in the case of bitumen coated tiles, when the backing is essential to ensure cleanliness of the pile during handling, transport and installation.

Plain weave jute fabrics of around 270 g/m^2 were the original secondary backing fabrics. Although they are technically satisfactory (except perhaps in terms of rot resistance) and have a 'green' image, they have lost the major market share to woven polypropylene on the grounds of availability, cost and product variety. A technical advantage is that they remain flexible during carpet laying in unheated buildings.

5.3.1 Woven Polypropylene

An important function of a secondary backing fabric is to upgrade the appearance of a tufted carpet and raise its status towards that of a woven carpet, and weavers of polypropylene have capitalised on this aspect. Secondary backing fabrics are now designed as much for appearance as for functional reasons. Some carpet manufacturers wish to associate their label with their own style of backing; and in Germany customised backing weaves/colours are used to identify the manufacturer when the carpet is to be recycled or otherwise disposed of.

A secondary backing fabric constructed like a primary fabric would be unsatisfactory in terms of appearance and, moreover, would not satisfy the key technical requirement of providing good adhesion in the laminate. Woven secondary backing fabrics have a more open structure than primary fabrics and are deliberately more hairy. Two-ply warp tapes and specialised weft yarns are used to lock an open leno structure and to allow the latex or other adhesive to penetrate and bond with the structure. By using yarns in white (to simulate cotton) and beige (to simulate jute) the backing has a superficial resemblance to the backing of woven carpet. Patterns of stripes and stripes of other colours can be introduced to customise products.

An example of material specially engineered to provide adhesion is Don and Low's standard range of Lobac™ secondary fabrics which are composed of a two-ply tape warp at 31 ends/dm and an air textured weft introduced at 22–46

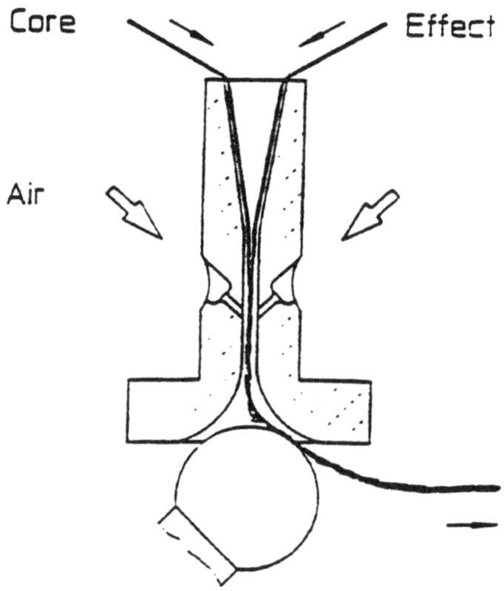

Figure 5.2. Production of core / effect yarns by air-jet texturing. (Courtesy of Heberlein Fibre Technology Inc)

picks/dm to give fabrics in the range 53–91 g/m^2. Such textured wefts are made from fully drawn polypropylene yarn by using a cold air-jet system [9]. The yarn is produced with core and effect components, as shown in Figure 5.2. The core material is fed 10% faster than yarn uptake, while the effect material is fed 30% faster, and the filaments are co-mingled to give a tightly bonded loop structure that has 70% of the strength of the feeder yarn. Processing speeds are 300–400 m/min.

An alternative weft yarn that has proved successful is spun on the DREF 2 system. Polypropylene fibre (usually 2.8 dtex, 60 mm staple) is combined with a core of fibrillated polypropylene tape on this friction spinning system to produce yarns with 14–20 tex fibre plus 45–64 tex core at delivery speeds of 180–200 m/min. Package weights are 8-10 kg and are ready for weaving. DREF yarns with a jute core have also been made.

Most manufacturers of woven secondary fabrics can supply versions of their products having conductive yarns spaced across the warp, as components of antistatic carpets.

A version of Amoco's Actionbac™ secondary fabric has a layer of melt-bonding fibre needled to one surface. This is intended to be used in conjunction with a primary fabric also bearing a layer of melt-bonding fibre, so that the two can be laminated simply by the use of heat and pressure [6].

5.3.2 Nonwoven Secondary Backing Fabrics

Nonwoven secondary backing fabrics have historically rarely been used on residential tufted carpets, because their appearance lacks consumer appeal. They are sometimes used on tufted carpets for technical reasons (eg, to meet a flammability specification or enhance dimensional stability). Because of their non-fray characteristics and excellent dimensional stability, nonwoven secondaries are, however, favoured when it is necessary to apply a backing to tiles (usually bitumen coated products).

Nonwoven secondary backing fabrics are usually much lighter than woven secondaries, for example, Lutradur™ polyester secondary fabrics cover the weight range 20–50 g/m^2.

The most common contract application for polyester spunbonded secondary fabrics is in automobile carpets. Use of fusible polyethylene in the laminate requires the thermal stability of polyester; and the extreme climate variations inside automobiles lead to a requirement for resistance to both shrinkage and stretching. Even a very light secondary fabric provides a significant increase in strength to the laminate, reducing the risk of damage to the carpet when making intricate mouldings.

Nonwoven secondary backings for carpet tiles are commonly made from a blend of polyester and polypropylene. When such products are used for tufted carpets they are often brushed to raise the surface and improve adhesion.

The Two-in-One system, offered by Fibertex AS, is a patented backing system using nonwoven fabrics which can be produced on a broadloom line for both carpets and tiles, particularly for contract installations. It incorporates two branded products, Fibertex's Comfortback, a three-dimensional mechanically cross-bonded thermoplastic fibre structure in which the fibres are bonded by heat, and Dow Chemical's Pecoback, a range of organically filled dispersion binders. The resultant textile backing, which can be used with traditional latex, hot-melt extrusion and powder scattering processes, is inherently dimensionally stable and highly resilient. Other claimed advantages include easy installation, since no permanent gluing is required, flexibility in design options and superior comfort and performance in use. Environmental benefits arise from the fact that it offers a single-process alternative to traditional latex foam.

5.3.3 Knitted

Raschel knitted backing fabrics are produced in Hungary. Striped effects to provide individuality of appearance can also be produced by knitting.

5.3.4 Needled

An interesting concept for tufted carpets that has received rapid acceptance in some countries is bulky nonwoven backing that may be regarded as integral underlay, competing with foam backings.

5.4 ENVIRONMENTAL CONSIDERATIONS

In some countries, most notably Germany, serious consideration is given to disposing of used carpets in a way that is friendly to the environment. Examples of German legislation that force the carpet manufacturer to take responsibility for the disposal of worn carpets are the Recycling Management Act, Product Liability Act and Technical Requirement for Municipal Refuse [10]. The carpet industries in the USA and other countries are also taking this issue seriously.

Although considerable progress has been made in recovering fibres, polymers and monomers, the most practical current proposal for disposing of carpets is to use them as a source of fuel. The presence of high proportions of latex and inorganic filler (which generate smoke and ash) is an obstacle to the use of carpets as fuel, and this has provided important motivation for the development of new backing materials (see Section 18.4). As fuel, the best carpet would be one with pile, primary backing, secondary backing, binder and filler all made from polyolefins having high calorific value.

A very different proposal is to use old carpets for preventing soil erosion (ie, as geotextiles). Synthetic products designed for the purpose are used in the construction of roads, embankments and levees, but there are cases where biologically degradable products are required to provide soil stability for a limited time until trees and shrubs become established (eg, afforestation in hill country). Used woven carpets having wool pile, cotton chain, jute weft and jute stuffer warp have been proposed for this purpose.

Other options being considered for reclaiming or recycling used carpets include:

- Filler materials: the carpet is chopped, granulated and pulverised for filling latex as an alternative to chalk.
- Thermoplastic resins: products such as moulded engineering components are being made from chopped and resinated materials.
- Nonwoven textiles, including floorcoverings. Edge trimmings from needled floorcoverings are, in fact, routinely recycled into this system of manufacture, and carpets with a high fibre content could possibly be treated similarly.
- Feedstock reclamation: hydrogenation or pyrolysis of carpet waste can give liquid and gaseous hydrocarbons usable as feedstock for some chemical processes.
- Recovery of monomers: the nylon from scrapped polyamide carpets can be physically separated from the other constituents and caprolactam (nylon 6) or hexamethylenediamine (nylon 6.6) recovered.
- Blast furnace reducing agent: granulated carpet waste has been recommended as an alternative to oil or coke to remove the oxygen from iron ore to give pig iron [11].

REFERENCES

1. W Klein. *Short-staple Spinning*, Vols 1–5. Manual of Textile Technology, The Textile Institute, Manchester, 1987–1993.
2. H Balk. Machinery and processing equipment for the improvement of tape and film qualities and characteristics. In: *Proceedings Second International Conference Polypropylene Fibres and Textiles*, 26–28 September 1979, The Plastics and Rubber Institute, London, 1979.
3. K Nick and F Legler. Carpet backings – economically woven on projectile weaving machines. In: *Proceedings World Textile Congress on Floorcoverings*, University of Huddersfield, 12–13 July 1995, 105–117.
4. P G Moore. Survey of polypropylene lubrication – fibres, tapes and fabrics. In: *Proceedings Second International Conference Polypropylene Fibres and Textiles*, 26–28 September 1979, The Plastics and Rubber Institute, London, 1979.
5. B Ladeur. Factors affecting the structure of woven polypropylene fabrics for carpet backings. In: *Proceedings Second International Conference Polypropylene Fibres and Textiles*, 26–28 September 1979, The Plastics and Rubber Institute, London, 1979.
6. J Diggle. AdBac™ – a latex and filler free backing system. In: *Proceedings Tifcon '98*, The Textile Institute, Manchester, 1998.
7. E Holler and H Fischer. 20 Jahre Lutradur Spinnvlies – Träger aus Polyester. *Chemiefasern / Textilindustrie*, July/August 1994, 44/96, 487–488.
8. U Gurniak and O Kohlhaas. The effect of backing materials on the combustion behaviour of tufted carpets. *Melliand Textilberichte*, 1993, 74(7), 632–633 (E246).
9. D K Wilson. Use of texturing jets to produce carpet face and backing yarns. In: *Proceedings World Textile Congress on Floorcoverings*, University of Huddersfield, 12–13 July 1995, 215–219.
10. R Wolff. Environmentally-friendly-carpets – a backing producer's view. In: *Proceedings World Textile Congress on Floorcoverings*, University of Huddersfield, 12–13 July 1995, 202–208.
11. A von Frankenberg. Utilisation of textile carpet waste in blast furnaces. *International Textile Bulletin Dyeing / Printing / Finishing*, 1996, 2, 28.

6 Principles of Tufting

Tufted carpet manufacture consists essentially of stitching pile yarn into a backing fabric (cloth) and subsequently fixing the tufts in place by means of an adhesive, usually latex. The carpet is formed pile down. The mechanisms for producing loop-pile or cut-pile carpets are somewhat different and variations on these two basic themes may be used to produce patterned carpets (discussed in detail in Chapter 7). Multiple needles (typically more than 1000) threaded with pile yarn span the working width of the tufting machine and must all operate precisely in the same manner if good quality carpet is to be produced.

6.1 LOOP-PILE TUFTING

The principle of loop-pile tufting is illustrated in Figure 6.1. During the downstroke of the needle, pile yarn is inserted through the backing fabric. When the needle reaches its lowest position a looper moves between the yarn and the needle so that a loop of yarn is held by it when the needle is withdrawn. As the needle moves down for the next cycle, the looper rocks backwards to release the loop, and the backing fabric moves forward to create space between the loops of pile yarn.

6.2 CUT-PILE TUFTING

Figure 6.2 illustrates the mechanism of cut-pile tufting. Interaction between the needle and the hook creates loops, as in loop-pile tufting, but the design of the hook is different from that of a looper and it faces in the opposite direction to the movement of the backing fabric. As a result, loops accumulate and move towards the throat of the hook, which is ground to a shear angle. When the hook withdraws after completing a loop, a knife in the form of a steel blade moves upwards at an angle and the oldest loop is cut by the scissors action of the knife and the throat of the hook.

Imperfect cutting can result in the loops pulling a hole in the backing fabric, creating a carpet fault. This problem may result from the cutting edges losing their sharpness or from incorrect settings. If the uncut filaments or fibres do not break, the faults may be serious and the tufting elements may be damaged.

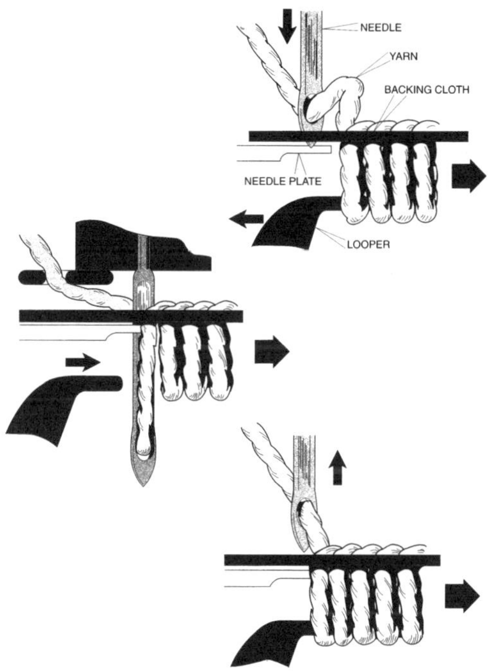

Figure 6.1. The principle of loop-pile tufting. (Courtesy of Cobble Blackburn Ltd)

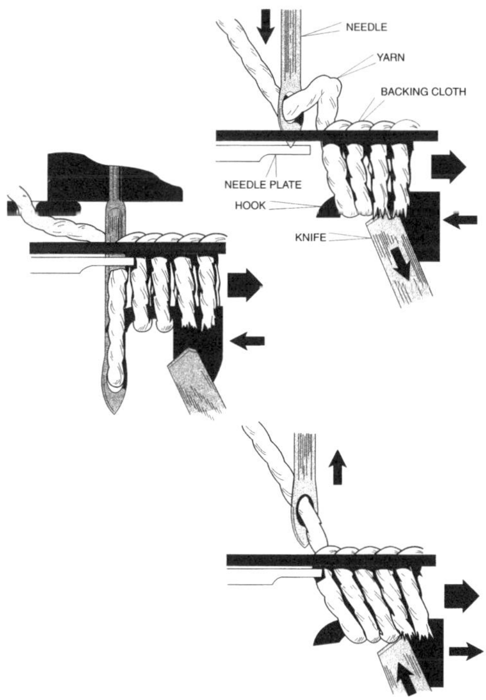

Figure 6.2. The principle of cut-pile tufting. (Courtesy of Cobble Blackburn Ltd)

6.3 CONFIGURATION OF TUFTING ELEMENTS

In Figure 6.1, the looper takes the yarn from the right-hand side of the needle, viewed from behind the looper or from the front of the machine (the side from which the primary backing and yarn are fed). This is the standard configuration of a tufting machine. The term right-hand take-off is usually used for loop pile, but in the case of cut-pile tufting left-hand to the hook point is right-hand to the operator, which could lead to confusion, so the term 'standard' take-off is preferable.

When a loop is formed, the yarn twist causes it to turn and it is desirable for the loop to turn on to the looper, rather than to turn off it. For this reason, S twist yarn should be specified on loop-pile machines with right-hand take-off. With Z twist yarn the yarn will tend to twist off the loopers and highly twisted yarns may not tuft smoothly because of this effect. In cut-pile tufting, with the hook pointing in the opposite direction, the effect is reversed. Yarns with balanced twist can sometimes be tufted on either cut- or loop-pile machines, but it is wise to apply the rule:

S twist for loop-pile with right-hand take-off,

Z twist for cut-pile with standard take-off.

Some carpet companies may prefer tufting machines with opposite-hand take-off (eg, to allow the same yarn to be used with equal facility on cut- or loop-pile machines) and must specify the twist direction of their yarns accordingly.

6.4 MACHINE LAYOUT AND MECHANISM

A unit for producing tufted carpet comprises a creel or other yarn supply system, backing fabric feed, the tufting machine itself, an inspection and mending frame and a carpet take-up system, which may be either a rolling or plaiting device. Figure 6.3 is a perspective sketch of a typical machine.

Figure 6.4 illustrates the general mechanism of a tufting machine. The backing fabric is fed at constant speed (for exceptions to this, see Chapter 7) to create spaces between the stitches (stitch length). Pile yarn is also fed at a constant rate, but is consumed intermittently as the bar of tufting needles is reciprocated by the eccentric mechanism. The temporary slack is taken up by changes in the geometry of the path from the eyes of the fixed jerker bar that moves with the needlebar relative to those in the adjustable jerker bar. Overall, the yarn feed rate is a little less than required by the spacing of the looper (or hook) relative to the backing fabric. The consequent pull-back tightens the backstitch. The bed plate, which incorporates fingers between which the needles pass, supports the backing fabric at the required height, and the presser foot prevents it from moving upwards as the needles are withdrawn.

Precise timing of the interaction between the tufting elements (needles, loopers/hooks, knives) is essential if tufting is to proceed at high speeds. The rocking-in of the loopers and upstroke of the knives can be adjusted to achieve the most

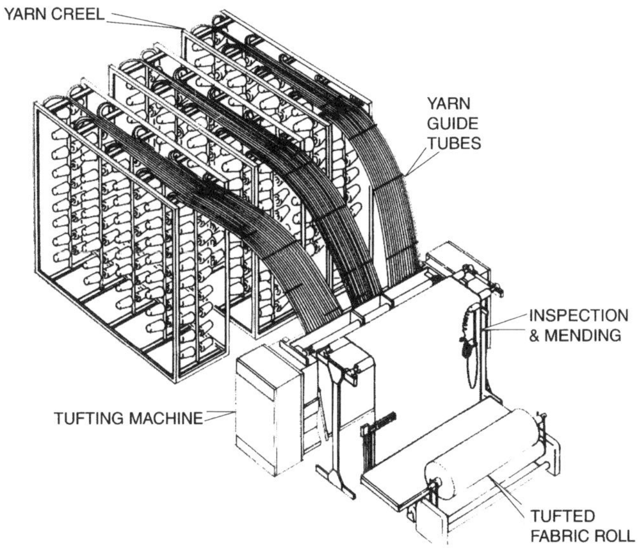

Figure 6.3. The principal components of a tufting machine (Courtesy of Cobble Blackburn Ltd)

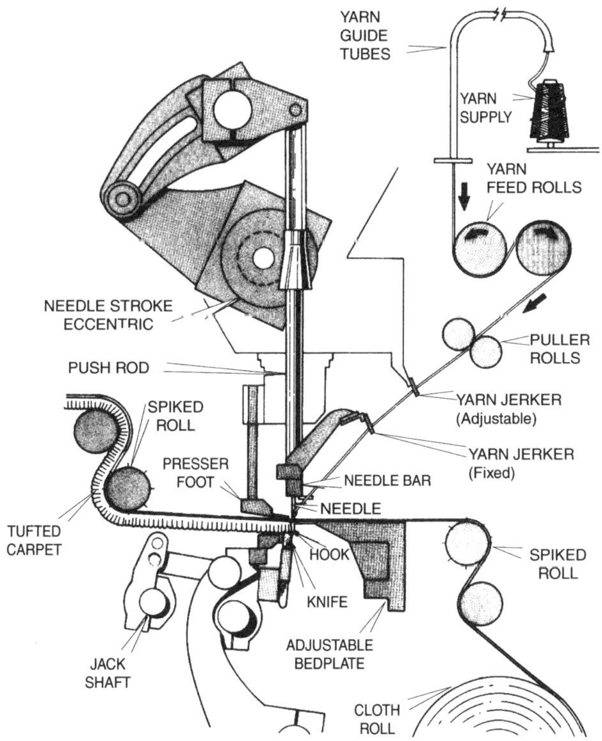

Figure 6.4. Basic mechanism of a cut-pile tufting machine. (Courtesy of Cobble Blackburn Ltd)

satisfactory running conditions. Tufting speeds for patterned carpets are usually limited by the patterning mechanism (see Chapter 7), so that there is a clear distinction between high-speed machines mass producing, for example, plain 'base cloth' for subsequent printing, and versatile machines producing patterned carpets. Machines designed for speed have lightweight mechanisms involving sophisticated materials and drives that accelerate progressively. They can operate at up to 2000 rpm on loop-pile and 1800 rpm on cut-pile constructions, generating as much as 15 000 m^2 of carpet (depending on construction) in a 10 hour shift [1].

Tufting machines are supplied in a variety of working widths. For broadloom carpets for residential applications, 4 and 5 m machines are common. Such widths are also used for the contract market, but a machine park with a selection of widths is useful, including, for example, 2 m wide machines for corridor carpets. Machines 2 m wide are commonly used for carpets to be cut into tiles (modular carpets). Sampling machines may be 1 m wide or narrower. In all cases, the actual working width is greater than the nominal width to allow for edge trimming and individual company requirements.

6.5 TUFTING ELEMENTS

The essential elements of the machine that create the carpet pile, often known as gauge parts, are needles, loopers (hooks) and knives. The reed fingers that support the backing fabric at the point of tufting are also considered to be tufting elements.

6.5.1 Needles

A tufting needle has a sharp point to penetrate the backing fabric, an eye through which the yarn is threaded, a groove to reduce the yarn/needle profile and a shank to link it to the needlebar. For obvious reasons, the finer the gauge of the machine, the smaller the needle. Other details of design required to optimise a needle for tufting a particular yarn in a particular carpet construction result in a wide variety of needles being available. For example, an exceptionally heavy count yarn will require a swollen eye, and a double sliding needlebar machine will require a different needle from a single bar (see Chapter 7). Needle points may be centrally located or biased, and the groove to the needle point can have different configurations. The needle must be profiled so as to facilitate yarn pick-up by the looper, but the rigidity of the needle must not be compromised.

Studies carried out by WRONZ [2,3] into the causes of stoppages in tufting revealed the principal cause to be joints or yarn faults jamming in the needle eye as the needle was withdrawn. This led to the design of the WRONZ-Eye Eisbär needle (Figure 6.5), made originally by Jos Zimmermann and now supplied by Groz-Beckert AG, which provides a straighter yarn path during the upstroke. WRONZ-Eye needles can accommodate bulky wool yarns, including yarns that have deliberate irregularities, such as tweed effects or flames.

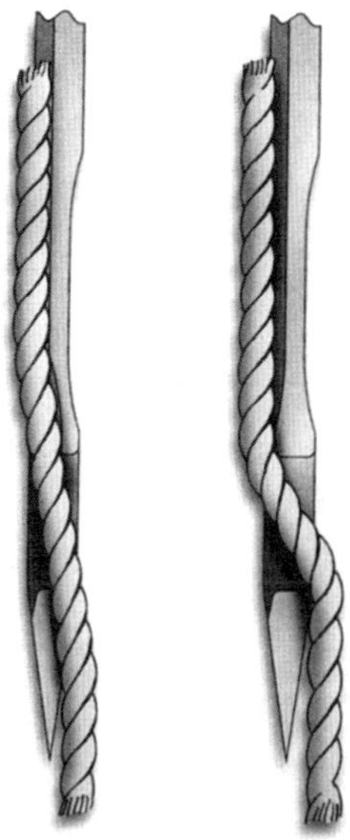

Figure 6.5. Cross-sections of tufting needles, demonstrating the straighter yarn path from WRONZ-Eye needle (left).

Singer's wool needles, developed in collaboration with Cobble (Figure 6.6), are claimed to provide the following benefits:

A. Extra-deep radiused groove to facilitate yarn flow.

B. Enlarged needle eye.

C. 'Maximum Target Area' for easier yarn pick-up, which also allows smaller rake angle settings to reduce needle wear and increase clearance between needles.

D. Deep radius groove along the total length for a smoother yarn flow.

E. Round shape of long groove to enhance needle stability.

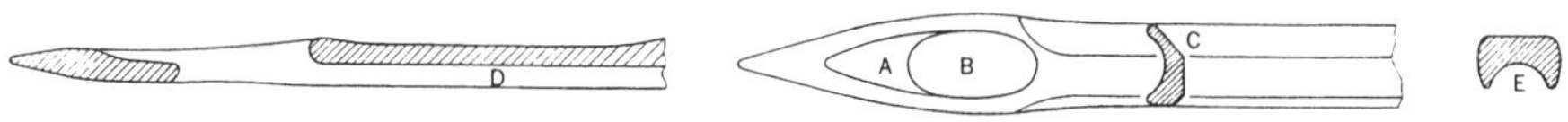

Figure 6.6. Aspects of needle design (see text). (Courtesy of Cobble Blackburn Ltd)

The Eisbär Fernmaster needle, developed by WRONZ and Groz-Beckert, has two eyes, as shown in Figure 6.7. The second eye holds the yarn more securely within the yarn protection groove of the needle, reducing stress on the yarn and producing significant improvements in tuft regularity and pile surface levelness. Insertion forces are said to be reduced by up to 70% compared with those for conventional needles, resulting in much better strength of the primary backing fabric after tufting.

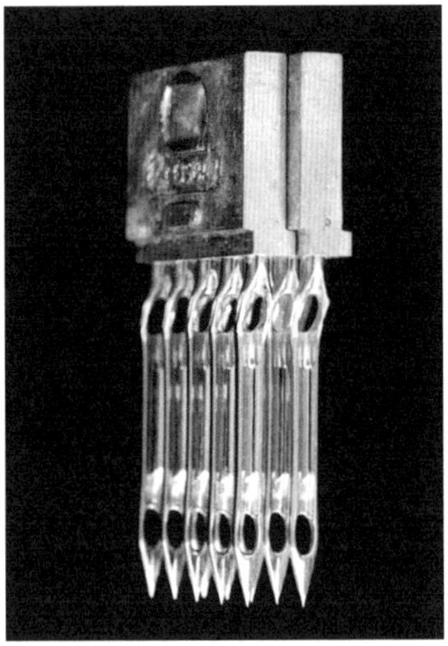

Figure 6.7 Fernmaster tufting needles, showing double-eye configuration.

6.5.2 Loopers/Hooks

It is the cut-pile hooks, rather than loopers, that set the more stringent engineering requirements. Both should be maintained in good alignment and free from burred metal, however.

Cut-pile tufting has, in theory, an inherent fault: the pile yarn is cut on one side of a hook of finite thickness, resulting in one leg of the tuft being a little longer than the other. In fine gauges the hook is quite thin so this 'J-tuft' effect is minor, but it can be very apparent in medium gauges. J-tufting can be minimised by avoiding high yarn tension. Another control is to avoid an excessive number of loops retained on the hook before cutting. The machine manufacturers' contribution is to provide hooks that are chamfered towards the throat, thereby improving the geometry of the loop in the cutting area.

Hooks are susceptible to wear and dulling of the cutting edge, especially with tough nylon yarns. High-speed steel is satisfactory for tufting wool, but hook life can be increased dramatically by incorporating a hard micro-grain tungsten insert around the cutting area. Tungsten inserts also reduce heat generation between hook and knife, minimising the risk of fusing when tufting polypropylene yarns. The face of a tungsten hook can be reground.

6.5.3 Knives

Knives are commonly made from high-speed steel. They should be reground for sharpness and precision of location at intervals specified by the machine manufacturer.

6.5.4 Reed Plate Fingers

For plain tufting, the principal requirement is to maintain reed fingers in good condition (ie, precisely aligned and undamaged). For crossover tufting (see Section 7.2) special reed fingers are used and alternative configurations are offered by machinery manufacturers.

6.5.5 Modular Gauge Parts

The development of fine gauge tufting was achieved through the improved accuracy of the interaction of the tufting elements and improved precision of their setting on the relevant bars or shafts. This requirement was met by precisely moulding needles, loopers, hooks, knives and reed fingers into modules, finished to overall length standards of half a thousandth of an inch. Examples of modular parts are shown in Figure 6.8. A further refinement is locator pegs enabling all the modules on a bar to be accurately aligned (Figure 6.9). Using such systems, tufting machines in gauges as fine as 1/20 in. have been made (although these are used for manufacturing pile furnishings rather than carpets).

Figure 6.8. Modular gauge parts. (Courtesy of Cobble Blackburn Ltd)

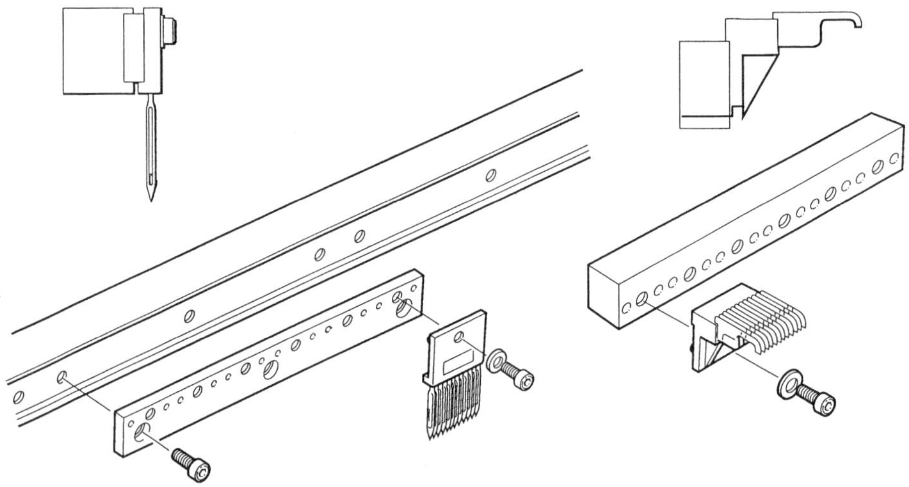

Figure 6.9. Locator modules for installing modular gauge parts on their relevant bars. (Courtesy of Cobble Blackburn Ltd)

Modular gauge parts provide a number of advantages that led to their application in medium, as well as fine gauge machines:

- improved carpet quality,
- increased wear life of components (because of improved interaction),
- rapid and accurate replacement of worn or damaged modules,
- regrinding of hooks and knives in modular form.

For medium gauges, semi-modular hook and knife blocks are often used. An individual knife block can hold up to 12 knives and is engineered so that the knives can be jigged to provide the required height, scissor angle, tension angle and looper throat position before clamping. The blocks are then located in keyways on the knife bar. Similar semi-modular systems are used for hooks.

Very coarse gauge machines may have their gauge parts located in the conventional (original) way, with each component clamped and set individually in the appropriate bar. Accuracy is achieved by hand and eye.

6.6 YARN CREELS AND BEAMS

Pile yarn may be fed to the tufting machine from creels or beams. Yarn creels are the more flexible option. Warp beams, onto which all yarns for a carpet are wound, occupy less space and are favoured by many companies for manufacturing plain carpet in loop or cut pile; however, they are unsuitable for patterned constructions where yarn consumption at all needles is not identical. Moreover, the machine must be shut down to change beams.

A yarn creel comprises a steel frame supporting tiers of cone holders, arranged in racks with aisles for access. Two cone holders are provided for each tufting needle: continuous running can be achieved by splicing the yarn tail from the

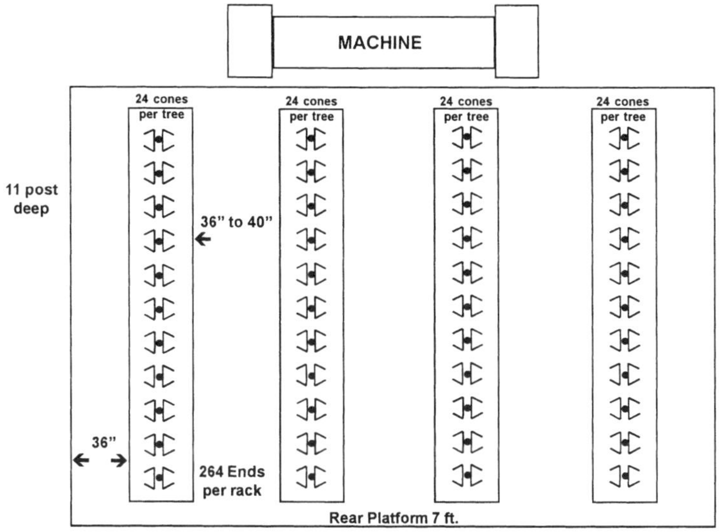

Figure 6.10. Plan of a creel for 5/32 gauge tufting machine with 1056 ends (6 + 6 layers). (Courtesy of Card-Monroe Corporation)

cone in use to the leading end of the standby cone. Creels can be supplied in a variety of configurations to comply with the space available or the operating system of the carpet company, but finer gauge tufting machines require larger creels to accommodate the greater number of yarn packages. Ladders and platforms are required to allow the operator access for creeling up a new carpet quality and replacing depleted cones. A plan for a typical creel is shown in Figure 6.10.

Each yarn end is led to the collection board associated with the yarn feed rollers via a tube, which provides a low friction path from cone to machine, that can conveniently be threaded using a compressed air gun. Tubes typically have 9.5 mm outside diameter with 0.7 mm wall and are made from antistatic plastic. Frictional drag from the different tubes must be about equal, despite their necessarily different lengths. This can be achieved by ensuring that all tubes have the same number of bends of equal radius. The tubing from a single post is bundled together and the bundle is bent from its vertical position in its natural radius to lie horizontal on the top of the creel.

As variations in tension can cause streaks in the carpet or, in the extreme case, yarn breakages, the creel must be designed and maintained so as to avoid surges in tension as the yarn is pulled off the cone. The distance from the cone to the eye of the tube should be at least 670 mm to allow for the yarn to swirl off the cone. The cones themselves (not the cone holders if they are the type that can sag) should be pointed at the eye of the tube.

Yarn is drawn from the creel by the yarn feed rollers at a rate related to yarn requirements at the needle. On modern machines the feed rollers are driven by servo-motors. For plain carpets, yarn feed is constant, but may be varied to compensate for needlebar movements on patterning machines. In the case of tension patterning systems, the patterning mechanisms control the feed rates of the individual yarn ends (see Section 7.1).

Residual yarns remaining on the creel when a tufting run is complete may conveniently be recovered by re-winding to large packages, using a special automatic winder/splicer. The Schlafhorst R-Coner may be stationary or mobile and can handle spun yarns or BCF yarns of 100 dtex or coarser. Typically the creel holds 15 feed packages that are wound into a full-sized cone with automatic air splicing (or fisherman's knotter if required). The machine stops and doffs the package when it reaches the required dimensions.

Yarn recovered in this way may be used as edge cones in future tufting lots or diverted for gun tufting. Otherwise, residual yarns are cut off the cones and shredded as waste to be used, for example, for needled floorcoverings.

6.7 CLOTH FEED

Backing fabric is drawn from the cloth roll by a spiked roller and the tufted cloth is similarly taken up by a spiked roller at a small differential to keep the backing taut at the point of tufting. The backing cloth feed stand can adjust the height of the roll and the roll can be centred by manual adjustment or automatically by edge sensing devices.

Feed rate and take-up rate can be controlled by variable pitch pulleys (flanges can be moved on the shaft to allow the belt to move higher or deeper), PIV gears or DC servo-motors. In the latter, the drive may be intermittent. Stopping the backing fabric while the needle is being inserted is advantageous for minimising the size of the needle hole when tufting very coarse gauges. Intermittent cloth feed is applied with certain patterning mechanisms.

6.8 INSPECTION AND MENDING

The system for taking up the tufted cloth at the back of the machine incorporates facilities for inspecting and mending. A common configuration is to pass the carpet underneath a walkway and then run it vertically to expose an area at least 2 m high for inspection. After a short (1 m) horizontal passage, and possibly a dancing roller to compensate during stoppages, the carpet is led downwards to a surface-driven roll-up unit or plaiter.

When commencing a new carpet quality or yarn delivery, the carpet should be inspected for faults using a check list, and any cones of yarn giving rise to stripes transferred to edge needles or rejected. If this is not done, the yarn supplier is unlikely to compensate the carpet manufacturer. Faults originating at the tufting machine should also be corrected at this stage.

During running, yarn breaks should be detected promptly, either by the operator or automatically, and missing stitches replaced with the aid of a mending gun.

The mending gun is a hand-held single-needle tufting machine, but operates on a different principle from the main machine. The mending tool is powered by compressed air. It has a pistol grip with trigger, yarn feed rollers, a reciprocating hollow needle and driven spiked rollers that assist the operator to guide it along the stitch line. Mending is performed from the back. The yarn is blown by air through the hollow needle as it penetrates the backing and forms a loop on the face of the carpet, the loop length being adjusted to that of the body of the carpet by the yarn feed rollers. For cut pile, the loop is made greater than the pile height and the excess is trimmed manually with scissors.

6.9 CONTROLLING CARPET CONSTRUCTION

The essential parameters controlling carpet construction are gauge, stitch length, pile height and yarn linear density (see Figure 6.11).

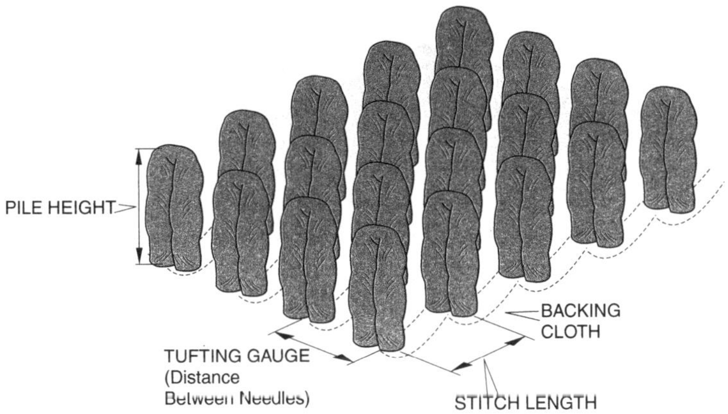

Figure 6.11. Parameters defining carpet construction. (Courtesy of Cobble Blackburn Ltd)

Tufting machine gauge is defined as the distance between needle centres in fractions of an inch (tufting machinery having originated in the USA before the metric system had been established there). Available gauges range from 1/16 to 1/4, although machines as fine as 1/20 and as coarse as 1/2 have been made. Gauge may be regarded as a fixed parameter. Conversion of a machine, including creel, to another gauge is a major operation undertaken perhaps to upgrade a redundant machine to a gauge more in line with market requirements. Usually, a tufting plant comprises machinery in a variety of gauges to produce different product lines.

Stitch length is controlled by the feed rate of the backing fabric: the slower the fabric speed relative to the machine revolutions, the shorter the stitch length and the heavier the carpet.

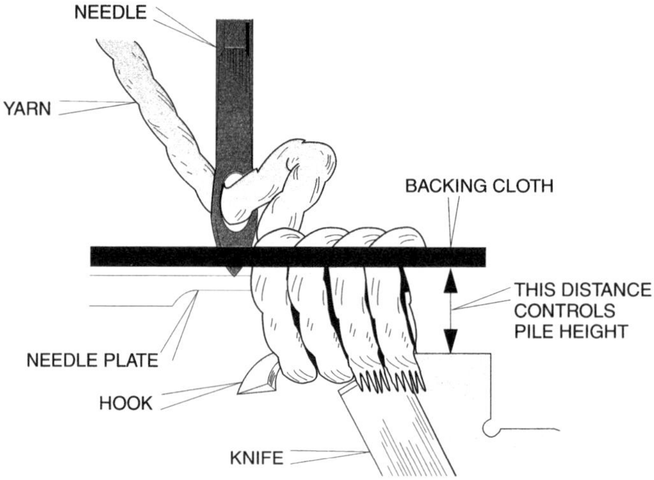

Figure 6.12. Control of pile height is by adjustment of the needle plate (reed plate). (Courtesy of Cobble Blackburn Ltd)

Pile height is controlled primarily by the distance between the backing fabric and the looper or hook (ie, by the height of the reed plate, see Figure 6.12). This can be adjusted manually or, in sophisticated machines, by servo-motor. If very large changes in pile height are made, it may be necessary to change the needle drive cam and push rods to ensure that the needles retract from the backing fabric by an appropriate distance.

Of course, the yarn feed rate must be adjusted in conjunction with any change in the geometry of the tufting elements to ensure that the exact amount of yarn required for the loop length and backstitch, is supplied. Yarn elasticity is a factor, too. In practice, slightly less yarn than is required by the loop length should be supplied in order to tighten the backstitch, which, in loop pile, may result in the slight robbing of yarn from the previously formed loop.

A tension control system based on a microprocessor, developed jointly by Teppich Forschung International and A and P Maschinenbau, improves the regularity of pile height and enables different qualities of carpet to be reproduced more accurately. It can be supplied on new machines or retrofitted

Yarn linear density and other yarn parameters (ply twist, level of set, bulk) have a major effect on carpet construction. Clearly, heavy yarns cannot be tufted on fine gauge machinery, and recommended count or linear density ranges for particular gauges and needle designs are available from machinery manufacturers or other sources (eg, fibre or yarn suppliers). With spun yarns, yarn linear density may vary between lots and it may be necessary to adjust the stitch length slightly to compensate.

Reliable control of carpet construction can be achieved rapidly on machines having yarn and cloth feed rollers driven by servo-motors, and with motorised adjustment of the bed plate. Such mechanisms can be controlled from a section of the control panel (eg, Card-Monroe's Command Performance or Cobble's Recall automatic machine setting systems). A typical system holds the settings for 500–1000 carpet qualities. When the operator selects a new program, the machine inches forward slowly while it achieves the required stitch rate, pile height and yarn feed. There is also the facility to produce carpet to a measured length: once the pre-programmed length has been produced, the machine stops automatically.

In the absence of automatic setting systems, the manufacturer must keep careful records of carpet specifications, and the machine settings required to achieve them, as well as the specifications of the yarns and backing materials to be used.

Although machine parameters are normally specified in imperial units, carpet construction as produced is tested in terms of rows per dm, stitches per dm, pile height in mm (note that pile thickness as measured by a thickness gauge is somewhat different) and total pile weight in g/m^2 (this can only be measured on unbacked carpet since it is not possible to separate latex reliably from a carpet with secondary or foam backing).

6.10 MONITORING AND CONTROL SYSTEMS

In addition to the foregoing systems for setting machine controls from a central panel, systems for monitoring the quality of production and integrating the tufting machine with mill management systems are available.

Research by the electronics company Sick [4] into the possibilities for eliminating faults and improving carpet quality led, in the first instance, to the Broken End Detector, the Tuftstop Broken End Detector and the Tight End Detector, offered on Cobble machinery. Competitive equipment has since been developed.

The Broken End Detector is usually used in combination with the TB Blower system. When a yarn breaks at the tufting needle or yarn feed, the blower system deflects the end into the path of an infrared beam and a detector reports the defect promptly. The device compensates automatically for contamination of the machine by fibre and the threshold level of contamination is digitally adjusted.

The Tuftstop Broken End Detector operates on a different principle. Each yarn passes through a detector hook that is held upright when the yarn is running normally. If an end breaks, the hook swings down to interrupt the beam to a photocell to stop the machine.

The Tight End Detector checks the tension of each pile yarn end between the creel and the feed rollers. If the tension increases above a pre-selected threshold value the machine stops, but only after a response time of 0.5–8 seconds to allow for fluctuations in tension that may be considered normal for the particular yarn. The fault zone (there are 16 sections) is indicated by a flashing red light, enabling the operator to find and correct the fault before a break or a row of tight loops is created.

The Ziessenis FST and FNT systems detect broken ends and tight ends, respectively.

The IMS Tufting Scanner (Sick) monitors all yarns in the region of the needles using a precision laser scanner. It can detect a variety of faults and locates the fault within a 1 m zone.

Matrix Vision's camera-based inspection system monitors the quality of the tufted cloth as it is produced.

Tension regulated yarn feed is possible as a result of a project carried out at Teppich Forschung International [5]. On-line analysis of signals from across the width of the machine is used to control the yarn feed. The result is more uniform carpet thickness and weight, elimination of the effects of atmospheric changes, a more uniform appearance, documentation for quality control and reproducible production of loom-state carpet.

A system proposed by Autologic Plauen [6] links machine data to a computerised management system. A data terminal at the machine records both automatic events and those declared by the operative via a keyboard. Thus, data such as shift records, efficiency levels, frequency of faults and waste statistics are accumulated and networked with the production planning and control system. The system is intended to incorporate machinery at all stages of carpet production, not just tufting, and extends to stockkeeping and dispatch of finished product.

6.11 STITCH PLACEMENT MECHANISMS

When manufacturing plain carpets it is sometimes considered desirable to break up the regimented structure of the tuft lines. Zig-zag tuft lines can be produced by reciprocating the backing fabric feed and take-up rollers laterally (a backing mover mechanism, commonly known as a jute shifter even though jute primary backings are no longer used). More commonly, the needlebar is reciprocated by one or multiple gauge widths, so that needles interact with different loopers or hooks. Sliding needlebars are widely used to create patterns and are considered in more detail in Section 7.2.

A more subtle approach is to displace the stitches by only half a gauge width. In coarse gauge loop piles, this technique tends to turn the loops and gives an attractive random structure. In both loop- and cut-pile constructions the overall effect can be used to give a denser carpet than would normally be produced on the machine (ie, simulated fine gauge).

In one such mechanism (Cobble's dual stitch placement system) the needlebar and the reed plate move together. At top-dead-centre each needle is aligned with the corresponding looper or hook. At the start of the downstroke the needlebar and reed plate move slightly out of line with the loopers (less than half a gauge width) and the backing fabric is penetrated while they are in this configuration. At the later stage of the downstroke the needles and fingers move back into alignment, displacing the backing fabric to one side and the stitch is made with the backing fabric displaced. The needles are withdrawn and the sequence is repeated, but with the displacement in the opposite direction. Thus, two lines of stitches are produced by each needle. In more modern versions of dual stitch placement systems, where the needlebar is accurately controlled by a servo-motor, the reed plate is fixed and needles are realigned between the fingers after penetrating the backing fabric while it is slightly displaced.

Perhaps the simplest mechanism for achieving high density tufting by stitch placement is to displace the primary backing fabric by small lateral movements of either the cloth feed rollers or the cloth take-up rollers. A mechanism of this kind, which provides a simple zig-zag movement of usually half a gauge width to the front cloth roller only, is illustrated in Figure 6.13.

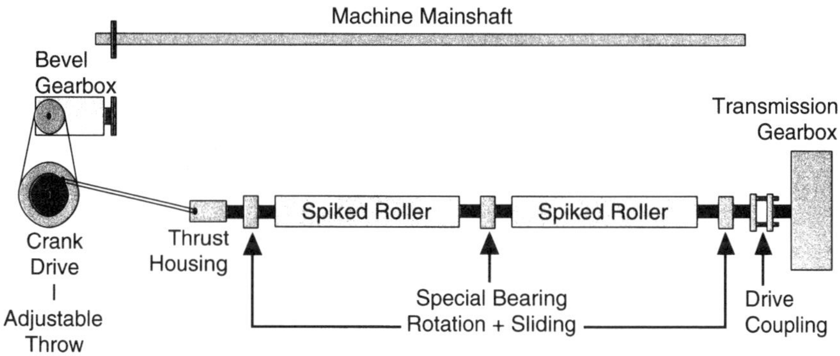

Figure 6.13. Cobble's Megashifter mechanism for creating a small (half to one gauge width) zig-zag movement of the cloth feed. (Courtesy of Cobble Blackburn Ltd)

REFERENCES

1. M T Allen. Advances in tufting technology. In: *Proceedings of World Textile Conference on Floorcoverings*, University of Huddersfield, 12-13 July 1995, 133–138.
2. G A Carnaby, G M Garden, M J Hammersley and D M Watts. Tufting efficiency. Part 1: Some causes of stoppages during tufting. *Textile Institute and Industry*, 1979, 17(8), 282–285.
3. G A Carnaby and D A Ross. Tufting efficiency. Part 2: Some improvements in tufting technology. *Textile Institute and Industry*, 1979, 17(9), 327–330.
4. G Katholing. Start of automation of tufting machines: state-of-the-art in fault monitoring. *Chemiefasern / Textilindustrie*, 1990, 40/92, 248–253.

5. H-J Schulz. Carpet tufting machines with tension-regulated pile yarn feed. *Melliand International*, 1996, 2, 85–86.
6. M Rudloff. Effective control and monitoring of the production of textile floorcoverings. *Melliand International*, 1997, 2, 79.

7 Patterning Systems for Tufted Carpets

Patterning systems on tufting machines are mostly in three general categories:

- Yarn tensioning systems that usually control pile height.
- Crossover systems that displace the relative lateral positions of the pile yarns.
- Colour interchange systems that allow one of a number of colours of yarn to be selected for insertion at a particular point in the design.

There are also systems that combine the first two technologies; and a few that do not strictly fall into any of these classifications but, for convenience, will be discussed together with yarn tensioning systems. Of course, printing of ecru carpet is an extremely important patterning system and there are many carpets that have design features (colour or texture effects) provided by the yarn – often known as 'semi-plains'.

7.1 YARN TENSIONING SYSTEMS

By varying the pile height in loop-pile tufting, a wide range of effects can be produced. In the simplest case, a sculptured pile in a single colour is produced. Two pile heights or three may be created. If a sculptured carpet is passed through a shearing machine (see Section 18.2), the tips of the higher pile are converted into cut pile to give a pattern of cut pile among the loops, known as a tip-sheared design.

With alternate needles threaded with different colours of yarn, very low loops in one colour can be totally hidden by high pile in the other colour, and reversing the situation at will gives a design in two colours. Of course, the density of the high pile is governed by the fact that it is only formed from alternate needles, and a bulky yarn is required to cover the buried pile satisfactorily.

Using a special hook having a spring clip to retain the loops (Figure 7.1) tufting can be carried out so that the high loop is cut while the tensioned low loop is pulled past the spring and off the looper, giving a cut/loop design. The buried-pile (buried-end) technique described in the previous paragraph can be combined with cut/loop tufting to give cut-pile designs in two colours.

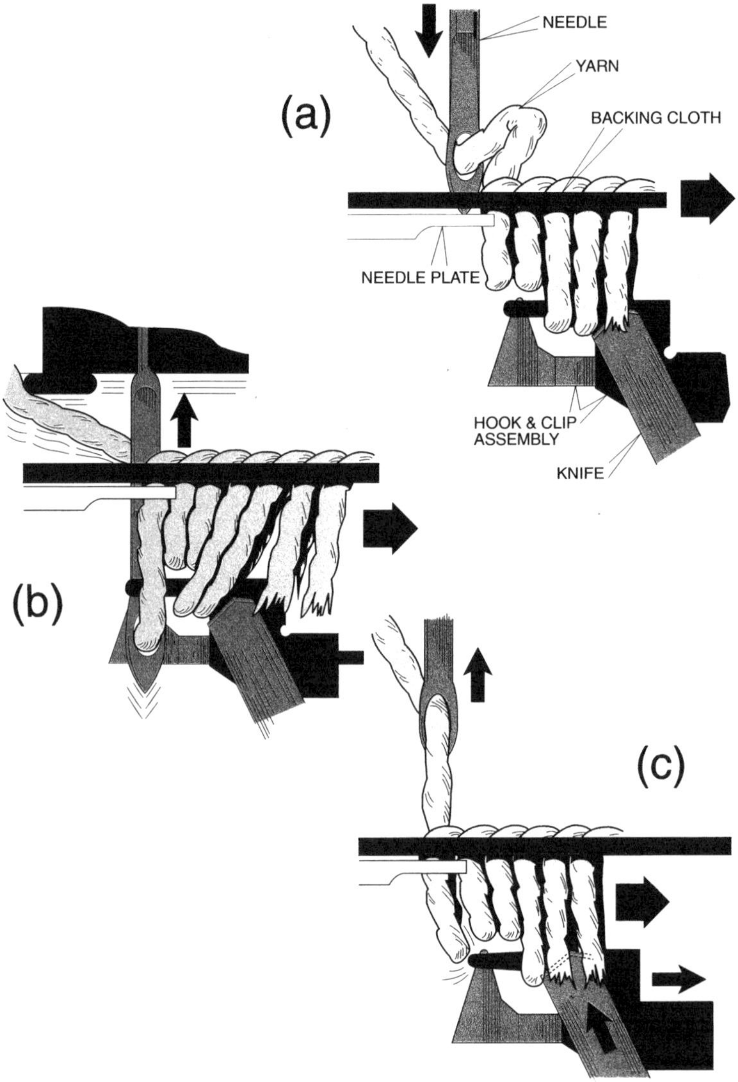

Figure 7.1. The principle of cut/loop tufting showing (a) the cutting action on long loops; (b) how a long loop is formed when yarn tension is low; (c) how high tension pulls the yarn past the spring clip to create a short loop. (Courtesy of Cobble Blackburn Ltd)

Various control devices can be used to create the pattern in tensioning systems, but the basic principle of tufting is the same – the yarn supply for selected ends is restricted, creating high tension in these ends. Whereas plain carpets are produced with the yarn supplied matching the geometry of the tufting elements at the point of loop formation, a patterning system provides a yarn deficit to some of the needles. This deficit is compensated by robbing back some yarn from the previously formed loop (ie, the loop being tufted is virtually unaffected by the yarn tension, but the loop that is a fraction of a second older is shortened).

7.1.1 Disc Cam

The simplest yarn tensioning system is illustrated in Figure 7.2. The pile yarns are threaded, often in pairs, over eccentric discs mounted in a pattern on a splined drive shaft, and the yarn path over each disc is alternately shortened and lengthened as a result of the eccentricity. By setting the various discs to different angular positions on the shaft, sculptured designs can be produced in graded pile heights. Quite small repeats are achieved, and the pile height change is gradual because of the sinusoidal action of the cams, so that the effect is often described as a textured surface rather than a design.

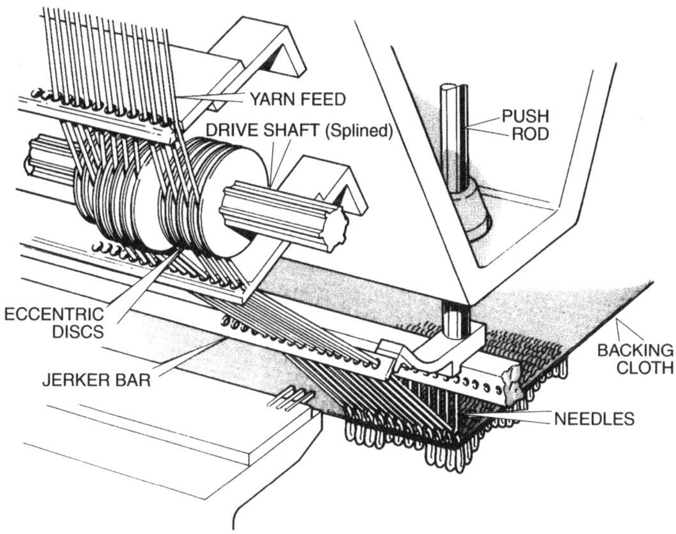

Figure 7.2. Production of surface effects by means of eccentric disc cams. (Courtesy of Cobble Blackburn Ltd)

The eccentric disc cam is in some circles colloquially known as the 'Mickey Mouse' attachment because adjacent discs create an impression of the cartoon character's ears.

7.1.2 Servo Yarn Feed Variator

A more precise system than the eccentric disc cam for achieving small repeat sculptured effects operates by adding or subtracting yarn from a continuous supply using oscillating guides (Figure 7.3). The guides can be rocked away from the tufting area by servo-motors that respond to a digital pattern. Because the servo-motors respond to the electronic impulse extremely rapidly, and the system is located quite close to the needles, the change from high to low pile is achieved in one stitch, so that sharply defined surface effects are obtained (see Plate 1 in the colour section).

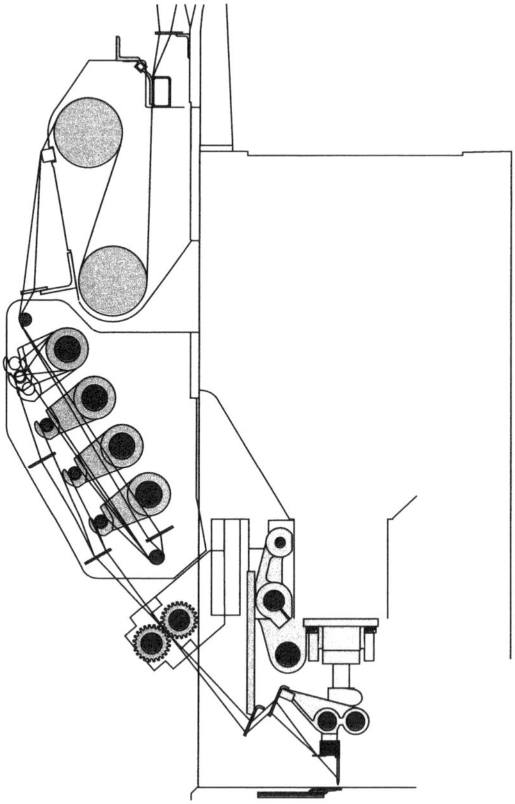

Figure 7.3. Cross-section of a yarn feed variator system in which pile yarn ends can be threaded over up to four guides. Each guide can be oscillated at will, as indicated by the alternative positions of the upper guide. (Courtesy of Cobble Blackburn Ltd)

7.1.3 Universal Type Patterning Attachment System (UTPA)

An older system for producing quite sharp surface effects is the UTPA, which has been made in two quite different forms.

The first is based on a system of slats having a standard configuration, as shown on the left in Figure 7.4. One chain carries plain slats and the other holds slats having regularly spaced cuts. Yarn guides can move pile yarns laterally according to the requirements of the design, so that the yarn follows a straight path or a zig-zag path, the latter delivering more yarn. Figure 7.4 shows the alternative version on the right, in which the guides transfer the yarn between high- and low-friction bands on slow and fast moving rollers, thus changing the yarn feed rate.

The guides may be controlled by various alternative patterning devices. A system having limited flexibility is based on a phenolic drum engraved with the design, working in conjunction with electrical contact fingers that actuate the movement

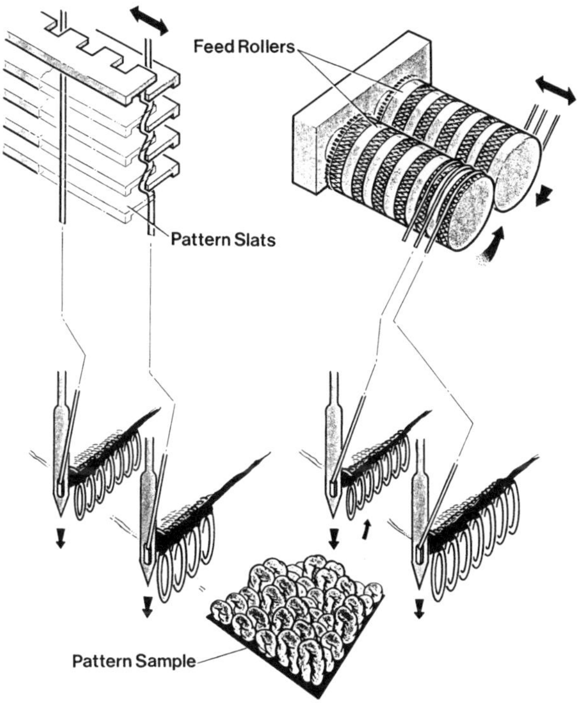

Figure 7.4. Alternative versions of the UTPA patterning mechanism (pattern slats and feed roller systems). (Courtesy of Cobble Blackburn Ltd)

of the yarn guides. A cruder system that is rather more flexible is to create the pattern in the form of insulating tape on a conductive brass drum, again using contact fingers to actuate the yarn guides.

UTPA systems were widely used for preparing samples and are still used in some companies to control cut/loop tufting of bath mats and small rugs (see Plate 2).

7.1.4 Slat Patterning

This early system, now used only for long runs of popular designs because of its limited flexibility, clearly demonstrates how the principle of yarn tensioning can be used to create full patterns of sculptured effects or buried-pile designs in two colours. The slat patterning system, developed by Mohasco in the USA, gives much sharper change in pile height than the disc cam system (and sharper than some of the older electronic systems that followed it) and is capable of large repeats.

Figure 7.5 illustrates how pile yarns are fed between two chains of intermeshing slats. The bottom set of slats has V-notches all of the same height, whereas the top set of slats is profiled in accordance with the design. As the two sets of slats

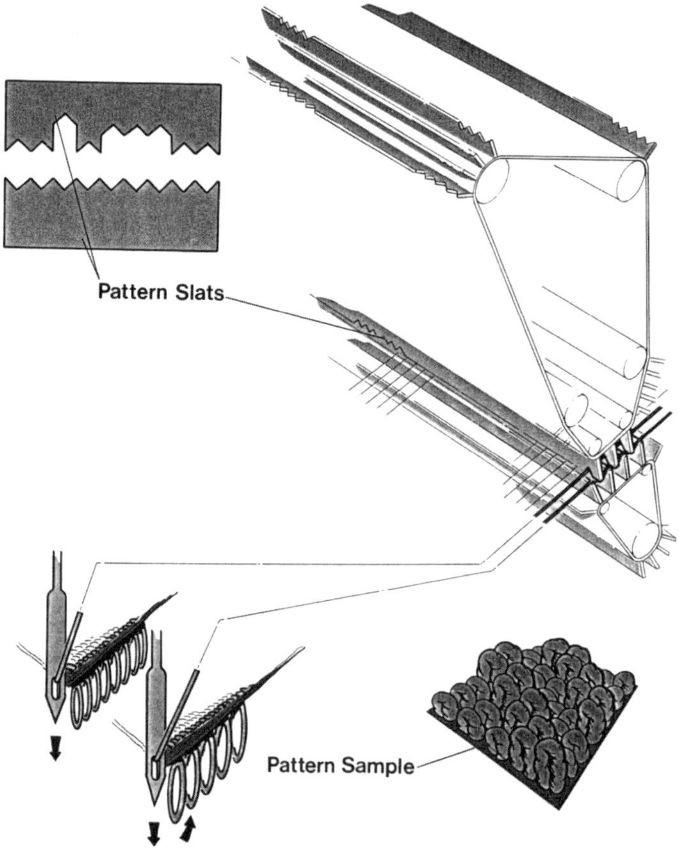

Figure 7.5. Mohasco slat patterning system showing how a deeper V-notch provides a smaller loop. (Courtesy of Cobble Blackburn Ltd)

intermesh, the yarns are deflected to an extent that depends on the depth of cut. A simple V-notch delivers sufficient yarn to allow a full-length loop to be tufted, whereas a deep cut delivers less yarn and creates a loop robbing situation. The slats may be profiled to give two or three pile heights. With alternate yarn ends of different colour producing high or low pile, buried-pile designs in two colours can be produced.

7.1.5 Scroll Patterning

'Scroll' patterning for sculptured, buried-pile or cut/loop (see Plate 3) carpet designs originated with a system of painting the designs in opaque black on scrolls of transparent film. The design film was mounted on a rotating transparent drum having an internal light source. Rows of photocells detected light passing through the transparent areas of the film and an electronic signal actuated a yarn feed roller to slow the delivery of yarn, producing low pile. No signal resulted in high pile. Separate drums were used for the alternate yarn ends in buried-pile designs.

Scroll patterning systems are now controlled from design data stored on computer disk. The design may be fed into the patterning system from a floppy disk originating in the design studio; or a laptop computer holding the design may be interfaced with the patterning system; or all the machines in the tufting plant may be networked with the CAD system in the design studio. Designing for the carpet industry using CAD systems is described in Section 16.3.

Signals from the electronic pattern are converted to the required pattern of yarn tensions via a multiplicity of yarn feed rollers, the speeds of which can be individually controlled. The 'clutch-in-roll' system, widely used for many years to control the surface speed of the feed rollers, has been improved in terms of precision and reliability of operation, ease of threading and compatibility with different tufting machines. Originally mounted on top of the machine, the scroll yarn feed rollers and associated equipment are now increasingly being mounted on the upper part of the side of the machine, making them more accessible and bringing them closer to the tufting elements to reduce loss of design definition due to yarn elasticity.

A typical clutch system has duplicate brushes (for reliability) that actuate a rotating-field clutch controlling the speed of the yarn feed roller. Typically, there are 120 or 126 yarn feed rollers, so that 120 or 126 ends of yarn may participate in a design width repeat. The actual width repeat depends on the machine gauge, too. For example, the width repeat on 5/32 gauge with a 126 roll scroll gives a repeat of $5/32 \times 126 = 19.7$ in. or 0.5 m. Therefore, on a 5 m wide machine, 10 repeats across the width of the carpet are necessary. To achieve this, 10 yarn ends are fed over each scroll roller and conducted to tufting needles spaced evenly across the machine via a 'scrambler box' or tube bank assembly. So that the distances between the scroll rollers and needles for the various yarns are approximately equal, the tubes in the scrambler box may follow complex paths – a necessity if the design definition is not to vary across the width of the carpet. Figure 7.6 shows the overall layout of Cobble's Low Profile Scroll LP III machine. Rollers for odd and even yarns in a buried-pile design are grouped at opposite ends of the row of rollers for simplicity of creeling.

Note that in practice, carpets are tufted a little wider than the nominal machine width and are trimmed to width after backcoating/laminating. To provide yarns for the additional needles, additional ends of pile yarn are threaded over a few of the scroll rollers (eg, for 5/32 gauge some of the rollers may carry 11 ends instead of 10).

Clearly, the repeat width is reduced as the gauge becomes finer, and some carpet manufacturers wish to compensate for this. For example, machines can be fitted with up to 256 yarn feed rollers to enable wider repeats to be achieved on finer gauges.

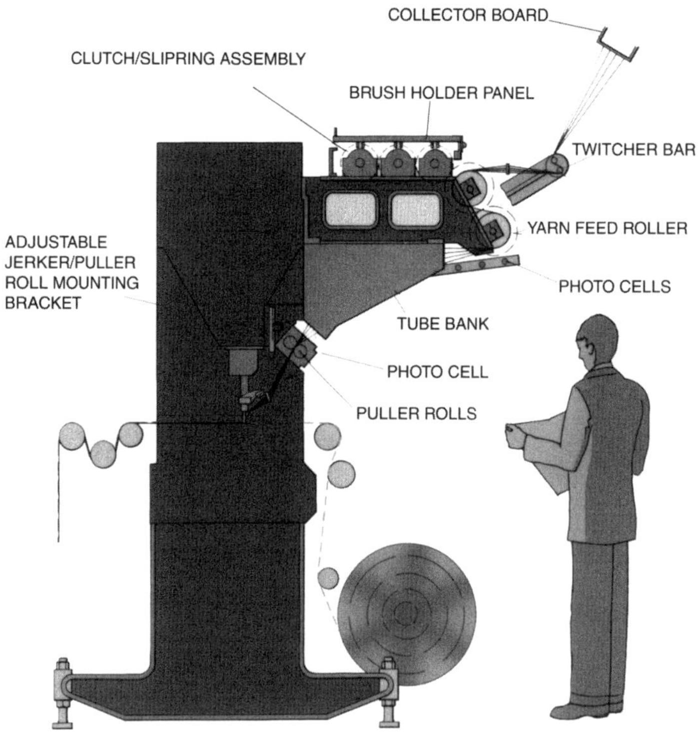

Figure 7.6. General layout of Cobble's Low Profile Scroll machine. (Courtesy of Cobble Blackburn Ltd)

Tuftco's Multi-Pile Servo Scroll operates on a somewhat different principle. Instead of the clutch controlled feed rollers, there are up to 120 servo-motors per pattern attachment, each controlling the feed rate of a group of yarns. The machine offers a wide range of pile heights, and crisp pattern definition owing to precise yarn control. The principle of the machine has been used as a basis for even more sophisticated machinery. For example, the Single End Servo Scroll, which has one servo-motor for every end of yarn, allowing single-end pattern control. The servo scroll system has a particular advantage when used in combination (see Section 7.3 for combination patterning) with a sliding needlebar: it can automatically compensate for the varying yarn requirement for the backstitch when the needlebar shifts.

7.1.6 Full Repeat Scroll

Cobble's unique full repeat scroll system (FRS) produces essentially two-colour buried-pile designs (Plate 4) by controlling each individual yarn (Figure 7.7). Four sets of yarn feed rollers span the width of the machine. There are slow (A) and fast (B) rollers for the odd and even (different coloured) ends of yarn and alternate colours by-pass the first pair, but are threaded over the second pair, and vice versa. However, the rollers only drive the yarn when it is pressed

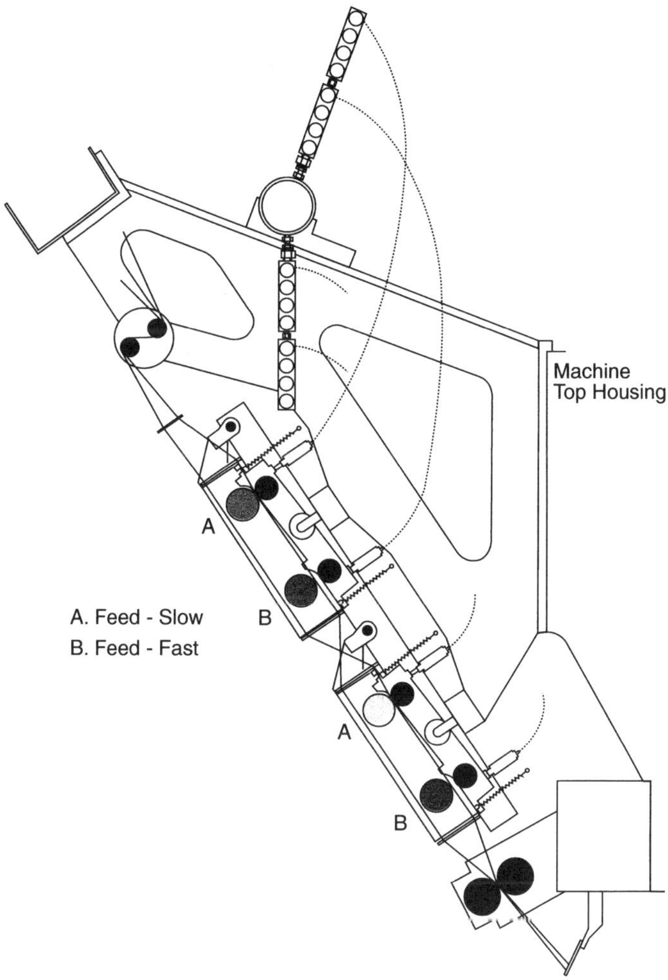

Figure 7.7. Principle of the Full Repeat Scroll patterning system showing the slow feed rollers (A) driving the two yarns. (Courtesy of Cobble Blackburn Ltd)

against them by very narrow upper rollers. Nipping of the yarn to the slow or fast rollers is activated according to the requirements of the digital design by means of an electro-pneumatic system.

FRS can be used with loop-pile, cut/loop or level cut/loop machines, and is used for the production of broadloom carpets (Plate 5), rugs and bath mats. More than one design of carpet can be manufactured simultaneously on the same machine (eg, rugs of different dimensions and designs can be tufted side-by-side).

A more sophisticated version of FRS has six sets of feed rollers, providing slow, medium and fast delivery of yarns for the alternate coloured ends. The possibilities are extended to include high cut, medium and low loops on a cut/loop machine, or three pile heights of loop-pile carpet.

7.1.7 Related Systems

Level Cut/Loop

Although listed under this general heading, the level cut/loop (LCL) system does not depend on varying the yarn tension. A mechanically actuated gate, rather than the spring clip shown in Figure 7.1, governs whether or not a loop is cut. Different machine manufacturers have different mechanisms for achieving this result. For example, Cobble's LCL system produces loop pile on just the tip of the looper if the gate is closed, but if the gate is open, the loop is admitted to the throat of the hook to be cut subsequently. Tags at the rear of the hooks are engaged with the patterning mechanism and open or close the gates as required.

This LCL system may be linked to a digital patterning system for large repeats, or small geometric patterns may be produced in response to a simpler digital design system. The machine is commonly used to produce cut/loop designs in a single colour, similar to some Wilton designs.

Velva-Loop

This too is a system not depending on yarn tension. Devised by Tuftco, it utilises a staggered needlebar, but has both loopers that rock away from the direction of the fabric feed and cut-pile hooks that move towards the oncoming fabric. The loopers interact only with the front row of needles and the cut-pile hooks interact only with the back row of needles.

Plain carpets are produced that might be expected to consist of alternate rows of cut and loop pile, but the stagger of the needles results in an appearance of each cut tuft being surrounded by four loops, and each loop being surrounded by four tufts.

The Velva-Loop system, too, provides the basis of some combination patterning systems (see Section 7.3).

7.2 CROSSOVER SYSTEMS

The concept of crossover tufting originated with the use of either a 'jute shifter' or cam actuated sliding needlebar to produce crude zig-zag or sinusoidal effects from a sequence of coloured pile yarns. Although mechanisms for reciprocating the backing fabric laterally are still used in some stitch placement systems (see Section 6.11), the emphasis on patterning systems that displace coloured pile yarns laterally has been firmly on sliding needlebar mechanisms. Very sophisticated effects have proved possible as a result of precise mechanisms and strong interest in the techniques from designers, aided by CAD systems.

When a needlebar moves laterally, it must do so quickly (within a fraction of the tufting cycle) and accurately so that the needles interact well with alternative loopers. Various mechanisms have been devised for achieving these requirements. Some of them additionally take into account the need to compensate for the longer backstitch that is created when the needles move laterally.

7.2.1 Single Straight Needlebar

If two colours of pile yarn are threaded in alternate needles of a straight needlebar, and the bar is reciprocated one gauge width for each revolution of the machine, the colours are interchanged and the linear effect is replaced by a more subtle mixture of colours. Note that the sequence of colours within a single row of tufts remains unchanged.

With more elaborate creeling sequences and more complex movements of the needlebar, variations on zig-zag or sinusoidal effects can be achieved. By applying creative skills, the basically simple nature of such effects can be made more acceptable aesthetically (Plate 6).

7.2.2 Single Staggered Needlebar

When using yarns of relatively high linear density on fine gauge machinery it can be advantageous to provide more space by staggering the needles into two rows, as shown in Figure 7.8. There are staggered loopers (short and long) to correspond with the staggered needles.

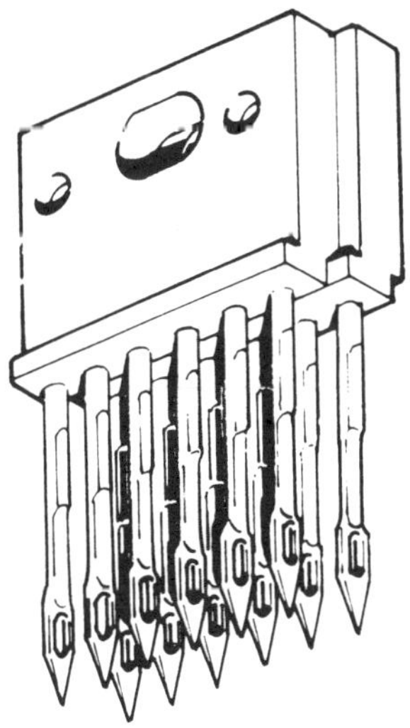

Figure 7.8. Module of a staggered needlebar. (Courtesy of Cobble Blackburn Ltd)

Needlebars of this kind may be used to create somewhat more interesting effects than those obtained with a straight row of needles. Because the front row and the back row of needles are essentially creating different rows of tufts during the same needle stroke (the front row of needles first tufts alternate stitches in a row and the gaps are filled by the back row during a later needle stroke), it is possible to interchange the positions of the colours within a row of stitches (the zig-zag backstitches overlap), as well as interchanging their positions in a line of stitches. Such reconfiguration of the colours enables simple chequered effects to be produced.

7.2.3 Double Sliding Needlebar

Clearly, if alternate needles are located on different bars, each capable of independent lateral movement, stitches from alternate ends can cross over each other more definitely than in the case of a single staggered bar. By having the facility to move the bars independently in opposite directions, together or singly, and by more than one gauge step, the patterning scope is increased tremendously. The possibilities are increased several fold if the number of colours of yarn is increased and their creeling sequence changed (Plate 7). Computer simulation is necessary to be able to visualise the alternatives, and this requirement was the initial motivation for introducing CAD systems into the tufted carpet industry.

Double sliding needlebar tufting machines are available for manufacturing both cut- and loop-pile carpets.

A key technical advantage of crossover tufting over tensioning systems is that the density of the pile is virtually the same as would be achieved by tufting plain carpet on the same machine gauge. In most buried-pile designs, the stitch density of the surface pile is only half of that obtained when tufting plain carpet.

A disadvantage of crossover tufting is that the diagonal backstitch is longer than a straight backstitch, especially if multiple jumps are made, and the extra yarn must be allowed for when costing the carpet. There is also a tendency for the long backstitches to be created at the expense of the pile, so that the carpet may be ribby. In order to compensate for this effect and to allow crossover tufting to be developed to its full potential, it was necessary to develop special mechanisms.

The simplest solution to the variable demand for yarn when tufting on the crossover principle is to provide a relaxed yarn feed, allowing the needles to take up the yarn they require. In practice, more sophisticated systems have been found to operate more reliably, especially when large needlebar movements are involved. Servo-motors are applied by the major tufting machine manufacturers to provide the yarn required for each individual stitch and are essential features of machine control systems that can store and reproduce the specifications for particular carpet qualities. Examples include Card-Monroe's Command Performance, Tuftco's Encore and Cobble's Recall systems. When the required pile height and shift pattern are entered into the computer control panel, yarn feed rates are calculated to include the requirements for the various backstitches.

Further refinements have been developed, such as Cobble's Recall system, which can also incorporate a servo cloth-feed system that can stop the cloth momentarily to facilitate needle penetration and removal. The servo cloth-feed is also important for minimising damage to the backing fabric when tufting coarse gauge products with over-sized needles.

When manufacturing crossover tufted products it is important to pay attention to the backcoating process (see Section 17.4). The backstitch pattern of a crossover fabric may be more difficult to penetrate with latex than a plain fabric, and a modified technique of application may be needed to achieve satisfactory tuft bind and adhesion of the backing fabric.

7.2.4 Double Sliding Needlebars with Needles In-line

The Vario-Design system, developed jointly by Teppich Forschung International and A and P Maschinenbau, was motivated in order to produce fine, clearly defined and correctly linear patterning, including small geometric designs, patterning options independent of stitch density, and minimise consumption of yarn in the backstitch [1].

Vario-Design utilises needles developed specially for the purpose by Groz-Beckert (see Figure 7.9). The needlebars are able to move both transversely to the direction of tufting and towards and away from each other (ie, in the direction of tufting). Transverse movements, as required by the design, take place at top-dead-centre when the two bars are spaced apart. The back bar then moves into contact with the front bar, so that the two sets of needles merge into a single row before penetrating the backing fabric. The loopers or hooks are, of course, in a single row. It is the absence of stagger at the time of stitch formation that accounts for the clarity of pattern and minimum backstitch. Fineness of design definition is achievable by the fact that colours can cross over by as little as one gauge width, whereas with a normal double needlebar system the displacement is by two gauge widths (a bar moves two gauge widths to allow its needles to interact with its particular row of loopers – front or rear).

7.2.5 High Definition Crossover Tufting

An alternative system for achieving crossover designs with definition as fine as one gauge width is Cobble's Hi-Fi system. The configuration of the gauge parts for cut-pile tufting is shown in Figure 7.10.

Instead of a bar of alternately short and long hooks, as in conventional crossover tufting, there is only one length of hook ($^1/_2$ in. throat). The stagger between the front and back row of needles is reduced from the conventional 1/4 to 1/8 in. This configuration allows either row of needles to interact with any hook, so that single gauge width movements, rather than the conventional two, are possible.

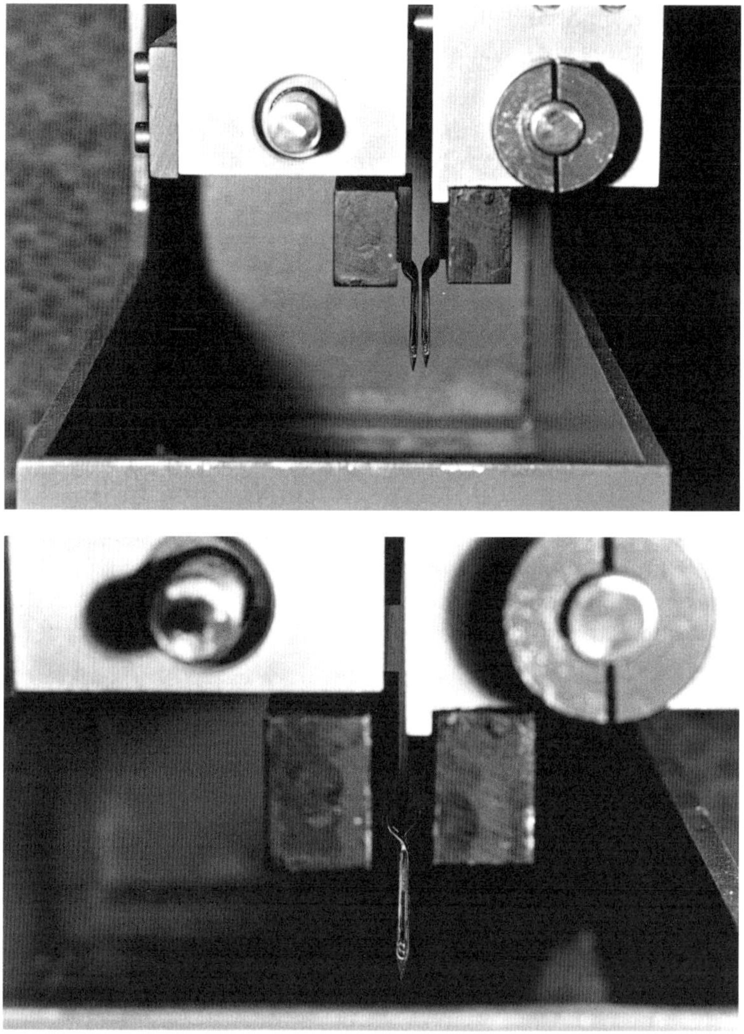

Figure 7.9. End view of needlebars in the Vario-Design system positioned for sliding (above); needles in line for tufting (below). (Courtesy of Teppich Forschung International)

7.2.6 Shifting Mechanisms

Several alternative principles have been used for actuating sliding needlebars: cam, hydraulic, pneumatic or electric. Movements must be rapid, accurate and reliable if the tufting elements are not to crash. The favoured systems are cam and electric positioning motor drives.

Cam mechanisms are used for both single and double sliding needlebar machines. They were originally used to create quite small movements but have been developed to provide large pattern repeats. A 420 mm cam allows pattern repeats of up to 70 stitches to be produced. The cams can be generated by

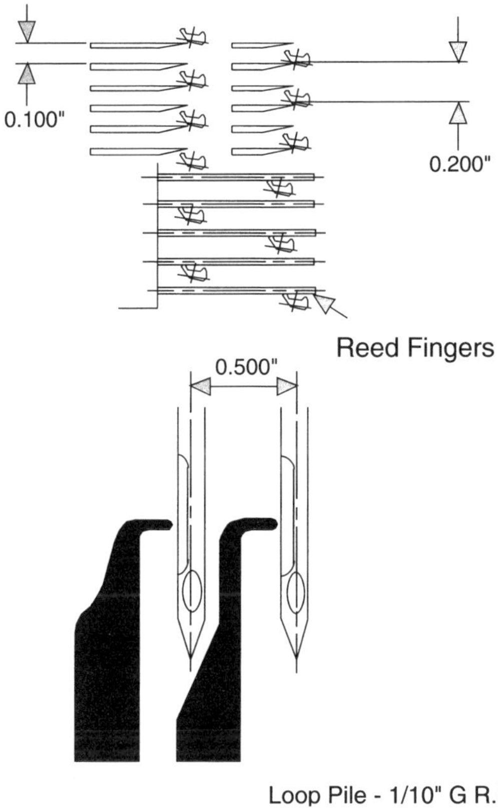

Figure 7.10. Arrangement of tufting elements in Cobble's Hi-Fi system for crossover tufting. (Courtesy of Cobble Blackburn Ltd)

computer according to the requirements of the design. Two such cams are required for a double needlebar machine.

Cam mechanisms are satisfactory for producing residential carpets, where designs are run regularly for several years, but are insufficiently versatile for those sectors of the contract carpet market that require carpets to be individually designed for customers.

More versatile systems are mostly based around servo-motors linked to computer patterning devices. For example, Cobble's Electroshifter has a hard disk drive that accepts and stores the patterns and (after creeling) the machine can be set up to run a particular design in less than a minute. The pattern, needlebar movement and creeling sequence are shown on a monitor and can be printed out if required. There is the facility to make small adjustments to the tufting elements to achieve the required carpet quality.

7.3 COMBINATIONS OF PATTERNING ATTACHMENTS

The principal patterning systems associated with tufted carpets each have fundamental limitations: the buried-pile technique utilises a small number of colours, while the sliding needlebar technique can have many more colours but all the colours are always visible in the pile. By combining variations on these and other techniques the patterning scope of tufting can be extended enormously.

Some combination systems involve quite modest changes to conventional tufting machines, but in many cases the combined machines have two distinct patterning attachments, one on each side of the machine, each feeding pulses of yarn to a sliding needlebar (ie, they are double sliding needlebar machines with a tensioning or other patterning attachment associated with each needlebar). The trend is for all the various systems to be based around servo-motor control and linked to control panels that can recall a carpet quality and design from stored digital information, although many existing machines are based on traditional mechanisms. The possible combinations, including machine gauges, are considerable and have not been exhausted at the time of writing. Perhaps the greatest potential is from combinations with the FRS or servo scroll systems. Many of the machines are not made in large numbers but tend to be purchased by particular companies to enable them to pursue their individual design themes. The following should be regarded as examples.

7.3.1 Mini-Level Cut/Loop

Cobble's Mini-LCL machine (Figure 7.11) has loopers facing the direction of manufacture and cut-pile hooks directed against the direction of manufacture, similar to the configuration of the Velva-Loop machine described in Section 7.1.7, but has a double sliding needlebar mechanism. The needle combinations on the bars are not necessarily alternately for cut or loop pile, but can be, say, in pairs or fours for cut or loop; and the needles in the bars can be staggered as well as straight.

Carpets from the Mini-LCL machine have the same kind of small repeat designs as are obtainable from conventional double sliding needlebar machines, but the pile is a combination of cut and loop.

7.3.2 Enhanced Graphics

Card-Monroe Corporation's first Enhanced Graphics machine (Graphics being their trade name for their double sliding needlebar machine) combined a scroll patterning machine with a cam operated double sliding needlebar machine, thus allowing the surface colours of a design area created by one bar to be changed by the buried-pile technique, as well as transposed laterally. The range of machinery under this heading now includes the following:

- Enhanced Graphics with Quick Thread attachment, utilising servo-shifter or computer generated cams.

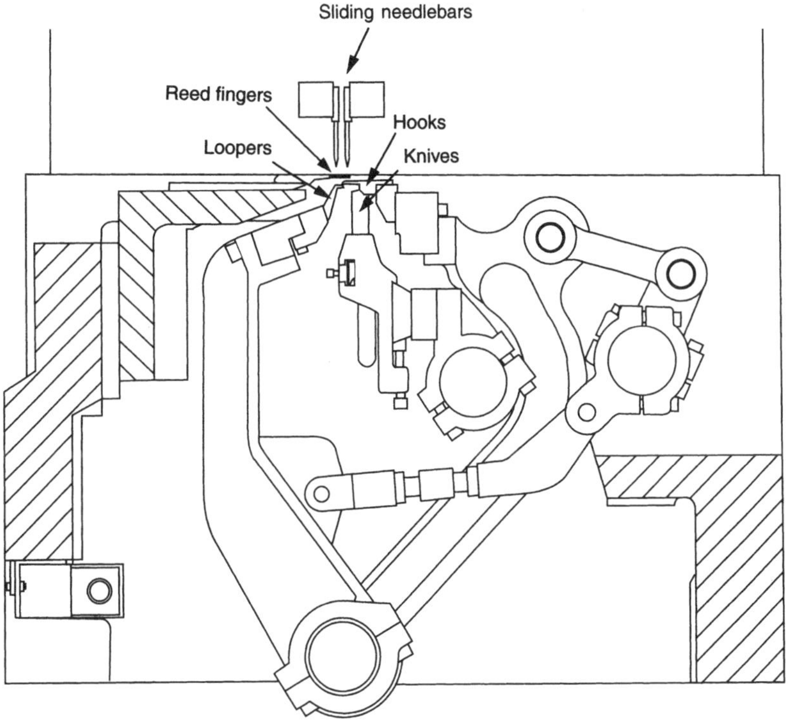

Figure 7.11. Essential components of the Mini-LCL machine, emphasising the mechanisms for the loopers (left) and cut-pile hooks (right). (Courtesy of Cobble Blackburn Ltd)

- Enhanced Precision cut/uncut, where the cut pile is the base, with designs on the loop-pile side. Patterning is by servo-drive roll or scroll systems, utilising computer generated cams.
- Enhanced Precision Loop-Pile Graphics with servo-drive four-roll or scroll patterning, utilising computer generated cams.

7.3.3 Split Rainbow Roll Pattern Attachment

Tuftco's Split Rainbow device can provide pulses of yarn to five ends of a design repeat (but six are said to be possible) from both sides of the machine to a double tufting area. The rollers are servo controlled so that the various yarns can be tensioned according to the requirements of the design stored in the Encore control system. The rollers are not continuous across the machine, but each is sectioned into six open-ended rollers for ease of threading.

The Split Rainbow system is intended to be combined with a variety of tufting techniques, including textured loop, Velva-Loop and Velva-Weve (high/low cut pile on one bar plus cut/loop on the other). Examples of carpets tufted using the Split Rainbow System are shown in Plate 8.

7.3.4 Multi Roll Attachment

Cobble's Multi Roll Attachment is designed to be combined with sliding needlebar and other patterning techniques (Plate 9). Figure 7.12 shows regular yarn feed on the left and a Multi Roll Attachment on the right feeding a double sliding needlebar tufting unit. Between four and six sets of servo controlled feed rollers can be provided. They span the width of the machine but are sectioned into snap-fit rollers for ease of threading, which can be done from floor level. As with the Split Rainbow system, the yarn feed rollers are close to the tufting area, providing positive response to the digital control.

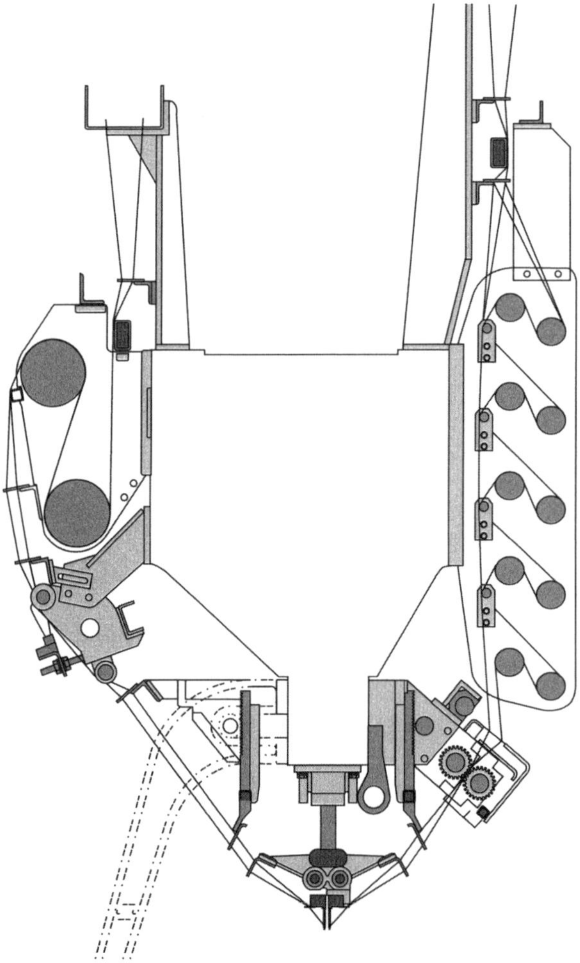

Figure 7.12. Threading diagram for double sliding needlebar machine with plain yarn feed on the left and Multi Roll patterning on the right. (Courtesy of Cobble Blackburn Ltd)

7.3.5 Combinations with Full Repeat Scroll

Using a double needlebar machine, FRS may be combined with other patterning systems (Plate 9). Figure 7.13 shows a combination of disc cam texturing with FRS.

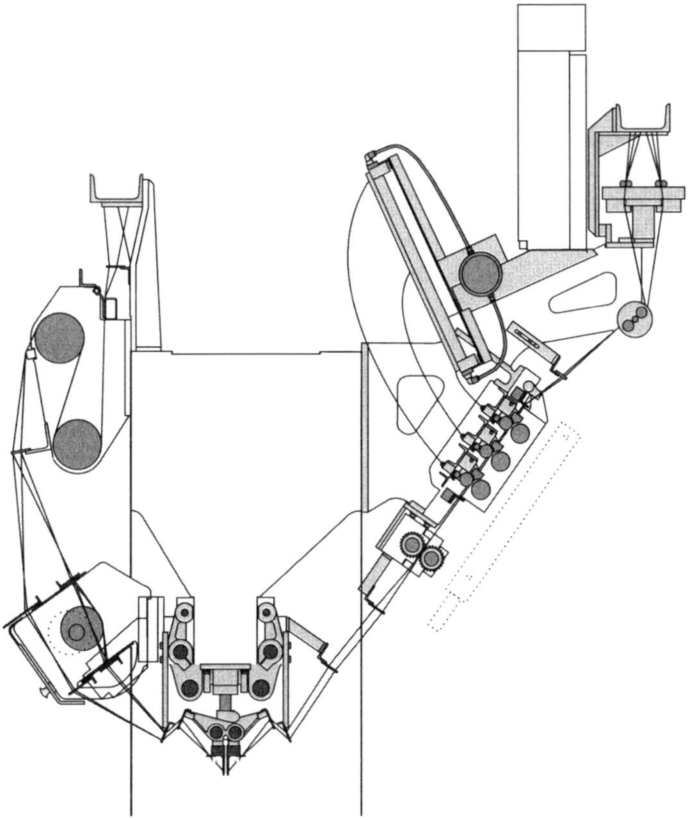

Figure 7.13. Threading diagram for a double sliding needlebar machine with disc cam patterning on the left and Full Repeat Scroll on the right. (Courtesy of Cobble Blackburn Ltd)

7.3.6 Texture Patterning

Cobble's Texturetec system provides patterns of level surface texture, rather than patterns of colour (Plate 10). It is based on FRS technology combined with a double needlebar: there is a FRS patterning system feeding each needlebar. The front bar is 1/5 gauge tufting straight. The rear bar also tufts 1/5 gauge but performs a fractional gauge zig-zag movement (see Section 6.11) to provide simulated 1/10 gauge pile in a non-linear texture. The two FRS systems operate in tandem so that when the 1/5 gauge straight texture is low, it is buried by the 1/10 gauge non-linear texture, and vice-versa.

The overall result is patterns similar to those produced by some wireloom weaving techniques (see Section 10.2).

7.3.7 Combinations of Gauges and Pile Type

Card-Monroe's Wilton Organic machine has a double sliding needlebar mechanism operating on two different gauges and pile types. The rear bar is 1/4 gauge and produces a soft cut-pile base. The front bar produces 1/8 gauge loop pile, and is coupled to an electronic scroll patterning system so that it can produce free-flowing designs that complement the geometric effects obtainable from the sliding needlebars. The needlebars may be actuated by either a servo-system or computer generated cams.

7.4 COLOUR INTERCHANGE SYSTEMS

A long-term goal in the carpet industry was to be able to associate the patterning scope of carpet weaving with the high production rate of tufting. Developments based on adhesive bonding have achieved only limited success (because of low production rates) and a yarn printing/tufting system that was, in effect, a more modern version of tapestry carpet weaving (the Crawford-Pickering system) had only ephemeral appeal because of the lack of design precision. Printing of tufted carpets has been the system that has enjoyed the most widespread success. However, weaving can provide a crispness of design that is lacking in printed carpets, so the tufting industry has felt a need for a system that can implant individually coloured tufts in cut pile to achieve Axminster style designs. Current systems achieve this objective and operate more rapidly than weaving looms. Of course, patterning systems never operate as fast as plain tufting and the more elaborate the patterning, the greater the restriction in speed.

Three examples are introduced below.

Individually controlled needles

Cobble's Individually Controlled Needle (ICN) provides the facility to tuft or not to tuft cut pile according to the requirements of the design, without the need for buried yarn in the backing. Each needle carries only one colour.

The mechanism of the ICN is shown in Figure 7.14. When not required for tufting, the needle holder is held in its upper position by the tension of the lift spring. The cylinder bar, which is analogous to the needlebar in a conventional machine, is reciprocated vertically by a system of push rods, and holds a row of pneumatically operated latch pins, one for each needle. When the pattern calls for a needle to sew, the latch pin engages the needle holder and the needle reciprocates to interact with a conventional cut-pile hook, generating cut-pile tufts, until the digital patterning system disengages the latch. A tail of yarn is left on the backing at each change of sequence.

Yarn feed is effected by the action of a 'tear-drop' yarn clamping device that grips the yarn and pulls it from the creel during the downstroke but slips over the yarn on the upstroke. Thus, only the needles that are sewing take up yarn. Yarn feed rollers are unnecessary.

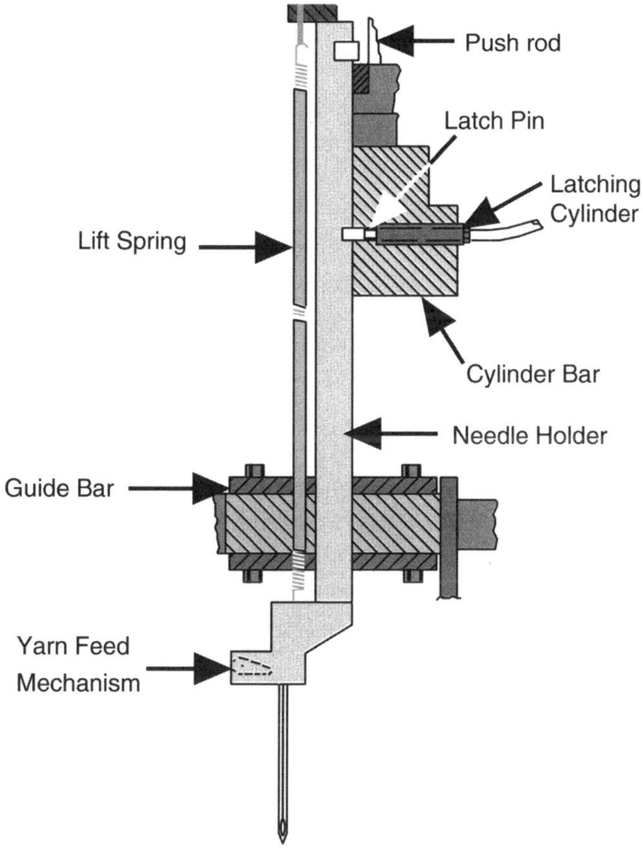

Figure 7.14. Components of the individually controlled needle. (Courtesy of Cobble Blackburn Ltd)

Because of the relative bulk of the needles, the finest (indeed only) gauge has been 3/16 in. The nominal speed is 450–500 rpm.

For many years, the limitations of the ICN machine meant that its use was confined to the production of bath sets and other small goods. Later, Bigelow, USA, initiated the use of the machine for overtufting pre-tufted carpet, and broadloom ICN machines were developed for this purpose. When overtufting, the pile height is a few millimetres higher than that of the base fabric so that only the colour of the overtufting yarn is visible. Although the overtufted area has denser pile than the base fabric, this is considered to be a less serious fault than the half-gauge surface pile in buried-pile and cut/loop designs.

Designs are held on disk and can include full-width repeats, as well as narrower repeats. Thus, although the overtufted designs are usually in a single colour or in space-dyed yarn (Plate 11), the overall possibilities are extremely interesting.

Colortec

The Colortec machine essentially combines ICN, sliding needlebar and servo cloth-feed technology, and allows individual implantation of up to six colours in carpets 5 m wide. Its design scope, therefore, approaches that of gripper Axminster weaving with electronic jacquard, including the possibility of exceptionally large repeats (Plates 12 and 13). Colortec was developed by Ohno, Japan, and Cobble, USA, and was first used for producing rugs. Outside Japan, the machine is widely used for producing broadloom carpets. Its versatility makes it particularly suitable for the hospitality sector of the contract trade.

The Colortec machine has a single sliding needlebar fitted with ICN needles, and the needles are threaded with up to six colours in the same repeating sequence across the machine. Its creel holds one cone of yarn per needle, so that a standard creel holds all the colours required, up to six.

Several cycles of the machine are required to produce one row of tufts, and the backing fabric remains stationary while the tufts are inserted. The number of cycles required to complete a row is the same as the number of colours in the design, but the colours are not simply tufted one per cycle. When producing a five colour design, for example, the sliding needlebar moves five times, one gauge width at a time, so that each needle in a group of five has the opportunity to interact with each of the five loopers in the sequence. The needles bearing the five colours are actuated by the digital patterning system as required by the design, so that the row of coloured tufts is filled progressively during the five cycles.

A particular feature of the product is that a tail of yarn is left on the backing after a colour change. These tails must be cushioned in some way at the stage of backcoating.

Clearly, the output of the machine is heavily dependent on the number of colours in the design. With six colours, 400 rpm and 70% efficiency the output of a machine is about 9 linear m/h, or about three times that of an Axminster loom (depending on which generation of Axminster loom is selected for comparison).

A common machine gauge for Colortec is 1/7 in., corresponding with 7-pitch (standard pitch) Axminster. Other available gauges are 3/16, 5/32 and 1/8.

As well as the machine's high rate of production relative to weaving and exceptional design scope relative to other tufting systems, the Colortec has two other particular advantages. The creel is no bigger than a conventional one (much smaller than an Axminster creel) and therefore relatively easy to creel up with coloured yarns; and the width of tufting can be easily changed (eg, from 5 to 4 or 2 m, important for the contract trade), without the need to change the threading.

Computerised yarn placement

The Computerised Yarn Placement (CYP) machine developed by Tapistron International (now CYP Technologies), USA, is an alternative, and very different system for producing Axminster style carpets by tufting. Its concept pre-dates that of the Colortec machine, but it was under development for a longer period so the Colortec came into general use earlier. The principle of operation of the CYP machine is quite different from that of any other tufting machine: it is based around hollow needles with a cut-pile facility and a system for conveying the backing fabric in two dimensions, permitting variable gauges to be produced.

The hollow needle is engineered with a tip that opens the backing fabric with minimum damage (woven backings that close to grip the tuft are recommended) and a groove that acts with a fixed knife to shear the pile yarn on the downstroke. Each needle has up to six yarn options running through tubes in the tube bar: a funnel channels the selected yarn colour into the hollow needle. Propulsion of the yarn from the tube bar to the needles is by airflow. The needles are located 2 in. (50 mm) apart so that each has a needle path 2 in. wide.

Figure 7.15 shows a cross-section of the tufting process. As the needles reciprocate vertically, the backing fabric is advanced as in conventional tufting, but more slowly, and is also shifted laterally by the pin rollers. After traversing the full width of the needle path (2 in.) the lateral movement of the backing fabric is reversed so that the backstitch has a zig-zag appearance. Thus, the stitch rate of the carpet is determined by the machine rpm and rate of advance

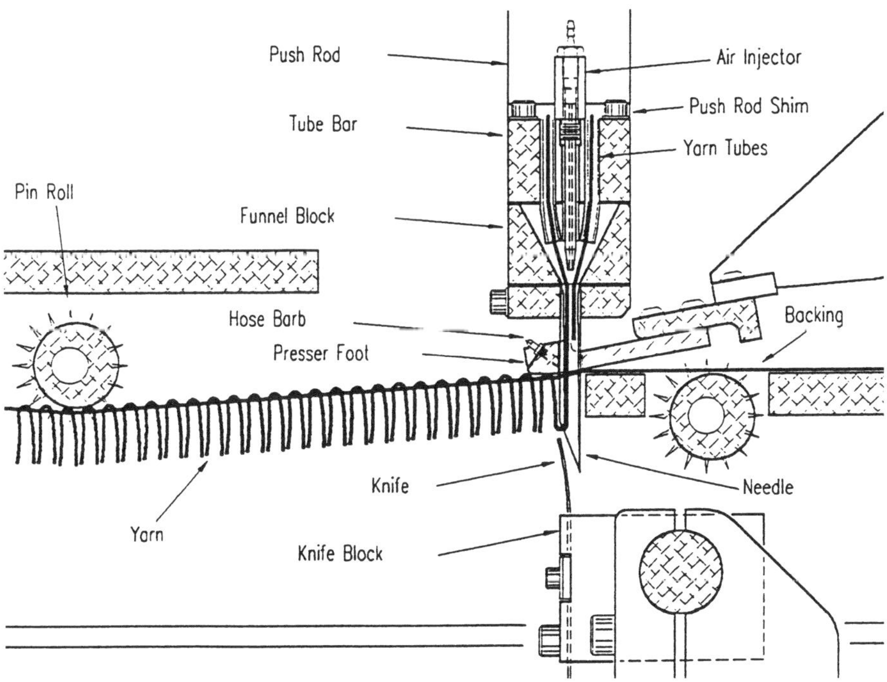

Figure 7.15. Cross-section showing the principle of the Computer Yarn Placement Machine. (Courtsey of Tapistron International Inc)

of the backing fabric, in the usual way, but the tufting 'gauge' is determined by the lateral movement of the backing fabric between the needle paths (ie, it is based on the number of stitches between reversals). On the one machine, gauges ranging between 1/6 and 1/20 can be produced.

Gauge and stitch rate, along with machine speed and pattern length, are selected via a touch screen on the central control panel. Pile height is controlled by pushrod shims on the tube bar plus, for large changes only, adjustment to the needle stroke. The knife rail must be adjusted accordingly. Pile heights in the range 8–16 mm can be obtained.

Pairs of feed gears pull the pile yarns from the creel as required by the design. The end of the gear assembly is linked to the gear air cylinder rod: when the air cylinder is activated the rod extends and meshes the gears, feeding the required colour of yarn to the pneumatic system for tufting. When the air cylinder is deactivated the yarn is held stationary.

The CYP machine operates in conjunction with a compatible NedGraphics CAD system that outputs designs on a diskette that is loaded into the CYP machine computer. Designs can be viewed and manipulated (compressed or extended, colours changed) on the main menu screen of the control panel. The digital design actuates the air cylinders that in turn control the yarns fed to the various needles.

A standard design of tufting creel is used, except that each yarn exits the creel through a simple ball-type tension device to maintain constant tension and prevent yarn from falling from the creel header. The unique aspect of the creel is its small size. As the CYP machine has only 20 needles/m (40 in.) only 120 cones of yarn/m of machine width are required. Two widths of machine are available: 2 and 4.4 m.

A remarkable feature of the CYP machine is the short down time required to re-creel it, contributing in a major way to its flexibility. Clearly, the output of a tufting machine is heavily dependent on the number of tufting positions available, as well as on the carpet construction and feasible rpm, and the CYP machine has only a small number of needles. Overall, the machine may be regarded as having the capability of efficiently producing short runs of designs and constructions in cut pile to customer's requirements from quite small lots of yarn (ie, one of outstanding flexibility, rather than high productivity).

REFERENCES

1. L Tezer. Vario-Design – a new tufting system. *Melliand International*, 1996, 2, 86–88.

8 Craft and Related Techniques

A variety of small-scale techniques for producing carpets and rugs are in use in different parts of the world which, however, produce mostly high value items and amount to industries of significant financial importance. Though they are as diverse as hand knotting and pass tufting, a uniting factor is a strong artistic element.

Note that there is no essential difference between a bordered carpet and a rug: both are covered by the term 'squares', often used in the European trade. However, a carpet occupies most or all of the area of a room, whereas a rug is a smaller item used to dignify some particular part of the room. 'Carpet' is the more general term.

Hand knotting is by no means a dying art. Demand is created by consumers interested in traditional arts and crafts and by 'western' companies wishing to have high quality products made to their particular requirements. Supply has been fostered as a means of creating employment in countries where labour is plentiful, cheap and skilled.

Artistic rugs and wall hangings can also be produced by tufting using hand-held 'guns'. The styles are quite different from those produced by knotting, but tufting and knotting actually compete in the market area of individually designed rugs.

Custom tufting companies make carpets to order, using small-scale equipment such as pass tufting and table-top machinery, as well as hand-held tufting guns. Despite the apparently small scale of the equipment, the value of the orders may be quite high: for example, it is common in South East Asia for a whole hotel to be furnished with heavyweight wool carpets, ranging from wall-to-wall carpets in corridors and lift lobbies to spectacular squares that form centrepieces for public areas, all made by hand-held equipment. At the other end of the volume scale, a bordered wall-to-wall carpet in company colours could be provided for an oval boardroom, or an artistic wall hanging could be designed for the chief executive's office.

Although not strictly custom carpet, the techniques described in this chapter are also used by artists/craftspersons who create individually designed rugs and wall hangings for residential sales, and by companies producing bath sets and other small goods.

The companies involved in custom (bespoke) tufting are service oriented and have individualistic styles. Their enterprise can result in their devising special equipment and techniques for their own use. Only equipment that is commercially available is described in this chapter, however.

8.1 TRADITIONAL HAND KNOTTING

Hand looms for carpet manufacture may be either horizontal or vertical. Nomads usually use horizontal looms, having components pegged to any convenient patch of ground (Figure 8.1). Only relatively small rugs can be made on such looms, the length being restricted by the reach of the weaver.

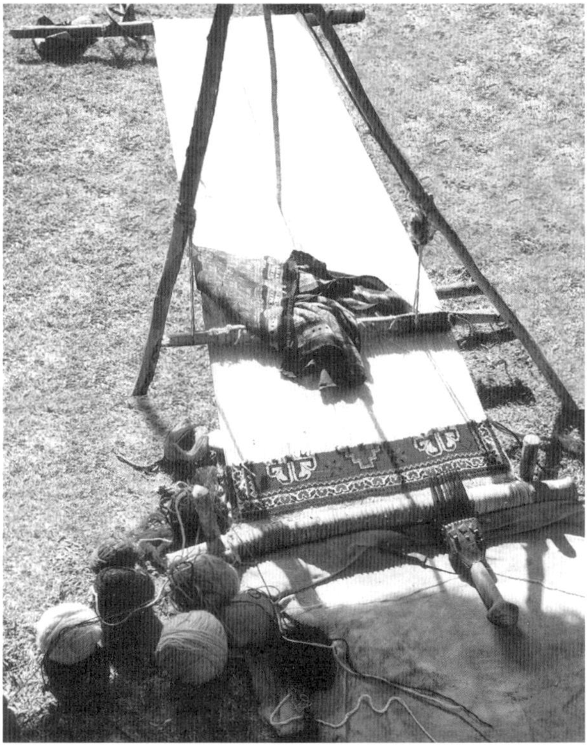

Figure 8.1. A portable carpet loom, as used by nomads. (Photograph: Warren Meade)

The upright loom (Figure 8.2) used domestically and in workshops consists of two posts supporting horizontal rollers. The upper roller holds warps which are stretched between it and the lower roller. Often, the warp is inclined from the vertical. As the carpet is formed, it is taken up on the lower roller and warp is let off from the upper beam.

Figure 8.2. Knotting on a vertical loom. Note the levers located behind the operative that can be used to reverse the shed of the warps. (Photograph: Feng Liu)

In factory operations, the warp may be fixed and the loom is tall to accommodate the full size of a large carpet. In such cases, several weavers may operate together on the same carpet and they move from the base of the loom to work on simple gantries as the carpet grows.

In addition to the frame holding the warp, there may be a reed maintaining a uniform spacing of the warp ends and preventing entanglement and, always, a system for creating a shed through which the weft can be inserted. The shedding mechanism consists of two wooden rods inserted horizontally through the warp. The warp ends pass alternately in front of or behind one rod and on the reverse side of the other. By moving one of the rods relative to the other, a shed is formed through which weft can be inserted to create the backing structure. A reverse shed is created by pulling harness yarns linked to the alternate warps that are at the back (away from the weaver) when the original shed is formed.

When two weft threads have been inserted, one through the shed and one through the reverse shed, the warp ends are levelled and a row of pile tufts, selected from bunches of coloured yarns available in the weaver's lap or hanging nearby, is knotted around the warps. The knots are tightened and cut to length. Another pair of wefts is inserted and a heavy metal comb is used to beat down the wefts, creating a dense backing fabric and locking the tufts firmly into place. The cycle is repeated. A skilled weaver may tie up to 10 000 knots a day.

Selection of the colours of tufts to create the designs may be made on the basis of point paper, a design sketch or, in the case of nomadic weavers, memorised traditional designs. Improvisation by nomads is considered acceptable, or even desirable, in some areas, whereas at the other extreme a design master may translate designs from 'western' artists into specifications for custom hand-knotted carpets.

When weaving is complete, adjacent warp ends are knotted to create a fringe and prevent unravelling. The long sides of the carpet are bound, usually with wool, to finish the edges. Carpets from the loom are shorn by hand to level the pile and washed to clean them and remove loose dyes. Washing may be in cold water, but the larger carpet companies often wash carpets in chemicals to age them artificially (see Section 17.7).

Current general practice is to use cotton for warp and weft and wool for the pile, but there are variations on this basic procedure. Traditional flax may be used for the backing material, and goat hair, camel hair or silk are used in the pile.

8.1.1 Types of Knot

There are two principal types of knot: the Turkish or Ghiordes knot, named after a weaving centre in Turkey, and the Persian or Sehna knot.

The Ghiordes knot is the most common and is widely used outside Turkey. It is formed by placing a length of pile yarn across two adjacent warp ends and turning its ends around the back of the warps to emerge between the centre of the pile and the warp. The two ends of the pile are located together between the warps (Figure 8.3).

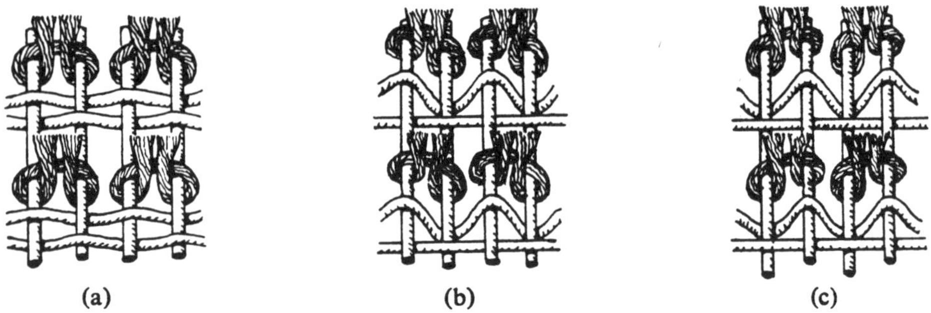

Figure 8.3. The Ghiordes or Turkish knot: (a) Tied with the warp on one level and two weft threads between each of the knots. (b) Tied with the warp on two levels, knots inclined to the left. (c) Tied with the warp on two levels, knots inclined to the right [1].

The Sehna knot is also tied around two adjacent warp ends, but the pile encircles one warp end and passes only around the back of the other. Thus, one pile end emerges between every warp end (Figure 8.4). The encircled warp may be on the left or the right, and the pile inclined accordingly.

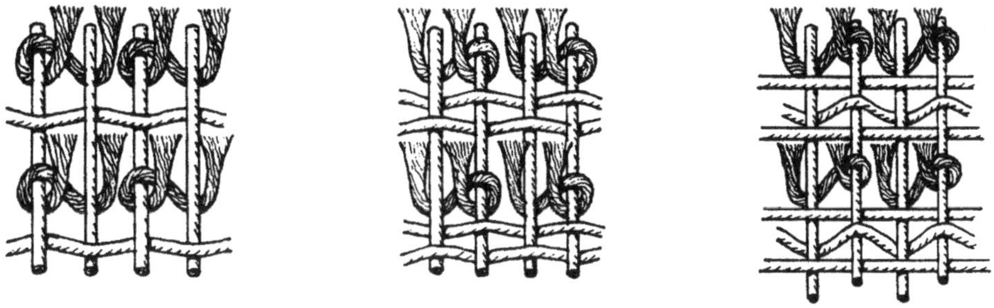

Figure 8.4. The Sehna or Persian knot: (a) Left-hand knot: warp on one level, one weft thread between each row. (b) Right-hand knot: warp on one level, two weft threads between each row. (c) Right-hand knot: warp on two levels, three weft threads between each row [1].

The Ghiordes knot appears to be more secure, but the Sehna knot is said to be more suitable for the finer constructions of carpet.

There are many variations on the two common knots. Most usually, knots are tied around the same pairs of warp ends right throughout the carpet. However, alternate rows of pile may be formed around alternate pairs of warp ends to create a twill effect.

To create fullness in the backing, one of a pair of wefts may be pulled very tight to straighten it and the other left crimped. This arrangement also changes the levels of alternate warp ends, so that a Ghiordes knot may be inclined to the left or the right, depending on which pair of warp ends is selected for knotting (Figure 8.3 (b) and (c)). It is also known for there to be one, two or three weft shots between each row of tufts, and this can affect the levels of warps and therefore the detailed geometry of the tufts (Figure 8.4).

The Jufti system of knotting, devised to accelerate production, involves looping the pile around four warp ends rather than two. It can be applied equally to Turkish or Persian style knots (Figure 8.5). Clearly, the Jufti knot is unsuitable for producing carpets with fine design definition, but the lack of quality in this sense is usually compensated by providing a thick pile.

The Tibetan knot incorporates a remarkable combination of Persian and Turkish knots and a simple U-tuft. It arises because of the system used by Tibetan settlers in Nepal of 'weaving with the iron rod'. An iron rod is placed in front of the warp and a piece of pile yarn, which may be fairly long depending on the requirements of the design, is looped around two warp ends above the rod, over the front of the rod and then around the same two warp ends, but below the rod. At the beginning of each row, and before each colour change, a special, more secure knot is tied which wraps one of the warp yarns once and the other, twice, as in the upper part of Figure 8.6.

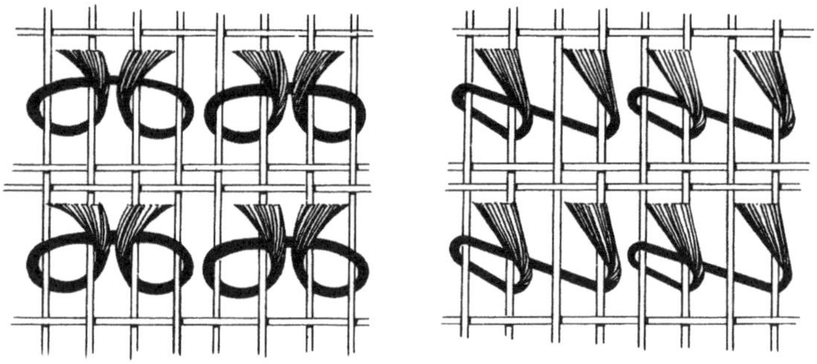

(a) (b)

Figure 8.5. (a) Jufti or double Turkish knot; (b) Jufti or double Persian knot.

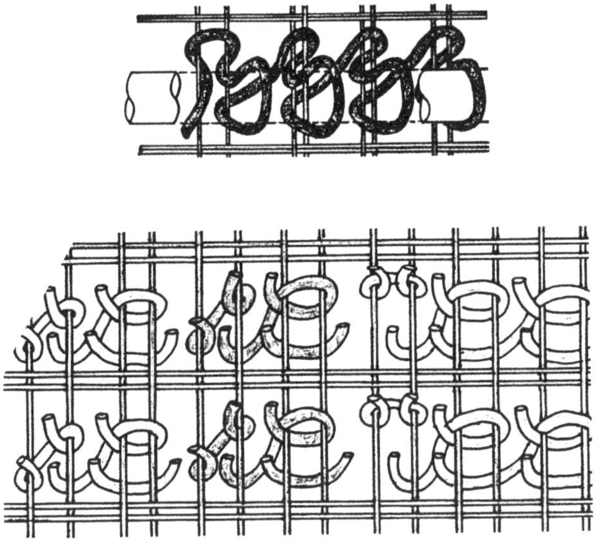

Figure 8.6. Weaving with the iron rod as practised in Nepal. Above: configuration of one colour section as first looped. Below: configuration of three colour sections in two rows after cutting.

When the rod is full, a knife is run along it to cut the pile, the rod is removedand the pile is beaten down to give the configuration shown in the lower part of Figure 8.6. The first knot in a colour becomes a Turkish knot and a sequence of overlapping Persian knots follows, with the final tuft an unknotted 'U' overlapping the last Persian knot. Nepalese carpets are typically thick and heavy.

The single warp knot (Figure 8.7) is used in some Spanish carpets. A length of pile yarn encircles one warp end, crosses at the back, and the ends are brought

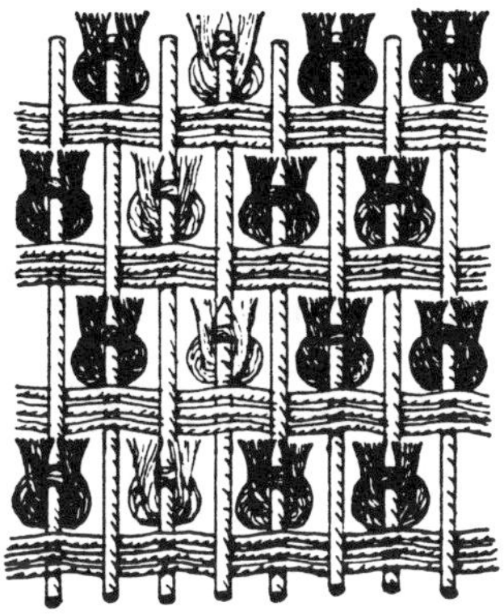

Figure 8.7. The single warp knot as used in some Spanish carpets, illustrating the smooth diagonals and serrated outlines in a pattern [1].

through a loop, one at either side of the warp yarn. The knots may be tied around each warp, but the illustration in Figure 8.7 shows knots on every other end, with the knotting staggered between rows. The single warp knot tied in this way gives designs with clear diagonal lines but serrated vertical lines.

WRONZ has developed a loom that produces a standard Persian knot more rapidly than conventional looms, using an ingenious knotting technique. Work on the WRONZ loom also stimulated the development of the 'Kiwi' knot structure that has tuft bind intermediate to that of the Persian and Turkish knots.

8.2 CUSTOM TUFTING

8.2.1 Pass Tufting

The technique of pass tufting was actually developed before broadloom tufting, as a means of producing full-width carpets on machines having the relatively small number of needles that were feasible in those days (see Section 1.3.4). Pass tufting remains important as a convenient means of providing carpets to room size with minimum waste of materials. Often, yarns for pass tufting are dyed in small lots to the customer's specified colours.

A typical pass machine includes a tufting head of relatively conventional design, usually for cut pile, but only 12-21 in. (300–530 mm) wide. The tufting head is cantilevered from the headstock, allowing space for backing fabric to be gathered partially into a rope, leaving only a flat strip for tufting. When one strip of carpet has been tufted, the material is returned to the front of the machine, a second strip is released from the gathered backing fabric, the needles are carefully aligned against the previously tufted strip with the aid of guides, and the material is again passed through the machine. Further passes are made until the required width has been tufted or the backing has been completely filled with pile.

An upper cloth feed roller is supported from the tufting head and can be raised to facilitate lateral movement of the backing fabric. The bed of the machine is mounted at one end of a long (up to 6 m) table that supports and optionally drives the portion of the backing fabric that holds pile. The operation of a pass machine is assisted by the use of motorised scrays, J-boxes or roll-up devices. Creels are similar in design to those for broadloom machines, but are smaller.

The GT Heavy Duty Pass Tufting Machine (GT Supplies and Engineering) exemplifies pass tufting machinery. It is made in a variety of sewing widths (eg, 12, 18 or 21 in.) and gauges from 1/10 to 1/2, cut pile or loop pile. The bed plate is adjustable by handwheel and the needle stroke is adjusted by screw action: pile heights from 6 to 50 mm and stitch rates from 17 to 50/dm can be obtained. Drives for the yarn and cloth are infinitely variable and speeds can be up to 650 rpm. Around 250 m^2 of carpet per shift can be produced, depending on the carpet construction.

8.2.2 Table-Top Tufting

Table-top tufting machines pre-date pass machines, and were originally developed from heavy duty sewing machines to manufacture candlewick bedspreads (see Section 1.3.2). They are now offered for tufting rugs and small goods such as bathroom sets and bedside sets. The tufting width is usually narrower than that of a pass machine and smaller areas of backing fabric are used. The backing fabric is guided by hand, and skilled operators can tuft curved as well as linear areas. Designs may be drawn on a piece of backing fabric and the various areas filled in with yarns of different colours.

The Venor cut-pile rug machine of David Almond, UK, is typical of table-top tufting machinery. In its standard form, it has 12 needles mounted in 3/8 gauge. The pile height is fixed during manufacture, but can be specified between 19 and 38 mm: small adjustments (±2 mm) can, however, be made by the user. Stitch rates are around 20/dm, and are adjusted by varying the cloth feed. Textures can be varied by tufting two or more ends of fine yarn, instead of one end of heavy yarn. Machine speeds at start-up are slow and are accelerated by a treadle control up to 1100 rpm. A guide is fitted to the machine so that each tufting run is exactly 3/8 in. from the next. An operative could be expected to produce 380 rugs 54×27 in. (1.4×0.7 m) in a 40 hour week.

Variants on this standard machine are available in tufting widths up to 6 in., and with more flexibility of control, ranging from 16 needle 3/8 gauge to 40 needle 1/8 gauge, with a standard pile height range of 5–11 mm, although non-standard machines can be provided to customers' specifications.

Rugs can be decorated with higher pile (20–35 mm) which may be cut or loop, using special overtufting machines. Overtufting machines are available in a variety of gauges and with 4, 6, 8 or 12 needles.

A cut/loop patterning machine is included in the Venor range. The patterns are formed by air operated yarn guides controlled by cams operating pneumatic valves. The standard machine is a 12 needle 3/16 gauge machine that produces 12 mm loop pile and 22 mm cut pile.

8.2.3 Weftwise Tufting

A prominent example of a weftwise tufting machine is the Sallee machine, developed by Ralph Sallee in the USA, that can produce full width carpets in loop pile. It is based on a unique hollow needle that has an eye in one side. As the needle penetrates the backing fabric (usually a 400 g/m^2 sateen weave jute fabric) the pile yarn is held by friction with the jute so that a tuft length equal to the needle penetration is withdrawn from the creel. When the needle retracts, this length is formed into a loop to give a pile height about half the needle stroke. The needle moves weftwise on an endless chain. When the end of a row is reached, the backing is indexed forwards to create the row spacing.

The working width of the machine is 4.6 m, but it can be programmed to produce intermediate widths and can also be programmed to produce a specific length and then stop, thus enabling it to make a carpet to fit a room or a carpet square. Finishing includes a shearing process. A specially designed unitary backing is applied.

Typically, Sallee machines are used to produce thick, heavy wool carpets in customer selected colours, including multiple colours, and in any width as required.

8.2.4 Gun Tufting

Single needle hand-held tufting guns are the most flexible of all the tufting machines available, since they can produce elaborate non-repeating designs in rugs (squares or other shapes) and broadloom carpets. They are labour intensive, but are used to produce goods in the upper market areas, and high prices can be obtained for the products. They are widely used by artists to produce individually designed wall hangings and squares and in the custom trade for luxury products for the hospitality and interior design markets (Plate 14).

There are two types of tufting gun: electrically operated and electro-pneumatic. The first type has the advantages of light weight and requiring only one service line, while the second also requires a supply of compressed air and therefore has two service lines. The electro-pneumatic type is usually more robustly engineered and is heavier, so it is supported by a balancing system as an aid to manoeuvrability. Figure 8.8 shows the production sequence for manufacturing rugs using an electro-pneumatic gun.

Electro-mechanical guns

The first tufting guns of this type were developed by the Tai Ping Carpet Company in Hong Kong for their own use. As the company expanded its activities into other countries in South East Asia, similar companies were set up and guns of a similar type became publicly available from engineering companies in Hong Kong and Japan [2].

A typical hand-held tufting gun is crank driven and the essential component is a combined tufting/cutting tool. The tool consists of a small scissors mechanism, spring-loaded in the opening direction, mounted on a shank and sliding in a

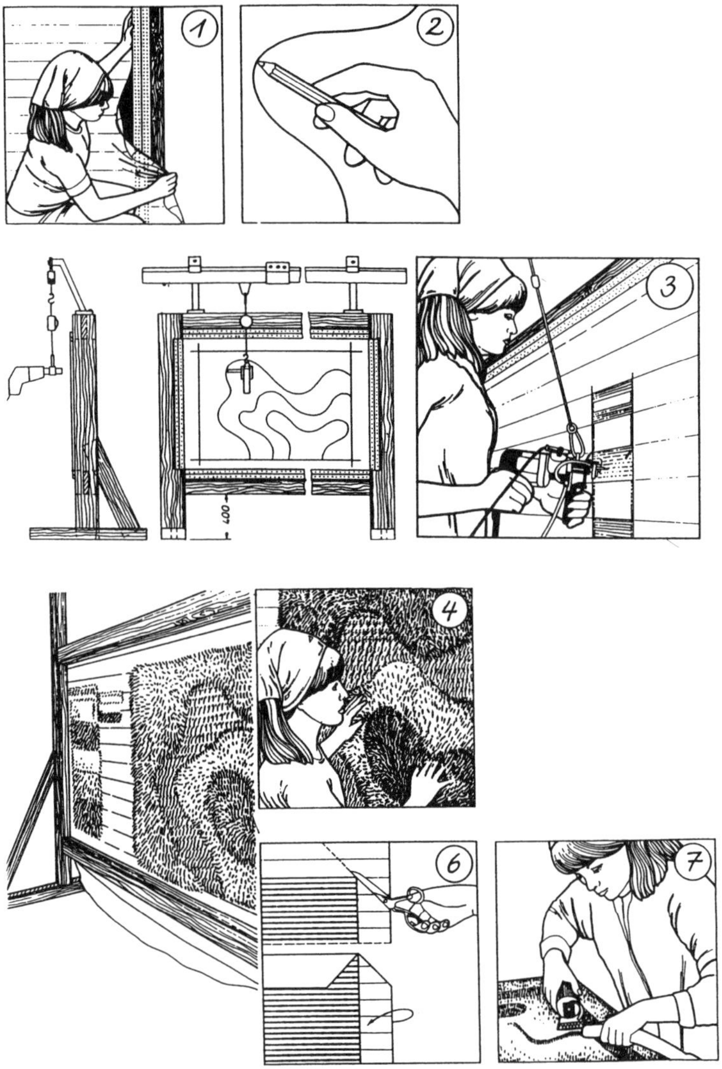

Figure 8.8. Production sequence for gun tufted rugs: 1. Mounting the primary fabric. 2. Sketching the design on to the backing. 3. The tufting step. 4. Inspection. 5. Latexing (not shown). 6. Finishing. 7. Carving, with dust extraction. (Courtesy of Hofmann Handtuft-Technik GmbH)

hollow needle and a tube that may be open at the side. The crank mechanism reciprocates the needle and scissors mechanism. At top-dead-centre the scissors are slightly open, creating a fork, and a pile yarn is laid across the fork so that a loop is formed when the scissors move forward into the needle. The scissors and the needle penetrate the backing fabric together, but the scissors move further and open up as they emerge from the needle. At this stage, the yarn is free to move back into the cutting position. Next, the scissors retract into the needle and the points are forced together, cutting the loop of yarn. Finally, the needle and scissors retract from the backing fabric leaving the two cut ends in place. One end completes the tuft partially formed in the previous cycle, while the other end provides one leg of the next tuft.

When tufting rugs, the backing fabric is mounted on a pin frame and the design is drawn on it. Tufting of the different colours then proceeds as for electro-pneumatic tufting (see Figure 8.8). When tufting roll goods, there are two possible techniques. In one alternative, a pin frame is used but has backing feed rollers at the top and a pinned delivery roller at the bottom and a batching system (Figure 8.9). Backing fabric is tensioned between the rollers and a section of carpet is tufted. The completed work is rolled on to the take-up roller and a further section is tufted. In the second method, it is possible to prepare carpets 5–10 m wide. Backing fabric is mounted on frames with the weft direction vertical and with a long length of fabric in the horizontal direction, thus making an enormous area of fabric available for work. Tufting takes place from gantries.

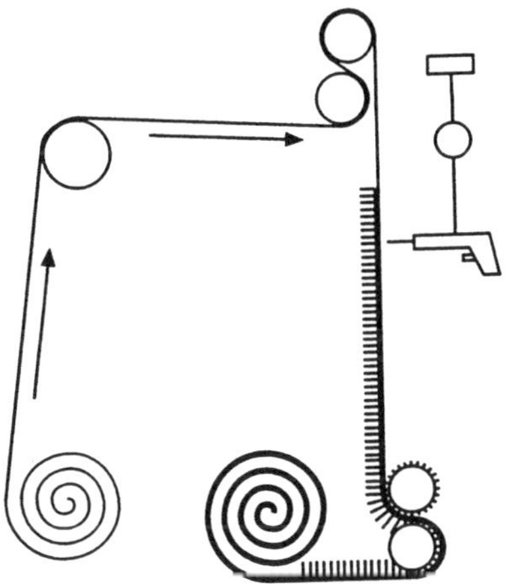

Figure 8.9. Arrangement for gun tufting of roll goods.

Electro-pneumatic guns

Electro-pneumatic guns are manufactured in Japan, Germany, the UK and USA. Most models are designed for producing cut pile, but there is also some demand for loop-pile tufting.

The principle of the cut-pile system is shown in Figure 8.10. Pile yarn is fed into a hollow needle by compressed air, and is cut by a rotary cutter, but the cut ends move together through the needle as it penetrates the backing fabric. The first part of the yarn completes the tuft partially formed in the previous cycle, while the second part provides one leg of the next tuft, as in electro-mechanical tufting. The leg lengths are controlled by a combination of needle length and yarn feed (the length of yarn fed per stitch is determined by the diameter of the driven roller in a pair of yarn feed rollers).

For level tufts, needles and feed wheels for the Hofmann VML machine are used in matched pairs, and can be changed in pairs to give pile heights in the range 16–45 mm. To create J-tufts, as shown in Figure 8.10 (d), the feed wheel only is changed so that, with a short leg of 16 mm for example, the longer leg can be adjusted between 20 and 74 mm. For lower pile heights, the Hofmann VML 12 machine is available, which gives a 12 mm U-tuft. The SML machine for loop pile does not have a cutter, and the standard version has a pile height range of 10–45 mm.

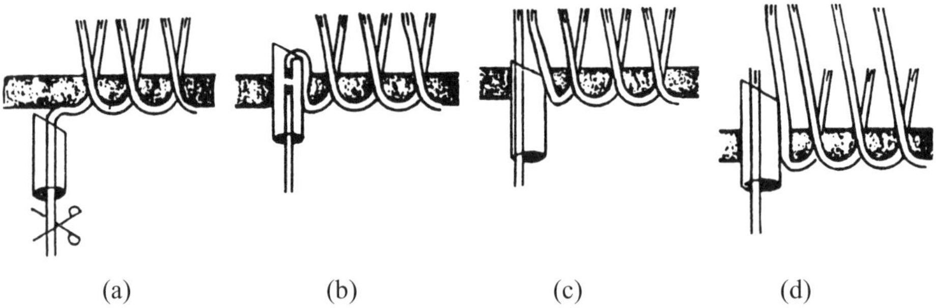

Figure 8.10. Principle of tufting with electro-pneumatic gun: (a) cutting the tuft length; (b) needle penetration; (c) needle withdrawal; (d) J-tufting. (Courtesy of Hofmann Handtuft-Technik GmbH)

The maximum yarn linear density that can be tufted depends on the internal diameter of the tufting needle. For example, a 6 mm needle of a Hoffman machine can accommodate yarns up to Nm 0.3. Needles of 4 or 5 mm diameter are available for tufting finer designs. Finer textures can be produced by tufting several fine yarns together.

The angled point of the needle drives the gun forward and, with a 6 mm needle, gives a stitch rate of 25/dm. Pile weights are in the range 2–10 kg/m^2. Machine speeds can be set between 50 and 2000 stitches/min, according to the intricacy of the design and skill of the operator. An experienced operator can produce 10 m^2 of carpet per shift.

8.2.5 Automatic Production of Rugs

There have been various proposals to automate gun tufting, or to devise other manufacturing techniques that automatically produce rugs in individual designs on a low output basis.

One possibility is to mount the backing fabric on a horizontal pin frame and have a vertical tufting gun moving on a pantograph over the fabric, under the control of a CAD system.

The Wilcom, Australia, AutoTuft™ machine has the backing fabric mounted on a vertical frame and the gun operates in vertical rows from a rail system that progressively traverses over the area, which may be up to 10 x 13 ft (3 x 4 m). A design may contain cut pile, loop pile and different pile heights, as well as multiple colours.

The Tapistron machine, described in Section 7.4, is a highly flexible machine that is suitable for short production runs of patterned carpet, and is an interesting alternative to gun tufting for the custom carpet industry.

8.2.6 Backing Fabrics

Backing fabrics of the types used for broadloom carpets (see Section 5.2) can be used for pass tufting, table-top tufting or gun tufting. For the more artistic ranges, cotton or spun polyester fabrics woven to about 70 picks and ends/dm tend to be preferred. Jute fabrics of 330–400 g/m^2 in a sateen weave are sometimes used.

8.2.7 Backcoating

In order to bind the tufts and improve dimensional stability, a unitary backing of about 1 kg/m^2 is typically used, applied by brushing or spraying. Formulations based on pre-vulcanised natural latex, air dried, require no further operation but, in large-scale production, drying may be assisted by infrared heaters. Pass tufted materials are sometimes backcoated and laminated as for broadloom products (see Section 17.4).

8.2.8 Finishing

Custom tufted, especially gun tufted, products are often sheared by machines (Figure 8.11) that are passed by hand over the carpet, rather than by passing them through a broadloom shearing machine (see Section 18.2). Such machines may be electrically or pneumatically driven and are desirably linked by a hose to a small dust-extractor unit. Machines that automatically traverse strips of clamped carpet are said to provide better control and a streak-free finish.

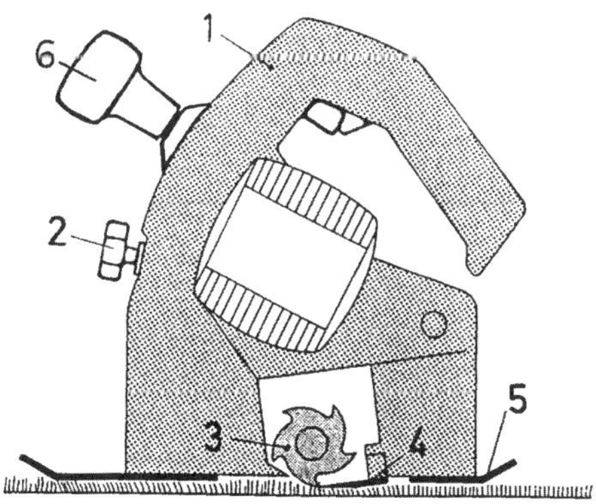

Figure 8.11. Components of a hand-held shearing machine for rugs and custom carpets: 1. Handle with trigger switch. 2. Adjustment for shearing height. 3. Shearing cylinder with helical blades. 4. Stationary blade. 5. Low friction baseplate. 6. Locking device for depth control. (Courtsey of Münsinger GmbH)

Carving (step 7 in Figure 8.8) to accentuate design areas is commonly carried out on gun tufted products, both rugs and wall-to-wall carpets. Hand-held electrically or pneumatically driven shears, similar to those used by hairdressers, but more robust, are manufactured specially for this purpose.

Edge finishing of rugs is an essential process. Excess backing fabric is trimmed to a standard width and mitred at the corners (step 6 in Figure 8.8) and the material is latexed to the underside of the rug. No further finishing is necessary for technical reasons, but a jute or other fabric is sometimes stitched to the backing, at the edges only, to improve the appearance.

REFERENCES

1. G Robinson. *Carpets*, 2nd edition, Textile Book Service, UK, 1972.
2. T Murdoch. Carpets in Asia. In: *Proceedings of 21st Annual Conference, Textile Institute (NZ Section)*, The Textile Institute, Manchester, 1993, 33–42.

Plate 1. Loop-pile tufted carpets with textured surfaces produced using a yarn feed variator (Section 7.1.2). Note the sharp change from low to high loop compared with the contoured product shown being produced by the cam-disc system in Figure 7.2. (Courtesy of Cobble Blackburn Ltd)

Plate 2. A bath set in cut/loop design produced by the UTPA mechanism (Section 7.1.3). (Courtesy of Cobble Blackburn Ltd)

Plate 3. Wall-to-wall cut / loop design produced using a clutch-in-roll patterning system (Section 7.1.5). (Courtesy of Cobble Blackburn Ltd)

Plate 4. A buried-pile design in two colours. The patterning system was the Full Repeat Scroll (Section 7.1.6), used in this instance to achieve precision of effect, rather than a large repeat. (Courtesy of Cobble Blackburn Ltd)

Plate 5. Carpet tufted from intermingled yarns using the Full Repeat Scroll patterning attachment (Section 7.1.6). (Courtesy of Cobble Blackburn Ltd)

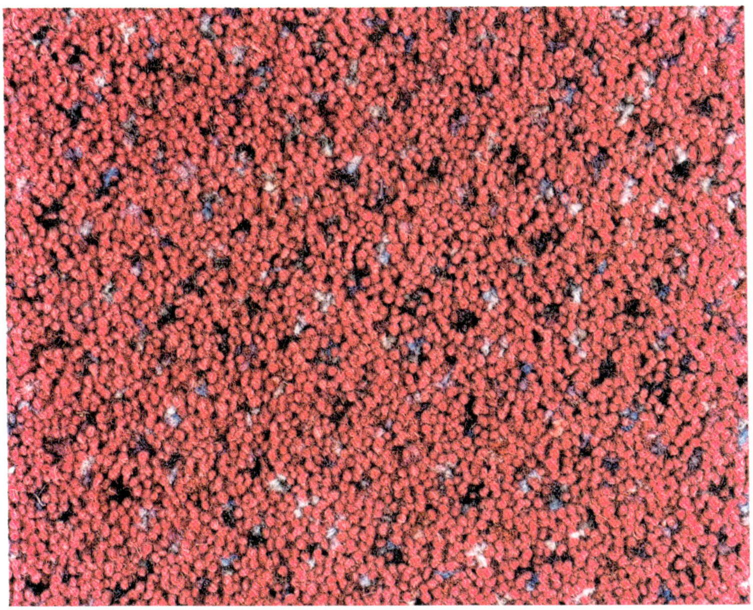

Plate 6. Cut-pile tufted carpet made using a single sliding needlebar (Section 7.2.1). One of the yarns is space-dyed and includes a colour matching the background colour, so that the simple zig-zag pattern is camouflaged. (Courtesy of Cobble Blackburn Ltd)

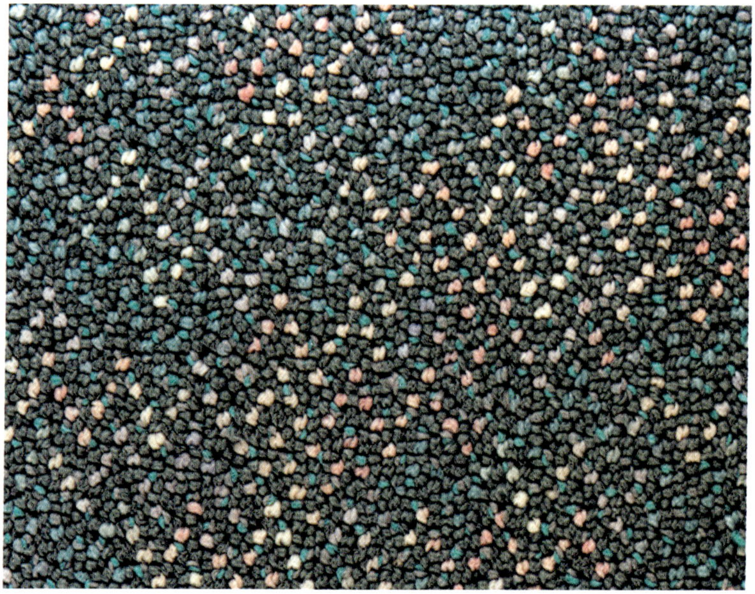

Plate 7. Loop-pile tufted carpet made from space-dyed yarn using a double sliding needlebar (Section 7.2.3). (Courtesy of Cobble Blackburn Ltd)

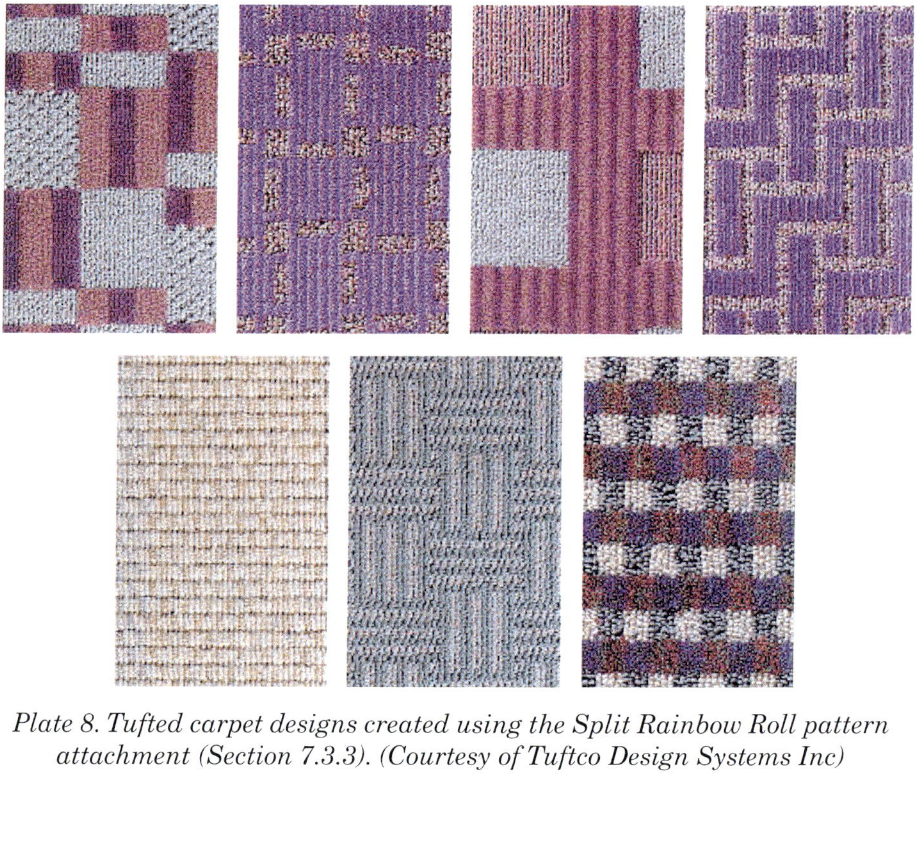

Plate 8. Tufted carpet designs created using the Split Rainbow Roll pattern attachment (Section 7.3.3). (Courtesy of Tuftco Design Systems Inc)

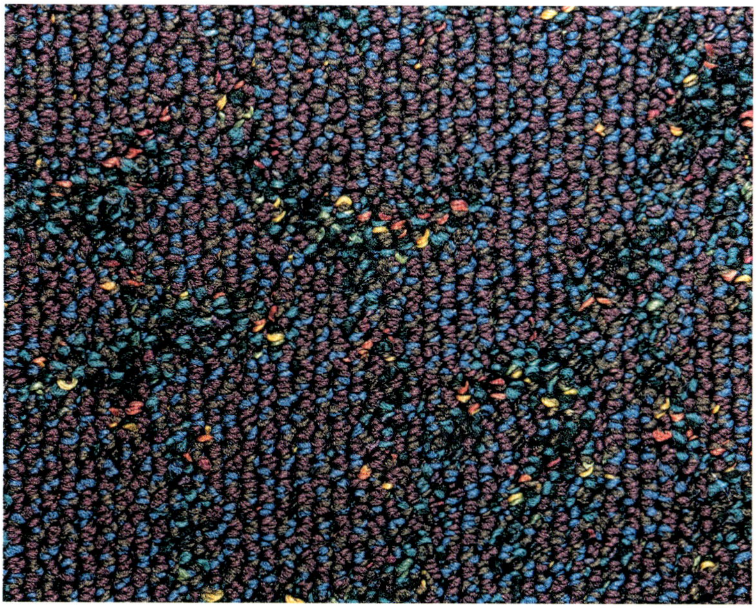

Plate 9. Combination of patterning systems (Section 7.3.4): Multi-roll and Full Repeat Scroll. (Courtesy of Cobble Blackburn Ltd)

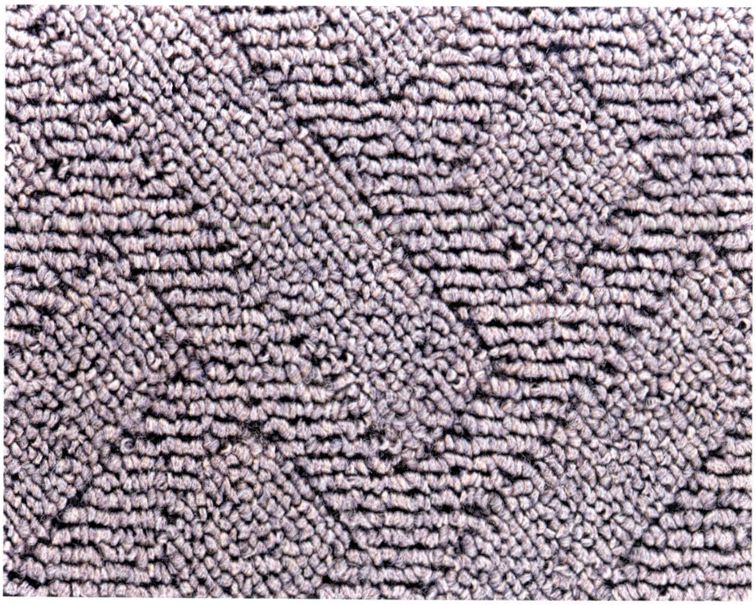

Plate 10. Loop-pile design in a single colour produced by the Texturetec system (Section 7.3.6). (Courtesy of Cobble Blackburn Ltd)

Plate 11. Overtuft design produced using a single colour ICN machine (Section 7.4) threaded with space-dyed yarn. (Courtesy of Wools of New Zealand)

Plate 12. Cut-pile carpet, Axminster style, produced on the Colortec machine (Section 7.4). (Courtesy of Cobble Blackburn Ltd)

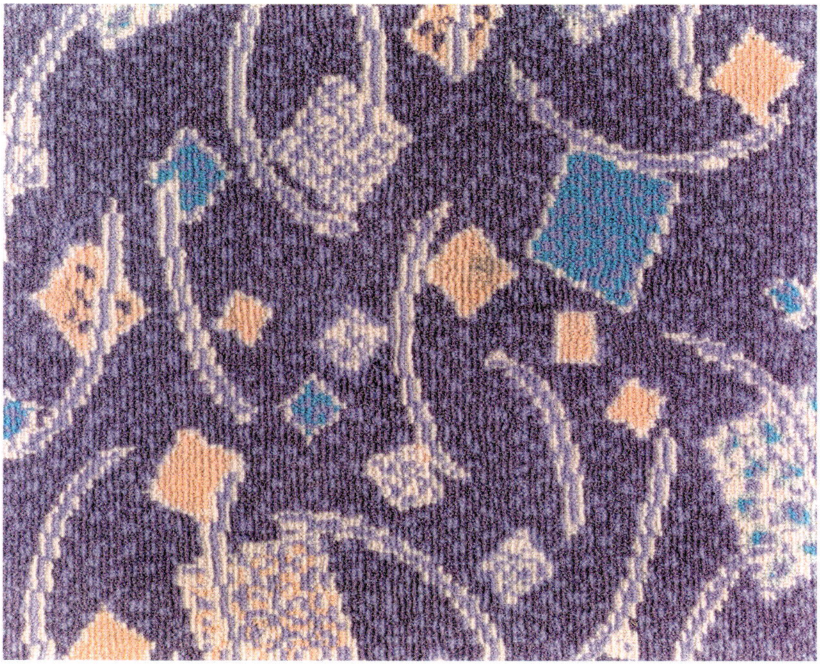

Plate 13. Loop-pile carpet, modern style, produced on the Colortec machine (Section 7.4). (Courtesy of Cobble Blackburn Ltd)

Plate 14. Gun-tufted rug (Section 8.2.4). Courtesy of Wools of New Zealand)

Plate 15. Axminster carpet in the P&O cruise liner Arcadia: a non-repeating design produced using an electronic jacquard (Section 16.2.2). (Courtesy Ulster Carpet Mills Ltd)

Plate 16. A long-repeat Axminster design at the Silver Star Casino, USA, demonstrating the capabilities of the electronic jacquard (Section 16.2.2). (Courtesy of Ulster Carpet Mills Ltd)

Plate 17. 5-frame Wilton carpet (Section 10.1) in a contract installation. (Courtesy of Michel Van de Wiele NV)

Plate 18. A collection of textured loop-pile carpets and a rug made on the LPM 32 loom (Section 10.3). (Courtesy of Michel Van de Wiele NV)

Plate 19. Carpet square in a modern design, produced by 6-frame face-to-face weaving (Section 11). (Courtesy of Michel Van de Wiele NV)

Plate 20. Wall-to-wall residential carpet produced by face-to-face weaving (Section 11) with 6 frames. (Courtesy of Michel Van de Wiele NV)

Plate 21. Carpet squares in a variety of designs and sizes from a production run of face-to-face weaving, demonstrating the flexibility of the CAD/electronic jacquard system (Section 11.3). (Courtesy of Michel Van de Wiele NV)

Plate 22. Sisal-look loop-pile texture produced on the SLC83 face-to-face triple-rapier loom (Section 11.4.4). Loops of pile yarn are formed alternately for the top and bottom carpets over thick weft inserted by the middle rapier. (Courtesy of Michel Van de Wiele NV)

Plate 23. Carpet square in a traditional design woven in a special 1+2/3 V-weave (Section 11.6) on an 8-frame face-to-face loom. (Courtesy of Michel Van de Wiele NV)

Plate 24. Royal Sovereign Isfahan, a rotary screen print (Section 15.10.3) on tufted nylon carpet. (Courtesy of Carpets International (UK) Ltd)

Plate 25. Richmond, a displacement screen print (Section 15.10.3) on tufted nylon carpet. (Courtesy of Carpets International (UK) Ltd)

Plate 26. A computer injection dyed (Section 15.10.4) broadloom carpet at the Charlotte City Club, South Carolina. (Courtesy of Milliken Carpet)

Plate 27. Musical Instruments, a repeating design (Section 16.2.2) for Axminster weaving. (Courtesy of Wools of New Zealand)

Plate 28. Use of die-cutting (Section 19.3.2) to create designs for carpet tiles. (Courtesy of Bonar Floors Ltd)

Plate 29. Flocked nylon carpet tiles in a Hong Kong play centre, demonstrating a non-repeating design produced by die cutting (Section19.3.2). (Courtesy of Bonar Floors Ltd)

Plate 30. Carpet tiles (Section 19) 36 x 36 in. from the Attitudes collection, with highlight colours. (Courtesy of Milliken Carpet)

Plate 31. Savanna tiles in repeating Millitron designs (Section 19.3.6). (Courtesy of Milliken Carpet)

Plate 32. Millitron tiles in a non-repeating design at the Osaki Electric Co, Tokyo. (Courtesy of Milliken Carpet)

9 Axminster Weaving

In theory, Axminster weaving comes close to the optimum method of manufacturing patterned carpets: the backing is woven at the same time as the pile is inserted, so that no major secondary backing process is required (as it is for tufting); the patterning scope is extremely wide and can include patterns of texture as well as colour; and a higher proportion of the pile is present in the use-surface than in other methods of creating patterned carpets (except printing).

In practice, the technique of tuft selection and insertion is relatively slow and weft insertion has also been slow, so that the rate of production of a traditional Axminster loom is lower than that of a face-to-face loom and much slower than that of a tufting machine.

However, the introduction of electronic jacquards has not only vastly improved the flexibility of Axminster weaving but has also enabled the pile selection step in the weaving cycle to be shortened; and more rapid weft insertion techniques are being introduced. A modern Axminster loom having projectile or rapier weft insertion and electronic jacquard operates at remarkably high efficiencies and can produce as much carpet as three or four of the older looms that were operating late in the twentieth century. Designers were quick to exploit the possibilities of the CAD/electronic jacquard combination and produced highly original non-repeating designs (Plate 15) and designs with long repeats (Plate 16) that stimulated sectors of the contract market. After a long period of decline, the new technology initiated a new period of growth in Axminster manufacture [1, 2]. For some years David Crabtree and Son Ltd (now Crabtree Textile Machines Ltd, a subsidiary of Cobble Blackburn Ltd) was the sole supplier of Axminster looms, but this situation was disturbed by competition from new technology. However, because most machinery existing in the industry was manufactured by Crabtree, and a high proportion has been modified to accommodate electronic jacquards, this chapter is largely based on information from them.

Two versions of Axminster weaving are in current use: jacquard-gripper and spool-gripper. Jacquard-gripper now means electronic jacquard and the flexibility of this modern version of the system (short runs, custom designs, non-repeating designs) and ability to weave dense constructions that perform well on the floor, place Axminster weavers in a favourable position in the contract carpet market, particularly the hospitality sector. The jacquard-gripper system is also used to weave rugs and carpet squares, and residential wall-to-wall carpets.

Spool-gripper weaving is faster than traditional jacquard-gripper weaving, because of the absence of a colour selection step in the weaving cycle and more rapid tuft insertion, and has unlimited scope in use of colour. However, it is only economically viable in long runs so that it tends to be used for mass producing popular residential qualities in designs involving subtle shadings of colour.

Designing for both traditional and modern looms is described in Chapter 16.

9.1 JACQUARD-GRIPPER

The key components of a traditional Crabtree gripper-jacquard Axminster loom are shown in Figure 9.1.

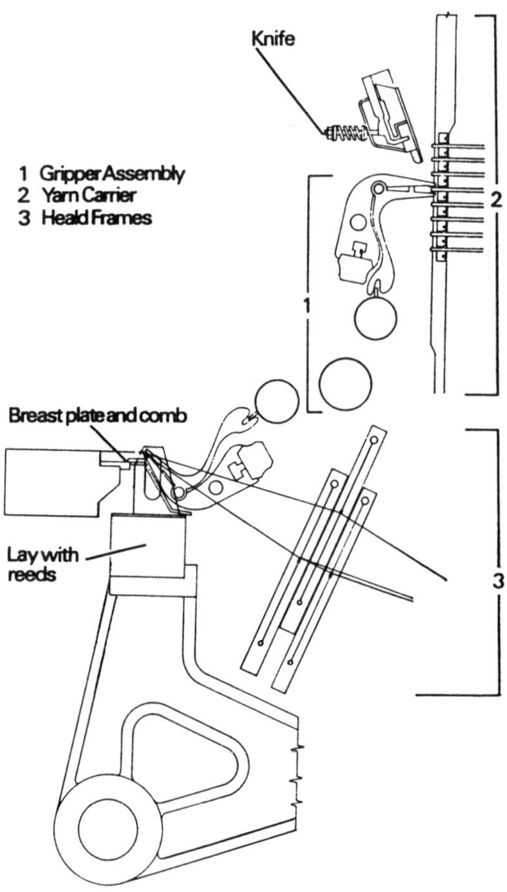

Figure 9.1. Key components of an 8-frame jacquard-gripper Axminster loom. (Courtesy of David Crabtree and Son Ltd)

9.1.1 Ground Weave

Weaving of the backing is controlled by three heald frames, two for the chain warps and one for the stuffer warp: these approach the point of fabric formation from below at an angle, leaving space for the pile mechanism. Weft is introduced to the shed by a rigid rapier having an eye near its free end, and weft yarn passes continuously through the eye. When the needle is fully inserted, a selvedge shuttle containing a cop of strong yarn (linen or polyester) is passed through the loop of weft and a catch cord from the shuttle holds the loop as the rapier is withdrawn, forming a selvedge. A double shot of weft is left in place and is beaten up by the sley (reed). The stuffer warp is moved by its heald shaft above and below the rapier, so that it lies straight between the weft shots in the finished carpet. Interlacing of the chain warp and the weft creates the basic structure of the backing. The principal function of the stuffer warp is to provide dimensional stability in the warp direction: it also fills the spaces between the pairs of chain warps. The weft also lies straight in the backing structure, providing stability in this direction. Alternative backing yarns are described in Section 5.1.

Chain and stuffer warps may be supplied from beam or magazine creel. In the beam system there is a minimum of three beams, two for the chain warps and one for the stuffer warp. Beams are up to 1000 mm in diameter. The beams may be full width, always the case in narrow-width looms, or may be half width for wide looms (ie, there are six beams rather than three).

Continuous warp feeding by the use of large creels may be provided as an alternative to beams. The upper ranges of the creel are for the pile yarns and the lower ranges for the ground warps. The warp yarns are led to the heald frames by means of rubber-clad feed rollers and tension rollers. Under this arrangement warp beaming is eliminated.

To facilitate smooth insertion of weft and provide consistent weft stability in the carpet, IRO weft feeder units are used in conjunction with the weft insertion system.

9.1.2 Pile Formation

In Figure 9.1 a gripper shaped like a bird's neck and head is shown in its two extreme positions. The gripper (one for each pair of chain warps) takes a tuft length of pile yarn (indirectly) from the creel to the point of fabric formation and forms it into a U-tuft, where it becomes locked in place by a double shot of weft.

Each gripper has two components pivoted at the eye. The outer jaw is fixed to a central shaft (fixing not shown) while the inner jaw has a fork that straddles a flat blade, mounted on an auxiliary shaft. This enables the whole row of inner jaws to be moved by rotating the auxiliary shaft slightly when caused to do so by a cam, opening and closing the beaks of the grippers at the appropriate phases of the cycle. The whole row of grippers can be rotated through an arc by the central shaft to take and grip ends of pile yarn presented by the carrier. Pile yarn is withdrawn to the required length by the grippers and cut off by a multiple knife system that is actuated by a rack and pinion mechanism. The grippers are then rotated to place the tuft lengths vertically in the point of fabric formation. A double shot of weft is inserted and the gripper moves upwards, tipping the second leg of the pile around the weft shot and finally releasing its grip to form a U-tuft, as shown in Figure 9.2

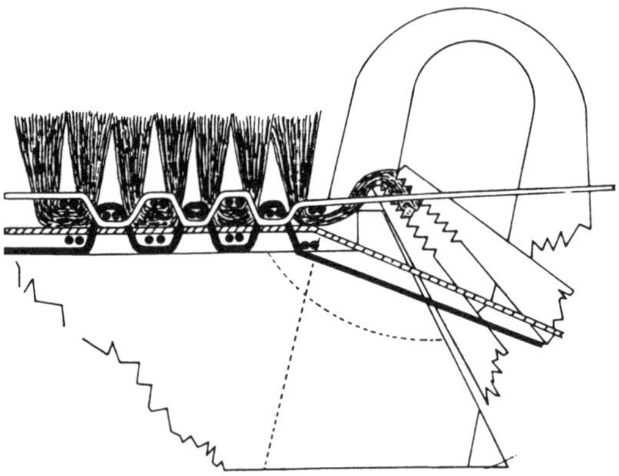

Figure 9.2. Tuft formation in gripper Axminster weaving. (Courtesy of David Crabtree and Son Ltd)

Four different weave structures can be produced by jacquard-gripper Axminster looms, as shown in Figure 9.3. The two essential styles are Kardax and Corinthian. Kardax, which takes the pile through to the back of the carpet, making the design visible on the back, is used for squares that are intended to simulate hand-knotted carpets as closely as possible. The Corinthian (ridgeback) weave embeds the tufts less deeply in the backing, economising on pile, and is used for wall-to-wall carpets.

Jacquard-gripper looms were traditionally set up to produce three-shot structures in both Kardax and Corinthian weaves (ie, three double weft shots were inserted for each row of pile). Recent models produce two-shot structures that grip the pile adequately and have higher output since only two revolutions of the loom are required to insert each row of tufts.

9.1.3 Pile Constructions

The pitch of an Axminster loom is fixed when the loom is made. The most common pitch ('universal pitch') is 7 dents per inch (3.65 mm between dents) although the list of pitches available is 4, 5, 6, 7, 8 and 9. Eight-pitch and 9-pitch

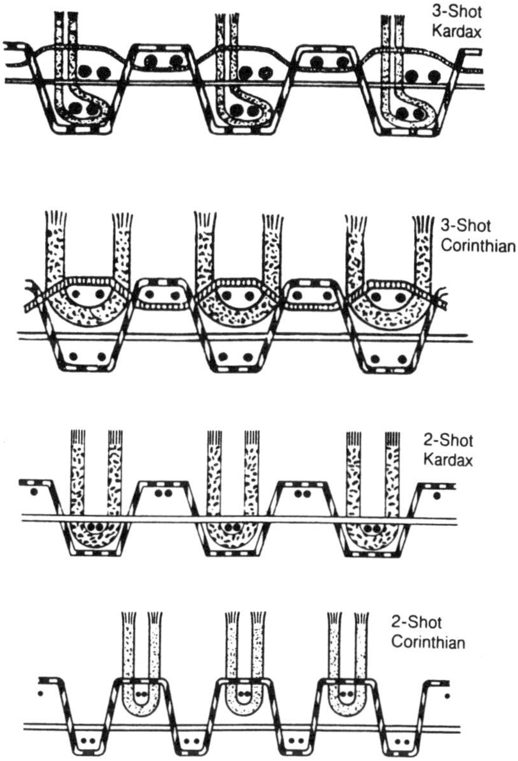

Figure 9.3. Alternative Axminster weave constructions. (Courtesy of David Crabtree and Son Ltd)

looms are used for dense contract qualities, and the very coarse pitches are used to weave carpets in special textures that can be made from mixed counts of yarn, including very heavy counts (eg, 2600 tex). Draw-off lengths are normally in the range 22–35 mm to give U-tufts with pile heights of 9–16 mm. Feasible linear densities (rows of pile per unit length) are between 16 and 48 rows/dm. Table 9.1 gives examples of some typical Axminster carpet constructions.

Table 9.1. Examples of Axminster carpet constructions[1].

Carpet quality[2]	Linear quality		Tuft length (at loom)		Pile count		Weft count		Warp count	
	Pile rows/in	Pile rows/m	in.	mm	Tex	Nm	Tex	Jute	Cotton	Tex
High	9	354	1	25.4	R560/2	3.57/2	344	10	3/9	R200/3
Medium	7	276	13/16	20.5	R620/2	3.2/2	482	14	3/9	R200/3
Low	5	196	13/16	20.5	R775/2	2.58/2	620	18	3/9	R200/3

1. Universal pitch: 7 dents/in (276 dents/m).
2. High = contract; medium = general domestic; Low = light domestic.

Source: David Crabtree and Son Ltd

One particular special texture is the J-tuft construction, which is achieved by changing the timing of the gripper mechanism. For this construction, tuft lengths up to 70 mm are drawn off and turned past their mid-points so that the long legs of the J may be up to 44 mm. Patterned J-tuft constructions are often used in Scandinavian style rugs.

9.1.4 Colour Selection

A creel for a jacquard-gripper Axminster loom has a number of frames (layers) each holding cheeses of yarn nominally in one colour. Creels are typically 8- or 12-frame (ie, up to 8 or 12 colours may be used in a design) and additional colours may be 'planted' in a frame if the designer wishes to use localised extra colour (see Section 16.2.3).

An alternative way of using a full creel is as a palette of colours for short-run weaving of contract carpets in a limited number of colours (eg, four to six). Both colour and pattern can then be changed simply by changing the design: no re-creeling is necessary.

Yarns are led from the creel over guide bars and through holes in a guide board to slots in moveable yarn carriers (see Figure 9.1). Each carrier has as many slots (one above the other) as there are frames in the creel; and there is a carrier for each reed dent. The function of the jacquard system is to position the yarn carrier so that the required colour of yarn is presented to the gripper when it arrives to draw off a tuft length.

Many existing looms have the traditional punched card jacquard mechanism (see Section 16.2.3) for selecting the yarn colour required at a particular point in the design, although the cards are now almost always punched by a Hardaker Electronic Jacquard Stamper operating on the basis of design data from a CAD system.

Modern looms are fitted with electronic jacquard mechanisms. Figure 9.4 shows the principle of operation of the electronic jacquard, although the configuration has since been developed to accelerate the weaving cycle. Movement of the yarn carriers is generated mechanically by the notched lifting bar that reciprocates in rhythm with the mechanism of the loom: but the positioning of the carriers is controlled by the electronics. A row of carriers, one for each dent, is positioned in front of the mechanism, and each carrier is operated by a solenoid. As the lifting bar moves upwards, the design computer causes the solenoid to fire a pin that pivots the carrier sharply forward so that its notch is placed in its required position on the lifting bar. The carrier, therefore, is lifted to the height required to present the required colour of pile yarn from the frame to the gripper. Note that pivoting the carrier also enables the top of the carrier to clear the spring that holds the carrier in its lower position. The row of springs then function as guides for the upward movement of the carriers. After the tuft lengths have been cut off, and whilst they are being woven into the carpet, the carriers are returned to their base position by the return stroke of the lifting bar.

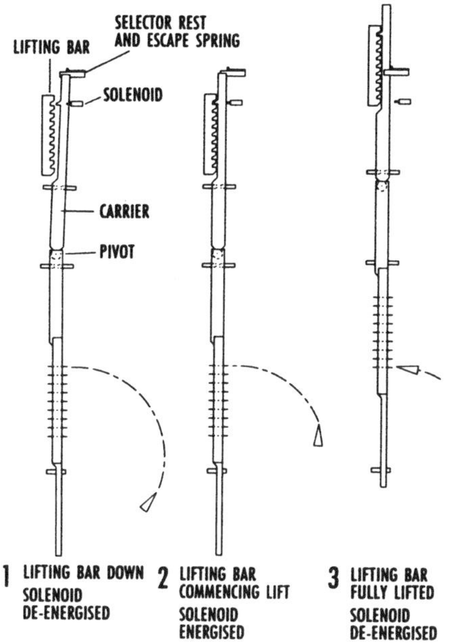

Figure 9.4. Principle of operation of the colour selector mechanism of the Crabtree electronic jacquard. (Courtesy of David Crabtree and Son Ltd)

Typically, the carriers are designed for 10 frames of pile yarn plus one frame for weaving backing cloth only (eg, headings of rugs or between batches of wall-to-wall carpet).

In most looms designed for electronic operation, the complete electronic jacquard unit is located at the front of the loom above the grippers, resulting in the overall height being about 2 m lower than that of a traditional loom, which has the mechanical jacquard on top. An important feature is that the mechanism does not have to be vertical, thus reducing the arc of travel of the grippers and shortening the loom cycle. The earlier electronic jacquards were mounted 30–45° from the vertical and, in later systems, the carriers move horizontally above the gripper, facilitating the shortest possible arc of travel.

Retrofit electronic jacquards for existing looms, originally supplied by Crabtrees themselves or by Griffith Textile Machines or Shuttle Systems, are mounted vertically on top of the looms. They provide all the versatility of operation and improved design scope of the integral units but do not reduce the arc of travel of the grippers and, therefore, give only a small increase in production rate. The most economic Griffith conversion retains the existing jacquard frame, selectors, wires and guides, but the lifting bars are converted to electronic units. A further modification can upgrade a loom from 8- to 12-frame weaving. The complete jacquard can be supplied for up to 16 colours.

Crabtree broadloom machines are fitted with computers having 32 Mb of memory as standard, which realises a pattern repeat sequence of 20 000 pile rows and a pattern width equal to the width of the loom. In a dense construction of 10 rows/inch, this translates into a design repeat of 50 m.

It is common operating practice for various electronic looms in an Axminster factory to be networked to the design studio. Designs from the CAD system are stored in a file store computer (server) that can be instructed to supply the required design data to the 'slave' computer in the selected loom. The loom computers do not, therefore, require hard disks for pattern storage. Designs may be viewed in full or in detail on the screen of the file store computer to verify their selection.

An auxiliary networking computer may be used to amalgamate a number of patterns in one file (ie, to organise a production run comprising a sequence of designs). A stop repeat may be set to end production at the end of a specified number of design repeats up to 999.

The file store computer in the design studio may transmit designs to a similar server computer in a distant factory by modem/telephone line or satellite.

Smaller factories may not justify a networking system, in which case the looms contain hard disk memory sufficient to store one file at a time. The design data are introduced from floppy disk. Designs may be viewed on a screen at the loom to confirm that the correct one has been loaded.

9.1.5 Loom Monitoring

Both jacquard-gripper and spool-gripper Axminster looms have traditionally been fitted with stop motions based on end-out detectors. Electronic devices are now available that reduce carpet faults and therefore mending costs.

Crabtree's Missing Tuft Detector incorporates camera scanning, as well as the original range of stop motions. The system detects missing tufts, uneven or short pile ends and 'fail-to-cut' and, on a screen, displays the identity of the fault to the weaver. There is the option of stopping the loom after one or several missed ends. The Missing Tuft Detector provides a production and quality reporting system that will interface with other monitoring systems.

The Modra Digital End Out Detection System incorporates fibre optic technology. Missing ends across the loom are detected immediately, and quality limits and alarm parameters can be adjusted. The system is said to facilitate multi-loom weaving (greater operator efficiency) and to reduce mending costs by up to 70%.

Boucher introduced an end-out detector based on lasers, with the claim that it was not susceptible to errors arising from vibration or bright light.

9.1.6 Production Calculations

The following calculations demonstrate the improvements in output that can be achieved by changing from a traditional 3-shot gripper loom with mechanical jacquard to a modern 2-shot loom with electronic jacquard. Further improvements can be achieved using the latest weft insertion techniques.

Traditional 3-shot gripper-jacquard

Width:	4 m
Speed:	50 ppm
Rows of pile:	80/m
Efficiency:	70% (varies)

Production per hour: $\dfrac{50\times60\times70}{3\times280\times100}$ = 2.5 linear m/h = 10 m²/h

Modern 2-shot electronic jacquard

Width:	4 m
Speed:	90 ppm
Rows of pile:	280/m
Efficiency:	70%

Production per hour: $\dfrac{90\times60\times70}{2\times280\times100}$ = 6.75 linear m/h = 27 m²/h

9.1.7 Projectile Weft Insertion

Griffith Textile Machines Ltd, involved with Axminster weaving through supplying electronic jacquards, became impressed with the general capabilities of Axminster weaving but believed that the picking rate could be improved significantly. Projectile weaving was selected as the fastest technique for inserting a heavy weft over a loom width of 4–5 m. A prototype machine was developed by combining the features of a Sulzer Type T projectile machine (as used for weaving jute secondary backing for tufted carpet) and a conventional gripper mechanism; and a loom speed of 180 ppm was achieved.

Production versions of the Griffith Fast Axminster Loom have lighter yarn carriers made from composite materials, and pile yarns are fed vertically, rather than horizontally, considerably reducing the arc of travel of the grippers. Speeds are in the region 180–200 ppm, so that over 50 m²/h can be produced. Griffith recommends larger capacity creels having packages of 130 mm, instead of the traditional 100 mm, to lengthen the times between changes. The Griffith loom is said to be able to weave a shorter and denser pile than has previously been possible, and the pile height is claimed to be very even, reducing shearing requirements.

The Griffith Fast Axminster Loom is supplied in nominal widths of 4 and 5 m and is fitted with a 12 colour electronic jacquard as standard. Backing warps are supplied from creel or large diameter beams.

9.1.8 Rapier Weft Insertion

The Crabtree Stage VI loom is a 3-shot Axminster loom utilising a modern hand-over rapier system for weft insertion. It is fitted with a 12 colour horizontal electronic jacquard and has a missing-end detector located in the knife box. The Stage VII loom is essentially an upgrade of the Stage VI loom, manufactured in 4 m or 15 ft widths.

Figure 9.5 (a) shows the rapier entering the shed and Figure 9.5 (b) shows the grippers taking yarns from the horizontal carriers. The arc of travel of the grippers from the carriers to the shed is beneficially short because of the move from vertical mechanical jacquards to horizontal electronic jacquards.

Figure 9.5. Stage VI Axminster loom: (a) Rapier weft insertion (above). (b) Gripper drawing yarn from the horizontal carriers (below). (Courtesy of David Crabtree and Son Ltd)

The Stage VI loom operates in the region of 120–135 ppm and the Stage VII loom 150+ ppm. The carpet take-up roller is servo driven, having beneficial effects on patterning and quality issues. A logic controller monitors and adjusts loom settings and displays loom and jacquard status on VDU and touch screens. The new looms are claimed to provide uniformity and consistency of draw-off, square-edge cuts and minimal gripper shaft deflection, maximising usage of pile yarn. Carpet constructions with lower pile are said to be possible.

9.2 SPOOL-GRIPPER

Spool-gripper Axminster weaving is a development of the obsolete spool Axminster system (see Section 1.2.6) in which the design is created by winding a parallel assembly of pile yarns, with colours in the sequence required in a row of tufts, on to a set of spools, each individual spool representing a different row in the design. Whereas the original spool Axminster system required a spool to be detached from its chain and taken to the weaving position when its turn for weaving came up, the spools in the spool-gripper system are transported close to the weaving position and tuft lengths are drawn off from them by grippers, as shown in Figure 9.6.

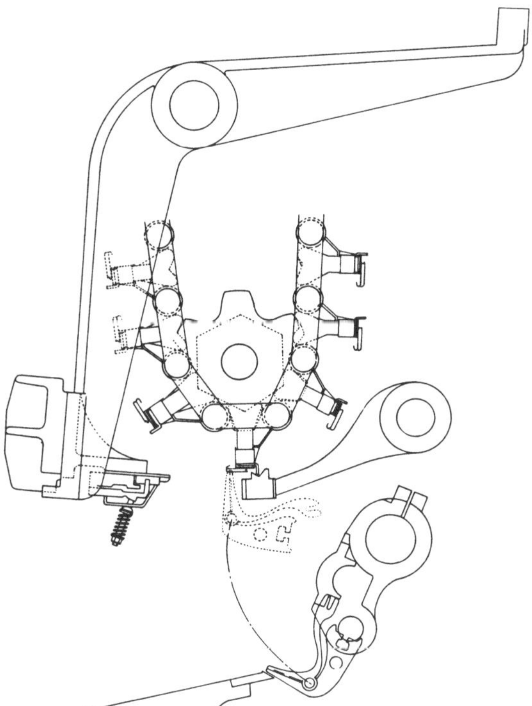

Figure 9.6. Gripper (dotted) taking tuft lengths from the spool of a spool-gripper Axminster loom to the weaving position (gripper outlined in solid line). The knife mechanism is on the left of the diagram. (Courtesy of David Crabtree and Son Ltd).

Any number of colours may be used in the design, and the design repeat, measured in rows, can be double the number of spools (the spools can be reversed to create a mirror image). However, the system is not easily associated with modern electronics so it is rather inflexible. Spool-gripper Axminster weaving is, therefore, associated with long runs, 400 m or more (ie, popular residential styles, rather than contract qualities designed to the customers' requirements). Design styles, such as floral and chintz, that require subtle shadings of colour, and elaborate rug designs are advantageously produced on spool-gripper looms.

9.2.1 Spool Setting

A spool setting machine consists of a flat table creel; an open reed having the same pitch as the loom; a device for rotating spools, typically 1 m wide, to wind lengths of yarn; and a board for holding the design in the form of point paper. The operative places a straight edge on the point paper to mark the first line of the design, selects bobbins of yarn in the required colours and leads the yarns through the reed in the required order. Lengths of yarn, typically 15 m, are wound on the spool, taking less than a minute. The full spool is replaced by an empty one: the operative moves the straight edge to the next row on the point paper, adjusts the creeling and winds another spool, and so on until the design repeat is complete. If the spools are intended for a 4 m wide loom and the width repeat of the design is 1 m or less, four spools must be wound for each creeling. To make this labour intensive process more economic, it is usual to wind duplicate spools (perhaps four times as many as the number of spools required for a linear design repeat).

If a spool holds 15.3 m of pile yarn, of which 0.3 m represents waste (end wrap, trimmings, etc), and the carpet requires 25 mm of yarn per tuft, then the spool will yield 15 000/25 = 600 tufts. As one spool represents one row in the design repeat, then one set of spools will weave 600 design repeats. If the design repeat is 0.5 m and there are 300 rows of tufts/m in the carpet, 150 spools will be required to weave a design repeat and the set of spools will produce $600 \times 0.5 =$ 300 m of carpet. For economic reasons, three sets of spools may be wound, so that the production run creates 900 m of carpet.

After the spools have been wound they are attached to a tufting frame and the pile yarns are threaded into an array of slots, one for each dent, so that they can be presented reliably to the grippers at weaving. The tufting frame contains a brake, so that tuft lengths may be let off smoothly, or there may be a pneumatic brake that is applied at the point of weaving.

A semi-automatic process for setting and winding spools from 3 kg cones has been developed, which is based on an electronic jacquard. There is a creel holding up to 24 colours of yarn and an assembly of yarn carriers, as for electronic jacquard weaving. A computer holding the design controls the firing sequence and the lifting bar lifts the carriers to a datum point so that the required colour in each carrier is adjacent to a gripper bar that spans the width of the machine (36 in. or 1 m).

Having closed on all the ends (252 or 276 for universal pitch) the gripper bar moves away from the carriers, pulling a sheet of pile yarn towards the spool winder. The ends are released over the spool winder and the operative introduces an empty spool. As the spool is wound, yarn is drawn from the creel through the carriers. When the spool is full, and another spool in the same set is required, the yarn is cut at the winder and another spool is loaded. When the last spool in the set has been wound, a knife cuts the yarn in front of the carriers, a new selection of colours is made, and the sequence is repeated.

Autosetting produces important increases in the rate of production and decreases in costs, and virtually eliminates setting errors. Spools are said to be wound tighter and are more consistently tensioned.

9.2.2 Gantry

Figure 9.6 shows only a small length of a chain of tufting frames with their spools that are mounted on a gantry. Taking the example of a design repeat of 150 rows in Section 9.2.1, the gantry would be required to hold 150 spools; and with a design repeat of 1 m in the same carpet construction, 300 spools would be required. A large rug would require more spools than this. Gantries may be specified to hold between 216 and 600 spools. The gantry is, therefore, a large construction with the chain of spools travelling in vertical loops to economise on space. The chain of spools is, of course, endless so that it goes around once for each design repeat (600 times to use up a set of spools in the example in Section 9.2.1). The chain can be reversed to create a mirror image of a large design repeat (eg, for a bordered square).

A spool-gripper loom has two gantries: the loom gantry from which weaving takes place and a loading gantry at the rear of the loom. A fresh set of spools is mounted in the loading gantry whilst weaving is in progress. When the production run is complete, the fresh set of spools is run on to the loom gantry and, at the same time, the empty spools are returned to the loading gantry. Operating in this way, a design change takes around 30 minutes.

9.2.3 Weaving

The tufting frames that hold the spools are conveyed accurately to the point of weaving by an hexagonal cog unit and are clamped briefly in locating blocks so that the projecting pile yarns can be taken by the grippers. Weaving then takes place as in jacquard-gripper looms. Note, however, that the arc of travel of the grippers is much shorter than in traditional gripper looms (64° compared to 131°), reducing the time for tuft insertion.

9.2.4 Spool-Gripper Carpet Constructions

Most spool-gripper looms are supplied in universal pitch (7 dents/in. or 28/dm), although they are also offered at 6 and 8 pitch. Change in linear rows is effected by the substitution of gears in the take-off gearbox and normal qualities range between 16 and 45 rows/dm.

Draw-off for U-tufts is in the range 22–35 mm, and for J-tufts 70 mm. Spool-gripper looms are most often designed for 2-shot weaves (for high output), although latexing of the back is required to achieve adequate tuft bind. Some weavers prefer to use 3-shot constructions.

9.2.5 Production Rates

Because of the shorter tuft insertion cycle, a spool-gripper loom can be operated at a faster picking rate than a corresponding jacquard-gripper loom (60 compared to 50 ppm) and the usual 2-shot version requires only two double weft shots compared with three for the 3-shot construction commonly used with the jacquard loom. Calculated on this basis, and with a loom efficiency of 65%, the hourly output of a spool-gripper loom would be 16.7 m^2, compared with 10.0 m^2 for the jacquard-gripper (Section 9.1.6).

Higher output has been quoted as an advantage of the spool-gripper system, but in this respect the system has been superseded by the electronic jacquard loom.

REFERENCES

1. D N Crabtree. Axminster: developments to reduce manufacturing costs. In: *Proceedings Tifcon '96 – Carpets: The Current Perspective*, The Textile Institute. Manchester, 1996.
2. J D Griffith. New developments in Axminster weaving. In: *Proceedings Tifcon '95 – 25 Years of Carpet Technology*, The Textile Institute, Manchester, 1995.

10 Wireloom Weaving

Wireloom weaving was the first mechanised system for producing carpets and is still used by the industry in two forms – Brussels weaving for loop-pile and Wilton for cut-pile constructions. There are many carpet constructions that contain both cut and loop pile so there is no clear technical distinction between the two, and they are conveniently considered together.

Essentially, the pile is introduced in the form of a warp running parallel and close to the warp yarns for the backing (chain warps). There is a frame holding packages of pile yarns for each of the colours required in the design. When a weft shot is to be introduced, half the chain warps are raised to create a shed through which the weft is inserted. When a colour of pile is required by the design, the pile warps of that colour are raised and a metal rod or 'wire' is inserted weftwise under them. The pile warps are lowered and the loops formed are secured by further wefts. The loops may or may not be cut, as required, when the wire is removed.

Most currently installed looms have weft insertion by shuttles, and patterning is often by punched card jaquards. New looms are equipped with rapier insertion and electronic jaquards, however.

For the production of multi-coloured carpets wireloom weaving has, to a large extent, been displaced by face-to-face and Axminster weaving, principally because of the high cost of the pile that lies 'dead' in the backing when that colour is not required in the design. Nevertheless, the traditional character of the products is valued in some areas of the trade. Densely constructed carpets woven on wirelooms perform well in use, and for many years the system was favoured for the production of heavy duty contract carpets: the cost of manufacture was kept low by restricting the number of colours to two or three, and small lots were produced to order on narrow width looms. The flexibility of wirelooms for this purpose (and other market areas) has been enhanced by the introduction of electronic jacquards and CAD systems, extending the future of this sector of the contract market.

Wirelooms are unique in their flexibility for producing textured surface effects in both cut pile and loop pile and designs that incorporate different textures. No other system, tufting or weaving, can compete in this respect. Continuing demand for textured loop-pile carpets has stimulated the development of a high-speed loom that can produce wireloom constructions by forming the loops over false picks instead of over wires. As wirelooms have been used principally for producing carpets with wool or wool-rich pile, because of the risk of fusing synthetic fibres by frictional heating when the wires are withdrawn, this development further extends the scope of wireloom constructions.

A modern wireloom 4 m wide operates at 60-70 rpm and efficiencies around 85%, depending on the number of frames and the quality of the yarns. The output is then related to the density of construction. A standard quality of 60 picks/dm should be produced at about 22 m^2/hour.

In a few companies, looms based around wires configured warpwise, rather than in the conventional weftwise direction, are used. Such warpwise looms operate at higher speeds than conventional wirelooms and can produce dense piles, but the patterning possibilities are not as great.

10.1 PRINCIPLES OF WIRELOOM WEAVING

Figure 10.1 is a schematic diagram of a modern jacquard wireloom. At the lower right there are beams and tensioning devices for the chain warps and stuffer warps. At the upper right, five colours of pile yarn are led in from the five frames of the creel and pass through heald-eyes that are linked with cords to the jacquard mechanism.

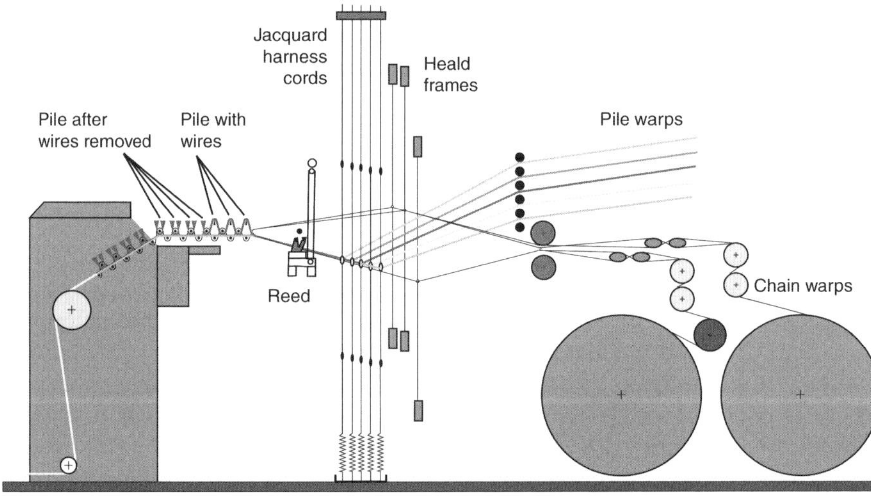

Figure 10.1. Principal components of a jacquard wireloom, showing shed for insertion of first pick. (Courtesy of Michel Van de Wiele NV)

Figure 10.1 shows the shed formation during the first revolution of the weaving cycle, which is for the backing only. A shot of weft is introduced (the dot in the shed) to be beaten up by the reed or sley (modern sleys have a parallelogram motion to minimise rubbing of the warps and to give optimal guiding of the wire during insertion). The section of carpet shows three rows of pile with wires in place (in reality there will be 20 or more wires in place) and the rows produced

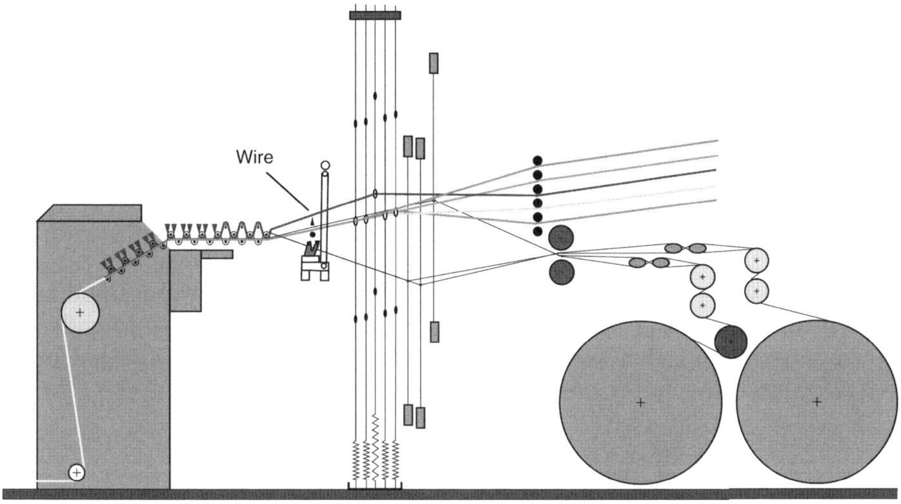

Figure 10.2. Wireloom with shed opened for insertion of second pick and for inserting a wire (triangle) under selected pile warps. (Courtesy of Michel Van de Wiele NV)

earlier as cut pile. At this stage of the cycle a wire is withdrawn from a pile row and is returned to the insertion mechanism.

Figure 10.2 shows the corresponding stage of the second revolution in the weaving cycle. Movement of the heald frames has alternated the positions of the chain and stuffer warps and one row of pile warps has been raised to create a double shed. The weft is shot above the chain warps but below the stuffer and all the pile warps; and a wire is thrown above the backing warps and all the pile warps except the selected ones. After beat up, the weft shot becomes located at the base of the backing and the wire plus selected pile loops are on the surface. Pile that has not been selected is always present in the carpet, but remains 'dead' in the backing. Although dead pile contributes in some way to the cushioning of the carpet, its presence contributes significantly to the cost of the product.

A cam mechanism is used to actuate the heald frames to create shedding of the backing warps, and a jacquard or electronic jacquard mechanism usually selects the pile, but simple designs may be executed by dobby. Looms for weaving plain carpets utilise cams and healds to lift the piles. Most installed wirelooms have shuttle systems for inserting the weft, but weft insertion for new looms is by flexible rapier. Wires may be 'round' for weaving low loops or 'rectangular' for higher loops. If the wire is bladed, as shown in the lower part of Figure 10.3, the pile is cut when the wire is withdrawn. Pile height is determined by the height of the wire. Flag wires (not illustrated) are sometimes used to create textures. Their function is to tension loops via unsharpened raised sections on withdrawal, rather as yarn tensioning patterning systems are used in tufting (see Section 7.1).

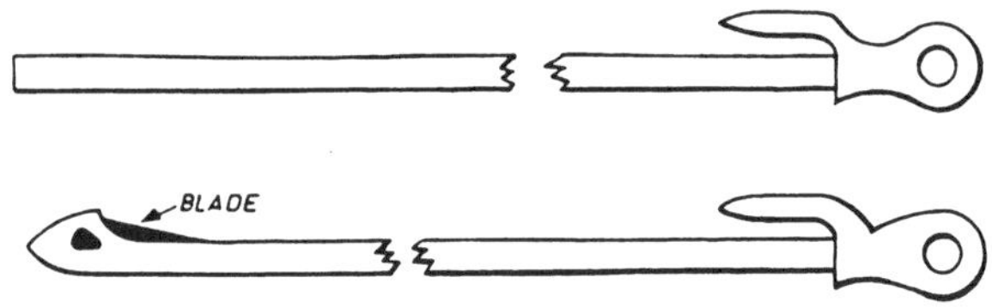

Figure 10.3. Pile wires: round wire for loop-pile carpet (above); bladed wire for cut-pile carpet (below).

Plain carpets in a 1/1 V-weave (Figure 10.4) are typically produced on single-frame looms without jacquard or dobby. Such constructions are considered to require latexing in order to improve tuft bind, and some companies prefer a W-weave, which gives better mechanical binding of the tufts.

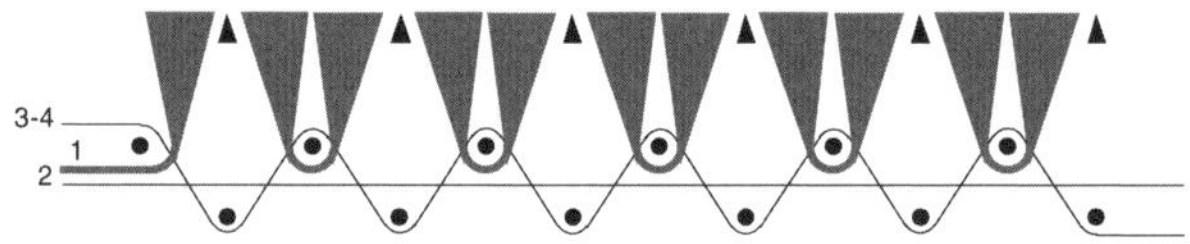

Figure 10.4. 1/1 V-weave for plain carpet (•= weft shots; ▲= points of cutting). (Courtesy of Michel Van de Wiele NV)

Most installed wirelooms are equipped with conventional jacquards, as described in Section 16.2.3. These may have a single lift, adequate for most purposes, or a double lift, required to produce some constructions, particularly textured effects. Looms recently supplied have electronic jacquards, and some of the older looms are retrofitted with electronic jacquards.

An electronic jacquard system provides three possible positions for the pile warp and may or may not have a driven rost. With a driven rost (a simple modification) the possible weave structures are extended to include designs visible on the back of the carpet, as in hand-knotted carpets. This possibility is important when squares are to be produced. Electronic jacquards also allow designs to extend over the full width of the carpet: it is not necessary to have repeats.

Figure 10.5 shows the operation of Van de Wiele's pile selection system with fixed rost for the Piletronic (PTR) electronic jacquard. Two spring hooks are reciprocated vertically in rhythm with the loom by knife rosts. When the digital pattern dictates, an electromagnet causes first one and then the other spring to latch on to its hook. When the electromagnet is switched off, the elements spring away from the hooks and are free to resume their rhythmic vertical movement. Both spring hooks are linked to an upper pulley system by a strong inextensible cord. When they reciprocate unlatched, the upper pulley does not move but latching causes the upper pulley to be raised by a distance equal to half the movement of the knife rost (L). Note that the left spring hook is linked to the lower pulley system which, in turn, is linked to the upper pulley so that this movement is amplified to a movement of 2L at the eye of the cord.

Selected pilewarp (fixed rost)

Pick 1	B	Kr1 bottom Kr2 top	No selection ○
Pick 2	T	Kr1 top Kr2 bottom	Selection ●

Incorporated pilewarp (fixed rost)

Pick 1	B	Kr1 bottom Kr2 top	No selection ○
Pick 2	C	Kr1 top Kr2 bottom	No selection ○

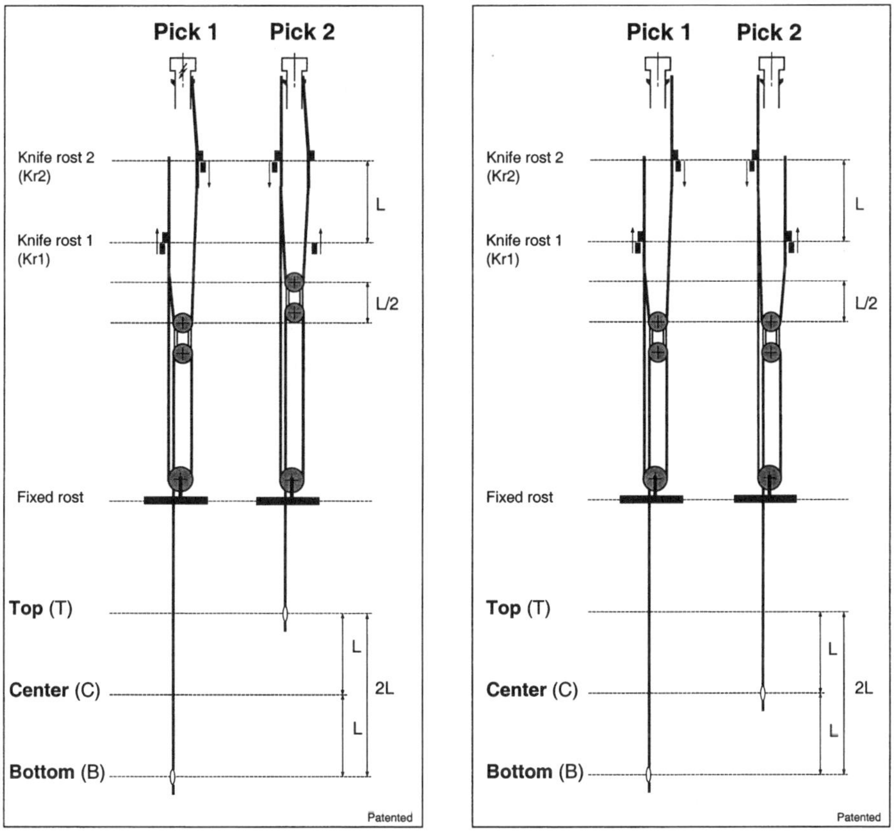

(a) (b)

Figure 10.5. Operation of the PTR electronic jacquard with a fixed rost. (Courtesy of Michel Van de Wiele NV)

Thus, Figure 10.5(a) shows that the eye holding the pile warp when it is not selected is positioned at its bottom position when pick 1 is inserted, placing the pick above the dead pile, but selection and latching moves the eye to its top position, placing the selected pile above both the pick and wire.

Figure 10.5(b) shows that the eye holding the pile is raised even when pile is not selected. This is because the harness cord is connected to the left spring hook via the lower pulley system so that when the left knife rost moves upwards by distance L, the eye in the harness cord also moves up by distance L. This places it in the intermediate position, laying it alongside the tight (stuffer) warp and above pick 2 (ie, incorporating it in the backing). Incorporated pile is the automatic result of the reciprocating mechanism without selection when the rost is fixed.

Figure 10.6 shows the weave structure of a classical 5-frame Wilton carpet (Plate 17) with incorporated dead pile, that can be produced using a jacquard with a fixed rost.

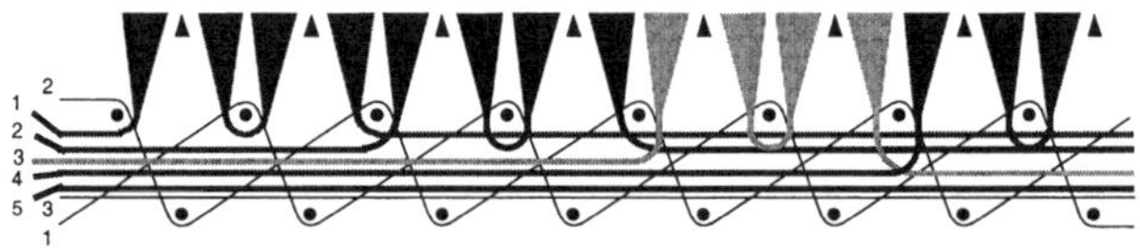

Figure 10.6. Weave structure for a 5-frame Wilton carpet with incorporated pile. (Courtesy of Michel Van de Wiele NV)

The operation of the PTR with a driven rost is shown in Figure 10.7: the driven rost reciprocates in line with the knife rost 1, but over only half the distance. Figure 10.7(a) shows that selection causes the pile warp to move from the bottom position at pick 1, which is visible on the back, to the top position above the wire at pick 2, just as with a fixed rost, but note that the point of actuation of the electromagnet is different. Figure 10.7(b) shows selection of pile for incorporation in the back. For pick 1, both spring hooks are raised, but the high position of the driven rost shortens the lower pulley system, allowing the pile to drop to the intermediate level, above pick 1, for incorporation in the backing. For pick 2, the left knife rost is down, but the pulley spacing is maintained by the low position of the driven rost and the pile moves into the bottom position, placing it below pick 2, which is the inside pick of the carpet back.

Figure 10.8 shows a Brussels carpet in four colours with the design showing through the back, woven using an electronic jacquard with driven rost.

Retrofit electronic jacquards for Wilton and Vincenszi looms are available from Shuttle Systems. Replacement of the loom cylinder allows the system to actuate the existing machine selector rods, making conversion extremely simple. The flexibility associated with electronics is achieved, but within the confines of the existing width repeat of the loom. Pattern files are provided on disk or CD-ROM, and the standard system supports NedGraphics and APSO CAD formats.

10.2 TEXTURED CARPETS FROM WIRELOOMS

The following are some examples of textures that can be created on wirelooms. Skilled technical designing is necessary for some of the effects.

10.2.1 Looms without Jacquard

Cut-pile friezé

A structure as shown in Figure 10.4 is woven, but using overtwisted yarns that have been relaxed and set, usually in hank dyeing, and a curled pile results.

Selected pilewarp (driven rost)

Pick 1	B	Kr1 and Dr top Kr2 bottom	No selection ○
Pick 2	T	Kr1 and Dr bottom Kr2 top	Selection ●

Incorporated pilewarp (driven rost)

Pick 1	C	Kr1 and Dr top Kr2 bottom	Selection ●
Pick 2	B	Kr1 and Dr bottom Kr2 top	No selection ○

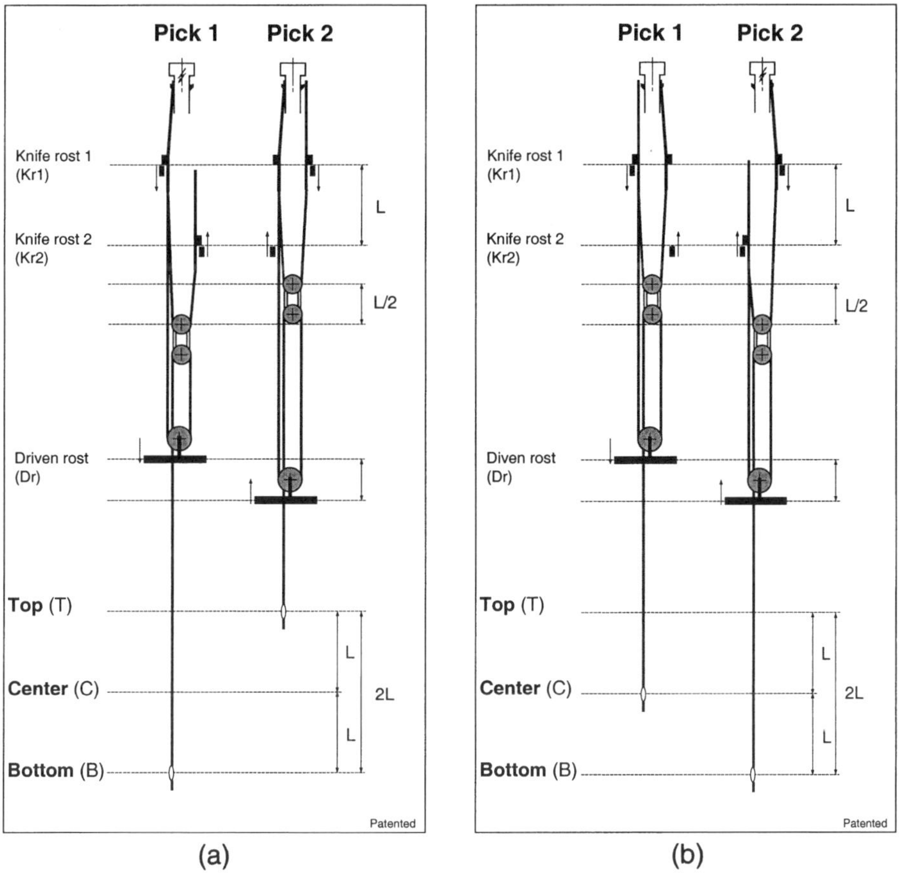

Figure 10.7. Operation of the PTR electronic jacquard with a driven rost. (Courtesy of Michel Van de Wiele NV)

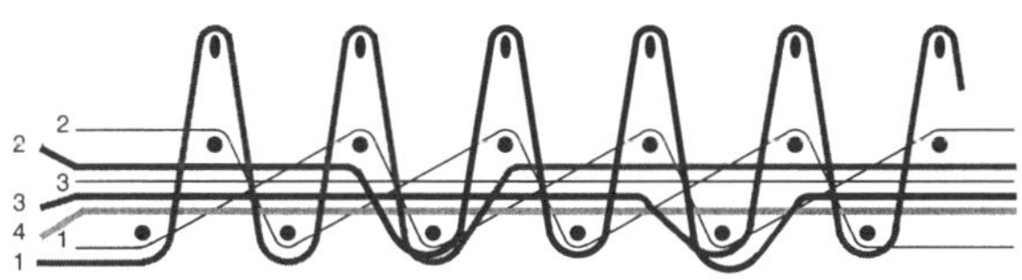

Figure 10.8. Weave structure for a 4-frame loop-pile carpet with design woven through to the back. (Courtesy of Michel Van de Wiele NV)

Cut-pile friezé carpets have been extremely popular in some countries over many years, because they do not show footmarks or pile reversal effects (shading, see Section 17.3) as easily as plain carpets from regular yarns. This texture is now most often produced by tufting, however.

Tonal chequered effects
By using groups of rounded wires followed by groups of bladed wires the side/end apparent colour difference of the tufts creates weft stripes. If the pile yarn is dyed in two fairly close tonal colours, creeled in groups, the combined effect is chequered.

Textured surface in loop pile
'Round' wires, engineered to a profile of shallow waves along their lengths, may be used to produce loops of different heights within one row. This structure does not persist in the same row, since the loops are all pulled to the same height when the wire is withdrawn (the wires carry a flag at the ends), but shortfall in yarn is made up by robbing the previously formed loops, so that the wave effect is transferred to that row. Wavy wires are used in groups having different wave configurations and are usually used alternated with plain round wires for low pile. The textures produced in this way are similar to those produced by the disc cam tufting system (see Section 7.1.1).

Turned loops
Plain loop-pile carpets tend to have a regimented pile, but it is often considered desirable to have a randomly oriented pile, especially with yarns of high linear density. The turning of the loops can be induced by taking advantage of the twist of the yarn. Using flag wires without wavy effects, loops are robbed to the same height, but turn in a semi-random way when the twist acts on the ground weave.

Shag pile
J-tufts can be created using special wires having both a flag and a blade. Such wires are used on all rows. The wire is selected to produce a fairly high loop (eg, 10 mm) and on withdrawal the flag will raise the loop further (eg, 5 mm) by robbing the previous row of pile. Immediately, the blade cuts the loop to give a row of 15 mm cut pile; but when the next row is formed, the nearest leg of pile is robbed to a height of 5 mm. The result is shag pile with a visible pile of 15 mm and hidden pile of 5 mm.

10.2.2 Single-lift Jacquard

Coarse cut-pile constructions
When very heavy yarns are used for visual effect in a plain colour, it is often necessary to reduce the number of tufts to avoid an excessively heavy carpet. One possibility is to use a simple jacquard design that selects pile so as to leave holes of various sizes in the pile in a semi-random array, with dead pile passing under one, two or three wires. Supplementing this procedure, the number of ends in the pile warps may be reduced.

Carved designs
Typically, cut-pile constructions are produced in dense matt pile in a single colour, but with designs created by the jacquard in the form of areas without pile. Clearly, these base areas should be narrow. Classical carpets in such constructions often have leaf designs and are green or beige.

Cut/loop designs
Using a plain colour, interesting and varied designs in two textures may be produced by using round wires and bladed wires in alternate rows. When cut pile is required, the pile lies dead under the round wire, and vice versa. Only one row of pile is, therefore, produced per two wires, so that the pick density must be exceptionally high. This is a case where skilled designing is required to avoid technical faults at the changeover from cut to loop pile.

10.2.3 Double-lift Jacquards

An extremely important area of loop-pile texture is provided by float weaves which may be based on floats over wires, floats over wefts, or floats over wires and wefts. Double-lift jacquards are required for most of these structures, allowing a pile yarn to be lifted to the intermediate height to float only over the weft or to be lifted higher to float over the wires.

Floats over wefts may be a component of a design (ie, there may be areas that are flat woven in the design), but in some carpets the use-surface consists entirely of floats over wefts. Pile may be held only by the weft shots, or may be woven through to the back. In reality, these are not wireloom carpets but are flat-woven carpets (see Chapter 12), although they are usually woven on wirelooms.

The principal reason for weaving such low, dense loop-pile structures is to create carpets that perform well but are light in weight (eg, for aircraft and some other areas of transportation). Pile may pass over one, two or three weft shots and simple designs can be produced by using different coloured yarns, floated in parallel or alternately as required.

One objective of floating over wires only is to create loops that are larger than normal. Typically, effect yarns in a fairly heavy count are floated over two round wires, and alternate pile ends are floated over different pairs of wires to create a diagonal arrangement of loops. A more random texture can be created by designing a carpet in a grainy pattern consisting of loops formed over one, two or three wires.

'Sisal' textures can be created by floating yarns in pairs over two or three wires in a simple pattern, and exposing some of the upper weft yarns.

Enormous loops can be formed from exceptionally heavy counts of yarn (eg, Nm 0.2) floated over two or three wires to give the required loop spacing. Such yarns are used only in alternate dents, and only half the usual number of chain and stuffer warps are used too.

The foregoing are only a few of the possibilities from wirelooms available to the designer. Figure 10.9 illustrates combinations of floats and cut pile. Using combinations of colours, yarn effects, yarn counts and weave structures, a wide range of possibilities is open to the designer, without the need to incorporate a high proportion of dead pile in the carpet. The skilled designer will not, however, neglect to ensure that an unusual structure of carpet is also technically satisfactory.

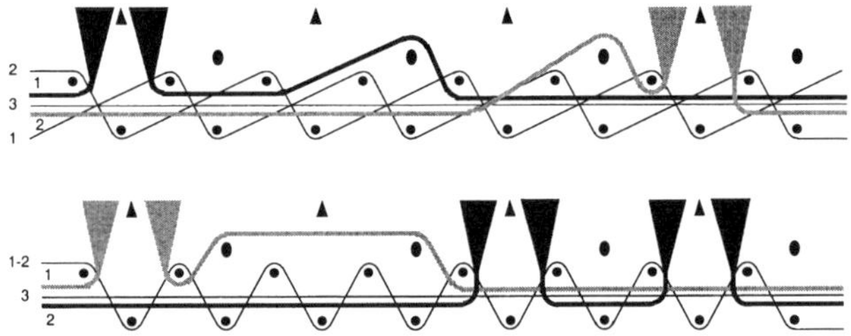

Figure 10.9. A 2-frame cut/loop structure with loops floated over one wire and one weft (above) and three wires and two wefts (below). (Courtesy of Michel Van de Wiele NV)

10.3 WEAVING LOOP PILE WITHOUT WIRES

Continuing interest in loop-pile constructions, coupled with consideration of the limitations of wirelooms, stimulated the development of Van de Wiele's high-speed Loop Pile Master LPM32, launched commercially in 1997. The LPM32 does not utilise wires, but the loops are formed over false weft picks supported by lancets.

A wide range of constructions are possible from the LPM32, especially when equipped with electronic jacquard. Products include both wall-to-wall carpets and rugs, having designs in texture, colour or both (Plate 18). Lightweight aircraft carpets having a low, dense pile are among the possible products.

An important technical advantage of the LPM32 over a wireloom is its ability to handle fusible synthetic piles, such as polypropylene, that are susceptible to damage by frictional heating when wires are withdrawn. The key production advantage of the loom is that it is capable of operating at 180 rpm, which is about three times the speed of a typical wireloom, by obviating the rate-determining mechanism for inserting and removing wires.

Figure 10.10 shows the basis of the weave structure of a 3-frame Brussels carpet as it is formed on the LPM32. The Loop Pile Master is equipped with lancets, thin metal plates, threaded on a metal bar that holds them in a warpwise orientation, one for each dent, through the point of fabric formation. The loom is a double-rapier weaving machine: the bottom rapier inserts the weft shots and the top rapier inserts false picks above the lancets. Pile selected by the jacquard mechanism (normally electronic) is lifted either above the false picks to create high loops, or above the upper weft shots to create low loops. Loops may be lengthened by floating over two or more false picks, if required. Figure 10.11 shows the formation of a 3-frame carpet with a sculptured pile containing low loops, high loops and loops floated over two false picks.

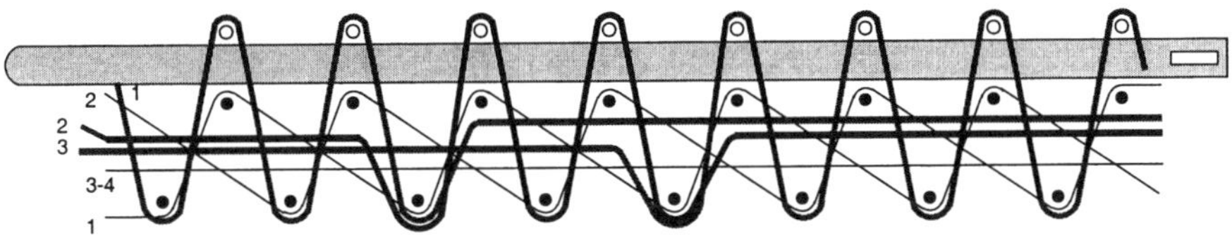

Figure 10.10. Formation of a 3-frame Brussels carpet on the Loop Pile Master: note the lancet and false picks supporting the selected pile warps. (Courtesy of Michel Van de Wiele NV)

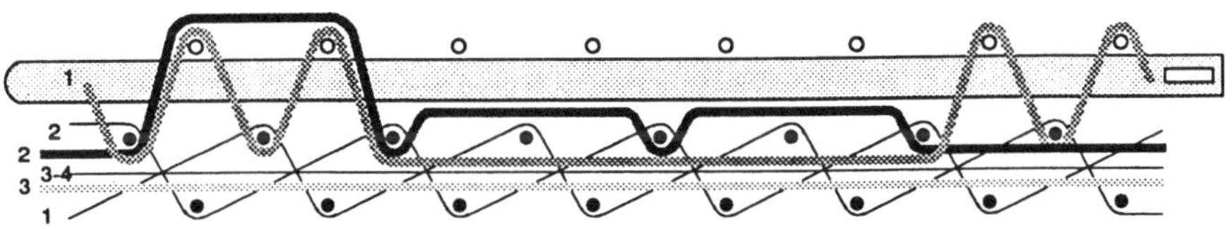

Figure 10.11. Formation of a 3-frame sculptured pile carpet on the Loop Pile Master. (Courtesy of Michel Van de Wiele NV)

The height of the high pile in this technique of weaving is determined by the height of the lancets. During weaving, the false picks are cut in the middle and are automatically removed. Most structures produced on the LPM32 cannot, therefore, be distinguished from those produced on wirelooms.

10.4 WARPWISE WEAVING

Looms having the wires permanently mounted in the warpwise direction were developed as a means of eliminating the time in the weaving cycle occupied in inserting a wire and removing another wire. For most constructions, the key to accelerating weaving has been the move to face-to-face constructions and, more recently, the system of forming loops of pile over temporary weft picks (see Section 10.3). However, the most successful of the warpwise looms, the Karaloc loom, is not on general sale but has been successfully operated by a small international club of companies over a period of many years. The production rate of a 4 m wide Karaloc loom is thought to be about 48 m^2/hour.

The loop-pile version of the Karaloc loom has a full set of looper wires, one for each dent, mounted rigidly across the loom with their open ends pointing towards the woven cloth (ie, in the direction of loop-pile loopers in carpet tufting). Pile yarn is fed through a series of guide needles that reciprocate in a weftwise direction, forming loops of pile over the wires. At the base of the loopers, the pile is incorporated in the ground weave. As the material moves forward, the loops slip off the open ends of the wires. In this way, plain loop-pile carpet is produced.

Simple designs, perhaps foreshadowing crossover tufting, can be produced on the Karaloc loom by creeling up different colours in a sequence (which would give a stripe design), together with the use of a mechanism for traversing the guide needles so that they can feed pile to alternative wires.

Cut pile can be produced on an alternative version of the machine that has bladed wires mounted parallel to the looper wires, but pointing in the opposite direction (ie, in the direction of cut-pile hooks in tufting). If yarn is looped over both wires, the loops are cut by the blade as the newly formed cloth moves forward. However, the design mechanism can cause the pile yarn to avoid the bladed wires, so that the loop is left uncut. In this way, simple cut/loop designs are formed.

10.5 HAND WIRE WEAVING

Although many flokati rugs are woven on hand looms, with the pile tufts being inserted manually, machine weaving is widely practised in this unique Greek industry, but with hand insertion of the wires. Industrial looms designed for producing flat fabrics are used, but are programmed to stop every few picks (5–12, depending on the quality) and a wooden 'shag bar' is inserted into the shed before the next pick. A high loop is created when weaving recommences. The operative runs a knife along a groove in the bar to create high cut pile.

Both the backing yarns and the pile are 100% wool, and total carpet weights range from 1.4 to 2.5 kg/m^2. As considerable shrinkage takes place in finishing (Section 17.8), two, three or four panels may be sewn together to create large rugs.

11 Face-to-Face Weaving

The face-to-face weaving system [1] was first applied in the carpet industry in the 1920s and has since benefited from a sustained programme of technical development. It has achieved particular success in the growing market for rugs and bordered carpet squares (Plate 19), and is used worldwide to create products ranging from fine and heavy constructions having 3-ply wool pile to cheaper products from BCF polypropylene. Increasingly, the system is being used for producing wall-to-wall carpets (Plate 20). Essentially, two backing fabrics are woven simultaneously and the pile yarn is interlaced between them to give a sandwich which is cut to create two carpets as shown on the left of Figure 11.1.

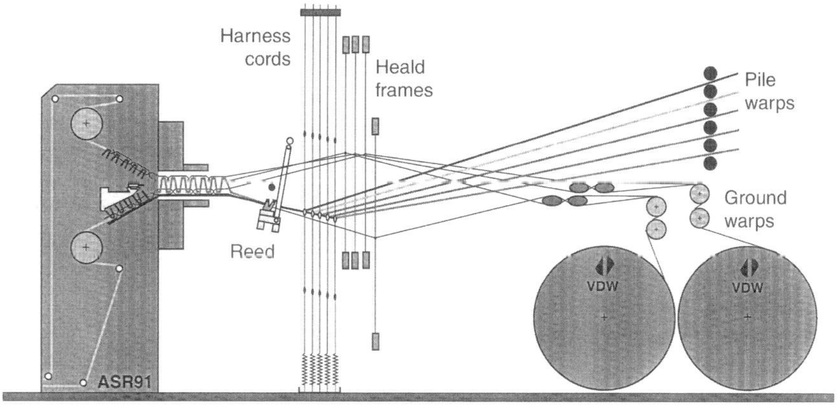

Figure 11.1. General layout of a face-to-face carpet weaving machine. (Courtesy of Michel Van de Wiele NV)

The advantages of face-to-face weaving compared with wireloom Wilton are:

- Higher rates of production.
- In jacquard constructions the dead pile is shared between two carpets so consumption of dead pile is roughly halved.
- The cutting mechanism gives a very level surface and is less liable to fuse synthetic fibres (wires may become hot during Wilton weaving).

The face-to-face principle has been adapted to produce textured loop-pile surfaces.

Wilton weaving has not been entirely displaced by face-to-face weaving, however. A Wilton loom is less sophisticated than a face-to-face loom, which in a reed dent has the required number of pile yarn colours, two pairs of binding warps and, possibly, two stuffer warps. Wilton looms are, therefore, easier to set up; and are more suitable for short production runs for custom contract work, especially in 2- and 3-frame designs.

The pile warps in face-to-face weaving are selected by a jacquard mechanism, traditionally based on punched-card jacquards: electronic jacquards are supplied on new looms. The weft insertion mechanism on modern looms is by handover rapier that operates from large packages of yarn. Weft insertion may be by single-, double- or triple-rapier systems: each has its particular advantage, depending on the quality and style of the carpet to be woven. Existing shuttle looms have mostly been upgraded by conversion to a gripper-shuttle system.

Rapier weft insertion methods have the following advantages over shuttles:

- Continuous weft insertion from large yarn packages, reducing stoppages and eliminating cop winding.
- Lower picking forces.
- The small cross-section of the rapiers allows smaller shed openings.
- Rapiers can be adjusted in height to match the shed and can be spaced in accordance with the required pile heights.
- Higher loom speeds and efficiencies result from a combination of the above factors.
- Energy consumption is lower as a consequence of the lower picking forces and shorter lifts.

Face-to-face looms are made up to 4.2 m wide. Modern patterning systems allow such wide looms to be used in a flexible way, weaving carpet squares of different widths and in different designs side-by-side.

11.1 WARP PREPARATION

Backing yarns (commonly cotton or polyester/cotton R165tex/2, Nm12/2) are prepared for weaving by direct beaming: a sectional warper can produce uneven carpets. A modern beaming machine incorporates a microprocessor to control winding speed, yarn tension (via a braking roller) and winding pressure; yarn length is also monitored. The beams may have barrel lengths between 1000 and 3000 mm and the maximum diameter is 1250 mm. Typically, there are two reeds, and the front one oscillates slightly to improve the regularity of beaming. Beaming speeds depend greatly on the yarn, but may be up to 100 m/min.

Pile yarns are normally wound on cheeses and are mounted on frames, one for each colour, and are led in parallel freely through the heald frames and into the relevant eyelets of the jacquard cords. To facilitate the mass production of some grades of polypropylene carpet, pile yarns may be supplied on large bobbins, and correspondingly larger creels are required (Figure 11.2).

Figure 11.2. CRM72 face-to-face weaving machine showing part of the creel that can accommodate large cones of polypropylene yarns. (Courtesy of Michel Van de Wiele NV)

11.2 GROUND WEAVES

Shedding for weaving the backing material (ground weave) is achieved by heald frames that are actuated by a cam mechanism. Cam boxes must be changed to produce different ground weaves. Beam let-off is controlled by electric motors, and warp tensions may be tight or slack, according to the requirements of the weave. The beam stand incorporates warp stop motions to eliminate weaving faults that would otherwise result from yarn breakages.

When designating a carpet stucture, it is necessary to define the backing weave and pile weave separately. When there is just a pair of chain warps, the ground weave is defined simply by the sequence of weft shots above or below a warp (eg, 1/1, 2/1). In the case of three warps (one tight and two slack) there are two sequences to be recorded, for the tight/slack pair and the slack pair, respectively (eg, 2/2 + 1/1).

Specific ground weaves may be advantageous for specific pile structures and weaving methods. Pile tufts may be shaped like the letters V, U or W, V-tufts being the most commonly used in carpets (U- and W-weaves are often used in face-to-face fabrics). There may be one, two or three shots of weft per tuft, so that pile structures are designated 1/1 V, 1/2 V or 1/3 V. In addition, a weave diagram is necessary to complete the detail, making it apparent how many colours (frames) of pile yarns are used and whether the dead pile is floated on the back of one of the carpets (see, for example, Figure 11.10) or incorporated within the backings (see, for example, Figure 11.12). Less commonly, dead pile is floated down the middle of the sandwich and is removed in finishing.

In weave diagrams the pile warps are usually numbered (eg, 1–6 for a 6-frame carpet) and the ground warps are also numbered, for reference to other aspects of carpet manufacture and design (eg, sequence of yarns in a dent or pile colours to be used).

Figure 11.3 shows three alternative ground weaves produced on single-rapier looms on the left. The upper 2/2 backing weave is used for 1/1 V-tuft structures having dead pile floating on the back of the bottom carpet (see Section 11.4.1). Its principal advantage is minimum consumption of warp yarns in low and medium grades of carpet. The middle diagram is a 1/1 weave with slack and tight warps: this structure is used for denser 1/1 V-weaves, either with floating dead pile or dead pile incorporated in the backing. The lower diagram shows a 2/2 + 1/1 structure having two slack and one tight warps. This structure, too, is used for carpets with high pile counts, with incorporated or floating dead pile.

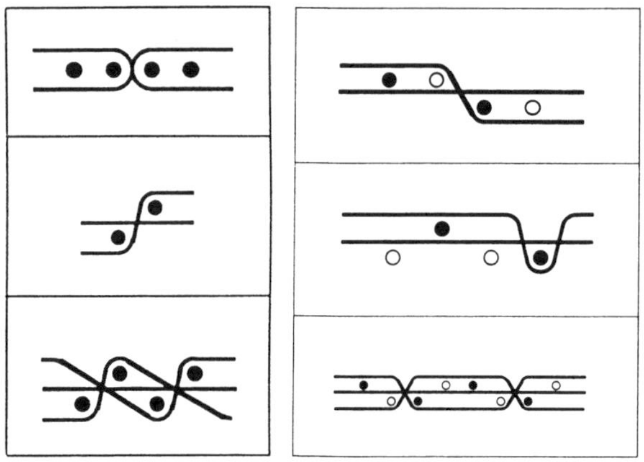

Figure 11.3. Ground weaves produced on single-rapier looms (left) and double-rapier looms (right). (Courtesy of Michel Van de Wiele NV)

The lower two weaves on the left in Figure 11.3 are also commonly used on double-rapier looms. They may also be produced using missed picks (circles show missed picks and solid discs actual wefts), as shown in the upper two diagrams on the right, to allow the proportion of pile to backing to be increased. The lower right diagram shows how a final 2/2 + 1/1 may be produced from a 4/4 + 2/2 structure with missed picks. Section 11.4.2 includes an outline of the rapier switch-off system that enables picks to be missed out, and the reasons for using it.

11.3 JACQUARD MECHANISMS

The pile design in many existing looms is formed by traditional jacquard mechanisms, as described in Section 16.2.3. All new looms are equipped with electronic jacquard mechanisms and older looms can be retrofitted with electronic jacquards.

The primary advantage of an electronic jacquard system is to allow a weaving company to respond quickly to customers' requirements, with minimum disruption to production. This is achieved largely as a result of close integration of CAD systems with the electronic jacquard, coupled with the ability to weave different designs one after another without stopping the loom (Plate 21). Indeed, designs of different sizes can be woven next to each other, and can be in different colourways, within the limits of the number of colours on the creel. Further advantages of electronic jacquards accrue from the facility to integrate them with production planning and process monitoring, as well as CAD.

A further advantage of electronic jacquards is enhanced weaving output. With the introduction of rapiers, the time required to select the pile with a punched card became the rate determining step in the weaving cycle.

The operation of an electronic jacquard machine for wirelooms has been described in Section 10.1: somewhat different mechanisms are required for double- and triple-rapier face-to-face looms.

Figure 11.4 shows the mechanism of the Piletronic (Van de Wiele) open-shed electronic jacquard machine with fixed lower roll (fixed rost) for use on a single-rapier loom. Quite simply, actuation of the solenoid causes pile yarn to be lifted from the bottom carpet, where it is floating on the back when not required in the pile, to the top carpet and to appear in the design.

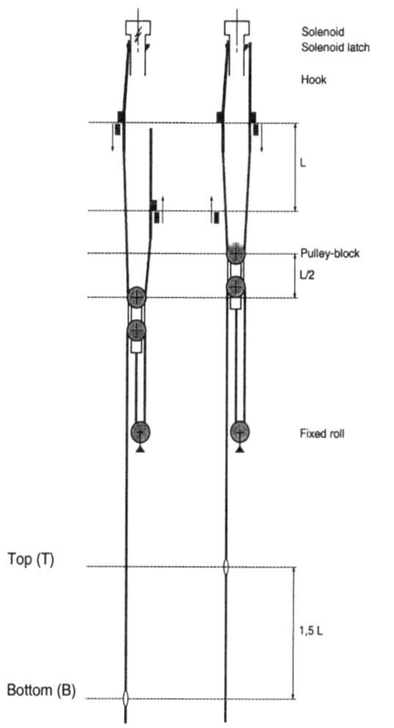

Figure 11.4. Mechanism of the Piletronic open-shed electronic jacquard machine suitable for single-rapier looms. (Courtesy of Michel Van de Wiele NV)

For double-rapier weaving, a three position electronic jacquard with both a fixed rost and a driven rost is used. The back part of the jacquard is equipped with a driven rost and selects pile or incorporates it in the bottom carpet, as shown in Figure 11.5. The front part of the jacquard has a fixed rost and can select the pile or incorporate it in the top carpet, as shown in Figure 11.6.

Selected pilewarp (driven rost)

Pick 1	B	Kr1 and Dr top Kr2 bottom	No selection ○
Pick 2	T	Kr1 and Dr bottom Kr2 top	Selection ●

Incorporated pilewarp (driven rost)

Pick 1	C	Kr1 and Dr top Kr2 bottom	Selection ●
Pick 2	B	Kr1 and Dr bottom Kr2 top	No selection ○

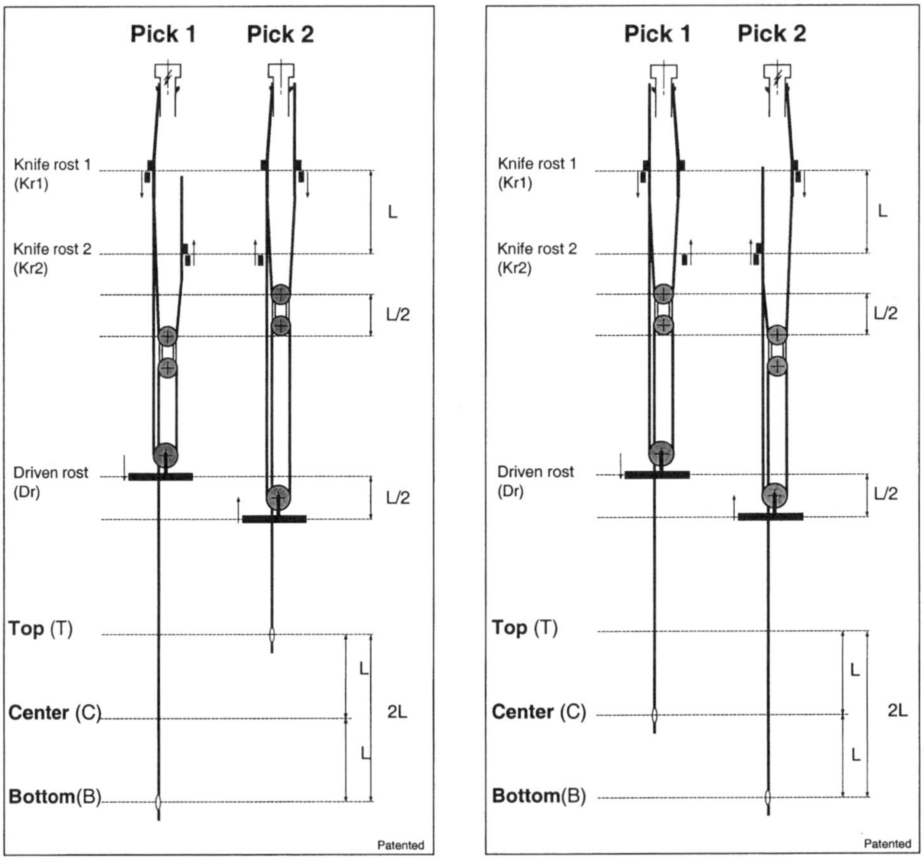

Figure 11.5. Back part of the Piletronic electronic jacquard machine for double-rapier looms. Driven rost for the bottom carpet. (Courtesy of Michel Van de Wiele NV)

Selected pilewarp (fixed rost)

Pick 1	B	Kr1 top Kr2 bottom	No selection ○
Pick 2	T	Kr1 bottom Kr2 top	Selection ●

Incorporated pilewarp (fixed rost)

Pick 1	T	Kr1 top Kr2 bottom	Selection ●
Pick 2	C	Kr1 bottom Kr2 top	No selection ○

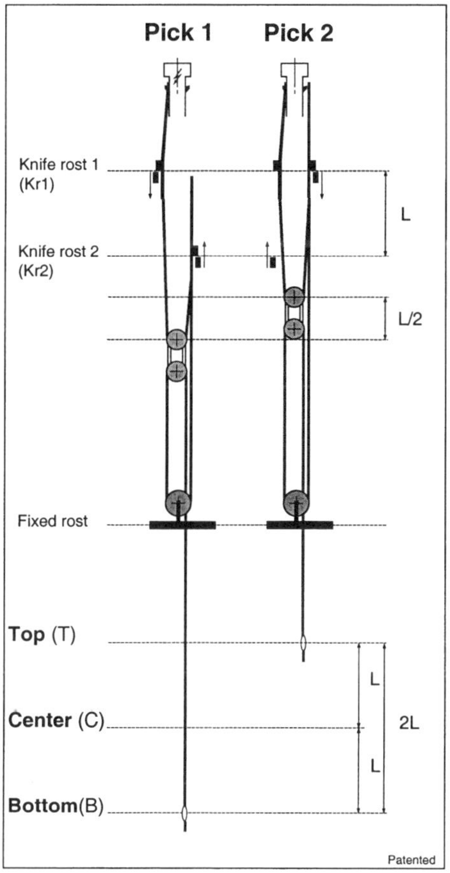

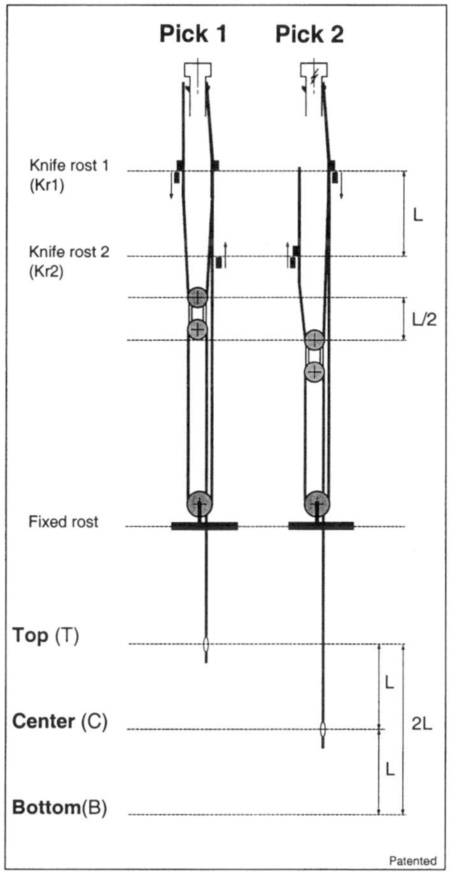

Figure 11.6. Front part of the Piletronic electronic jacquard machine for double-rapier looms. Fixed rost for the top carpet. (Courtesy of Michel Van de Wiele NV)

For triple-rapier weaving a four position electronic jacquard with driven rosts on both the front part and back part of the jacquard is used. The back part of the jacquard selects pile or incorporates it in the bottom carpet, as shown in Figure 11.7. The front part selects pile or incorporates it in the top carpet as shown in Figure 11.8. Note that the driven rost in the front part moves only half the distance of the reciprocating mechanism, whereas on the back part it moves the full distance.

Selected pilewarp (driven rost Dr1)

Pick 1	T	Kr1 and Dr1 bottom Kr2 top	Selection ●
Pick 2	B	Kr1 and Dr1 top Kr2 bottom	No selection ○

Incorporated pilewarp (driven rost Dr1)

Pick 1	CB	Kr1 and Dr1 bottom Kr2 top	No selection ○
Pick 2	CB	Kr1 and Dr1 top Kr2 bottom	Selection ●

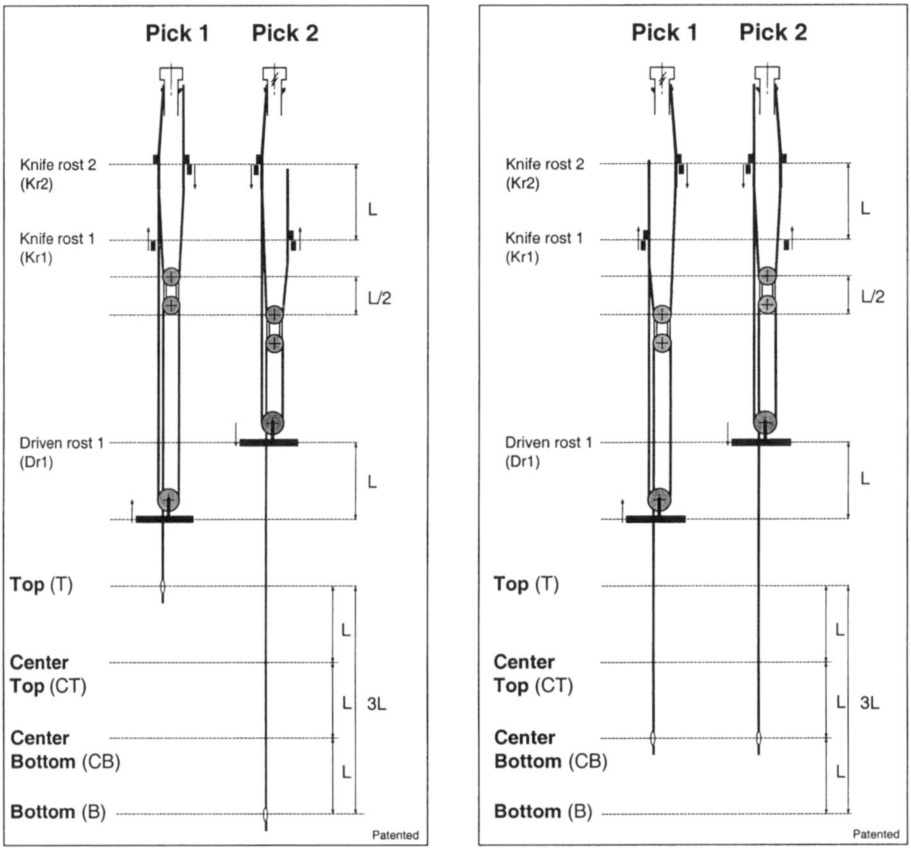

Figure 11.7. Back part of the Piletronic electronic jacquard machine for triple-rapier looms. Driven rost for the bottom carpet. (Courtesy of Michel Van de Wiele NV)

In some weave structures, a tuft of mixed colours is formed by some sequences of colour change. The greater capabilities of electronic jacquards provide better possibilities for eliminating such defects at the design stage. The machine manufacturer can supply a set of rules for designers to follow in order to achieve satisfactory results.

Selected pilewarp (driven rost Dr2)

Pick 1	T	Kr1 and Dr2 bottom Kr2 top	Selection ●
Pick 2	B	Kr1 and Dr2 top Kr2 bottom	No selection ○

Incorporated pilewarp (driven rost Dr2)

Pick 1	CT	Kr1 and Dr2 bottom Kr2 top	No selection ○
Pick 2	CT	Kr1 and Dr2 top Kr2 bottom	Selection ●

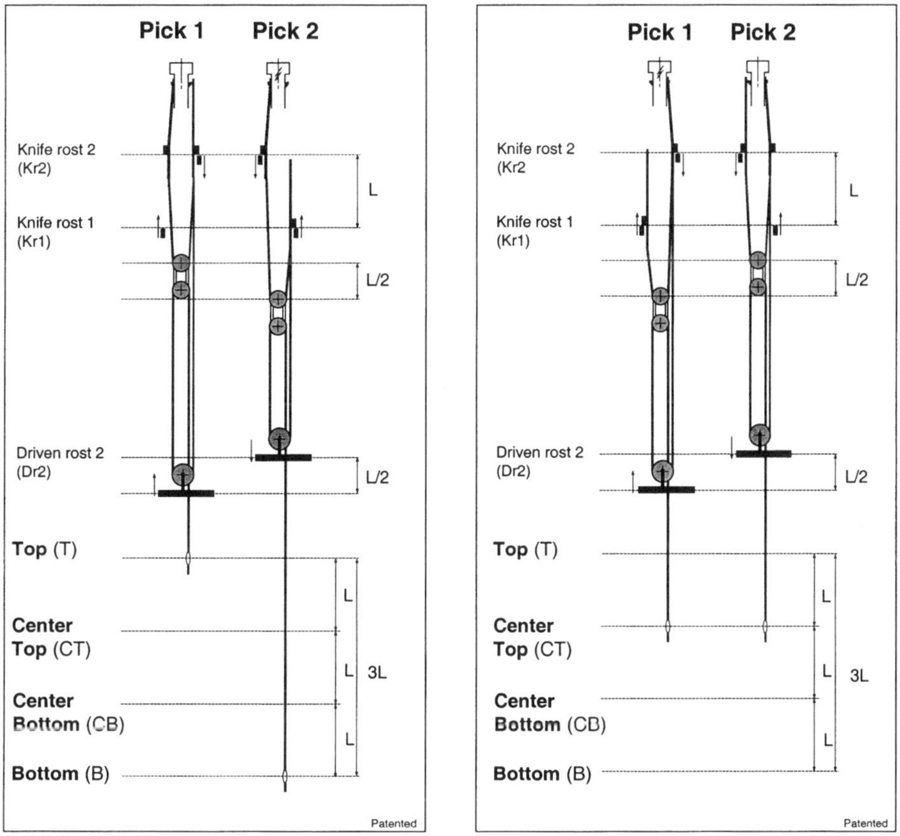

Figure 11.8. Front part of the Piletronic electronic jacquard machine for triple-rapier looms. Driven rost for the top carpet. (Courtesy of Michel Van de Wiele NV)

11.4 LOOM TYPES

11.4.1 Single-rapier Weaving

Figure 11.1 showed the first weaving cycle in single-rapier weaving, in which a shed is made for the bottom carpet. Weft shots are inserted alternately in the bottom and the top carpet and Figure 11.9 shows how a shed is made for the top carpet in the second weaving cycle. A selected pile end is moved across the shed whenever a shot of weft is introduced so that each time the weft is beaten up a new row of pile is produced (ie, a row of pile is introduced for every revolution

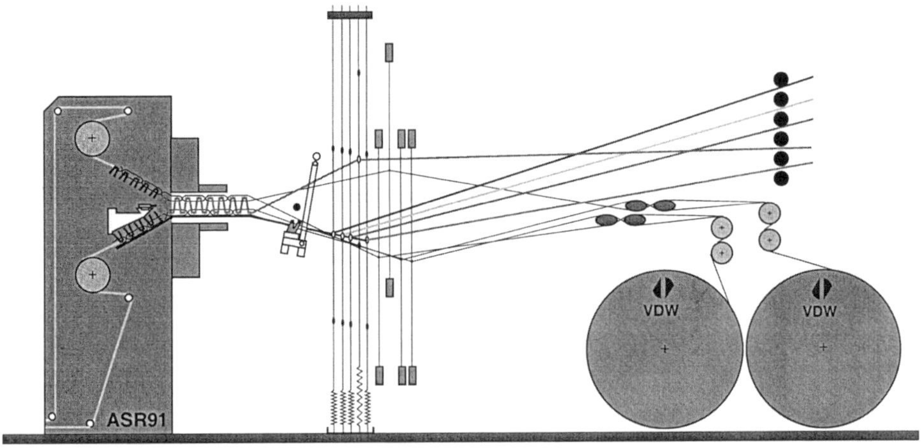

Figure 11.9. Second weaving cycle in single-rapier weaving, showing the shed for the top carpet. (Courtesy of Michel Van de Wiele NV)

of the machine). An important consequence is that very dense carpets can be woven in this structure.

Traditional 1/1 V-structures woven on single-rapier machines are shown in Figures 11.10 (for low pile densities) and 11.11 (for high pile densities). The two-level arrangement shown in Figure 11.11 can compact together more readily to create a denser pile. Note that these structures have all of the dead pile floating on the back of the lower carpet. A supplementary machine is required to scrape off this dead pile, which is then sold as waste. Another problem is that, when the dead pile is removed from the lower carpet, a tuft is removed where there is a change of colour. Corresponding tufts remain in the upper carpet, so the two carpets are not quite identical, creating a potential marketing problem.

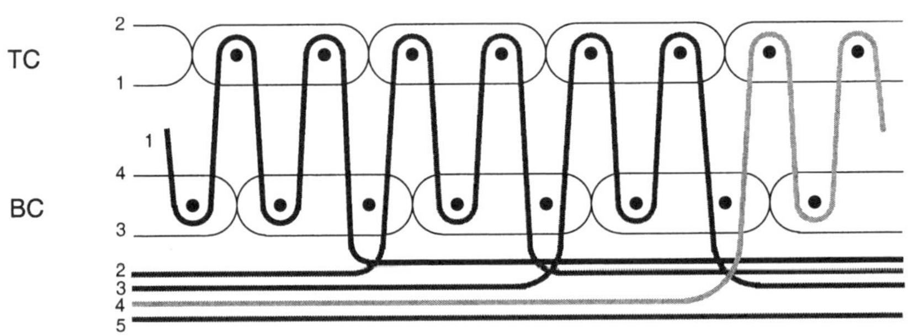

Figure 11.10. 1/1 V-structure with floating dead pile and 2/2 ground weave (for low pile densities). (Courtesy of Michel Van de Wiele NV)

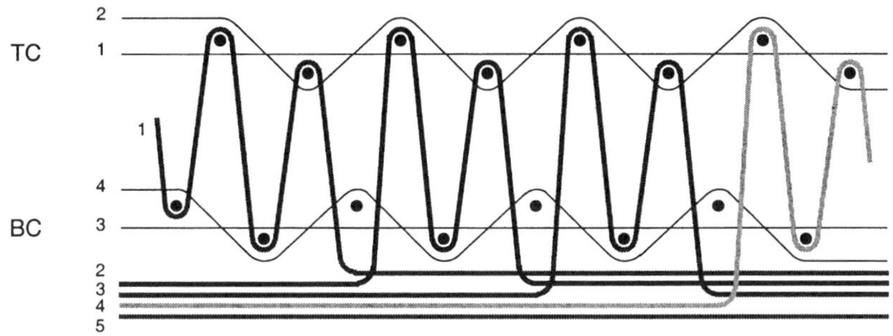

Figure 11.11. 1/1 V-structure with floating dead pile and 1/1 ground weave (for high pile densities). (Courtesy of Michel Van de Wiele NV)

A more satisfactory structure would be one incorporating the dead pile in the backing structure (and sharing it between the two backings) as illustrated in Figure 11.12. In this structure the two carpets are identical (except that the designs are a mirror image of each other – unimportant if the design is symmetrical) and the dead pile contributes positively to the mass and resilience of the carpet. Moreover, no finishing operation is required to remove the floating pile from the back.

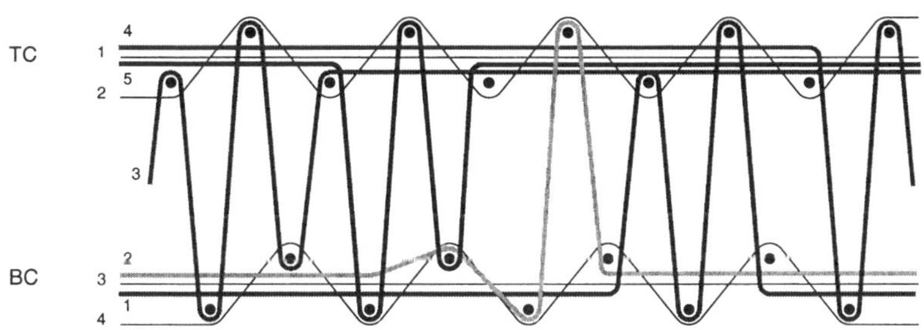

Figure 11.12. 1/1 V-structure with incorporated dead pile and 1/1 ground weave. (Courtesy of Michel Van de Wiele NV)

However, it is not easy to weave a carpet incorporating dead pile using the single-rapier technique. To weave good quality with the pile lying straight in the backing and because yarns cross over more, high tension has to be applied to the pile yarns. The extra tension tends to pull the two carpets together, making it difficult to control the pile height. The use of lancets (see Section 11.4.2) is not possible because of the high tension.

11.4.2 Double-rapier Weaving

The primary objective of introducing double-rapier weaving was to increase the rate of weft insertion, but many other benefits emerged.

In the double-rapier system, two separate sheds are formed – one for the top and one for the bottom carpet – and two weft shots are inserted simultaneously by rigid rapiers, one in each shed. The sheds are angled away from the centreline of the carpet sandwich, as shown in Figure 11.13, so that the resultant forces tend to pull the backings away from each other, assisting in controlling pile height.

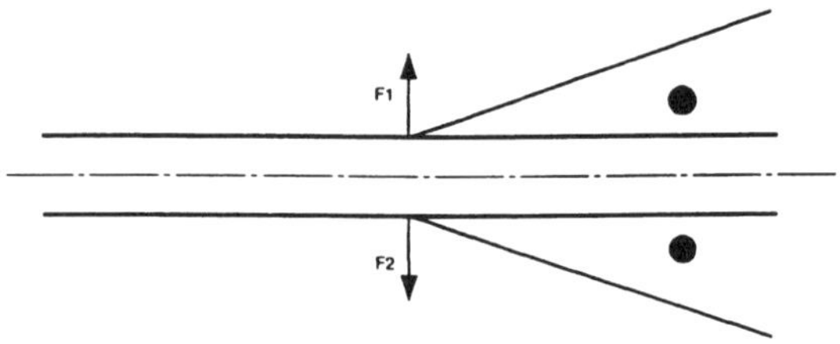

Figure 11.13. Geometry of shedding in double-rapier weaving (right) giving resultant vertical forces (F1 and F2) that assist in controlling pile height. (Courtesy of Michel Van de Wiele NV)

The use of separate sheds means that the ground warps do not need to cross from the top to the bottom carpet and vice versa; and the incorporated pile has only to move half a shed width from a central position between the two rapiers towards the top or bottom positions, so that crossings of pile yarns are limited. These factors contribute towards smooth weaving and facilitate the production of structures with incorporated dead pile. They also permit lancets to be used at the point of weaving.

Lancets are thin strips of metal, first used in velvet weaving, that are used, in effect, as gauges to control pile height (see Figure 10.10). They are threaded on a metal rod and are supported by it from a position behind the reed and otherwise move freely between the two sheds of chain warps, one for each reed dent, and project into the point of fabric formation. When first introduced for carpet manufacture, around 1985, they were simple parallel-sided strips and a set had to be changed when it was necessary to change the pile height. In the early 1990s, stepped lancets came into use, typically with gauges of 14, 16, 18 and 20 mm, so that to change pile height the assembly of lancets is simply advanced or retracted from the point of fabric formation.

As well as providing a convenient gauge for controlling pile height, lancets contribute to the uniformity of pile formation. Often, the need for shearing is eliminated, leaving brushing (see Section 17.5) as the only pile finishing process.

Figure 11.14 illustrates the general layout of a double-rapier loom and shows the top and bottom sheds with rapiers in place. Using this standard procedure one row of pile is created for every two picks of weft (ie, a 1/2 V-weave is produced).

In order to produce carpets with higher pile densities, a double-rapier machine may be adapted to produce 1/1 V-weave structures (pile on every pick) using a

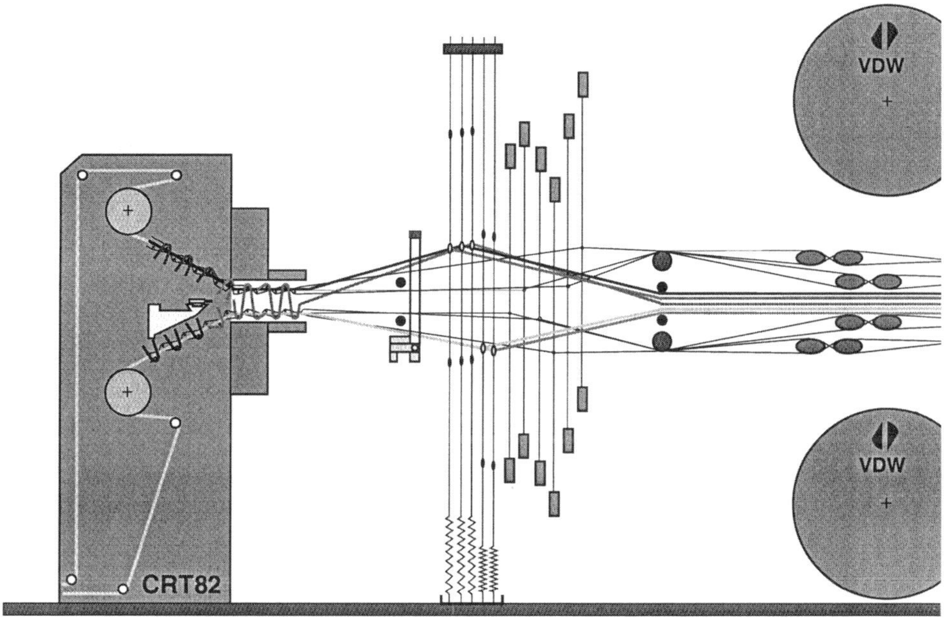

Figure 11.14. Shed formation in double-rapier weaving to produce 1/2 V-weave. (Courtesy of Michel Van de Wiele NV)

rapier switch-off system. The technique is illustrated in Figures 11.15 and 11.16. On the first cycle, no weft is introduced into the bottom carpet, and in the second cycle no weft is introduced into the top carpet. Although the rapier switch-off system negates the advantage of high rates of weft insertion, it does allow a 1/1 V-weave structure with incorporated dead pile to be produced much more easily than on single-rapier looms and also permits the benefits of lancets.

Figure 11.17 shows how an ephemeral construction produced by a double-rapier machine with switch-off becomes a 1/1 V-weave structure.

Most carpets manufactured on double-rapier looms have incorporated dead pile, but in some weave constructions the dead pile is floated down the middle of the sandwich and is removed in the subsequent brushing and shearing operations.

11.4.3 Triple-rapier Weaving

Machines are available with three rapiers. These have the ability to produce a wider range of weave structures than either a single- or double-rapier loom, and to improve quality and/or productivity in some constructions.

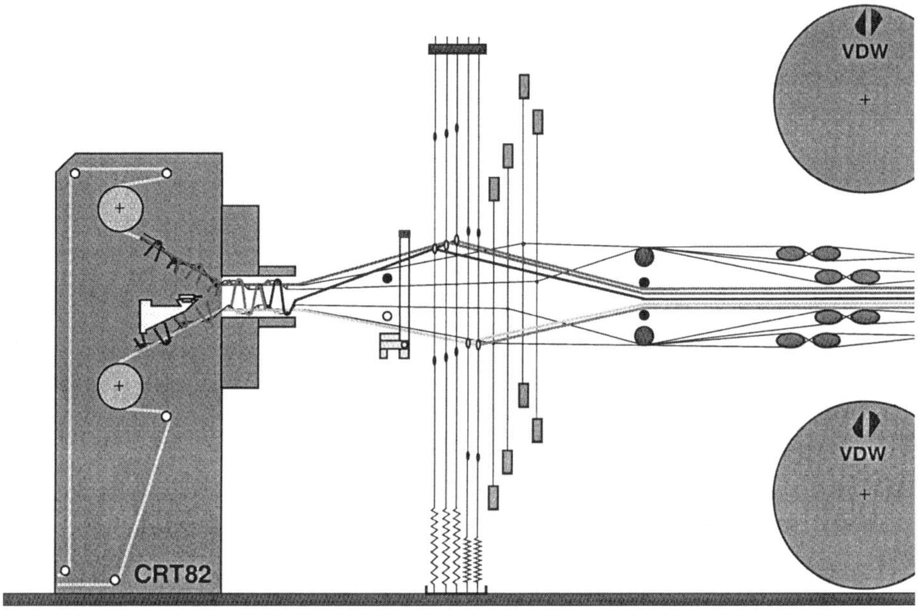

Figure 11.15. Use of the rapier switch-off system to produce 1/1 V-weave: first cycle, no weft inserted in bottom carpet. (Courtesy of Michel Van de Wiele NV)

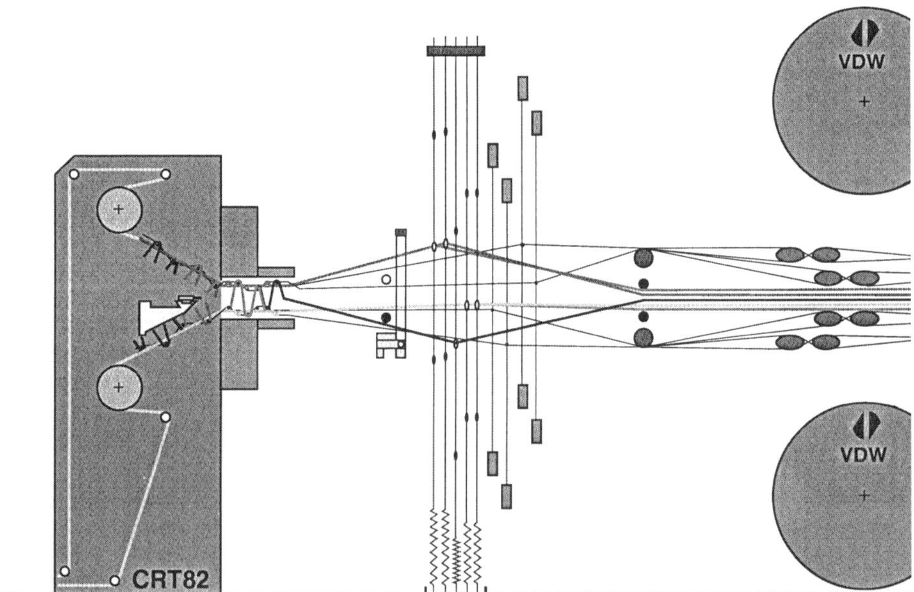

Figure 11.16. Use of the rapier switch-off system to produce 1/1 V-weave: second cycle, no weft inserted in top carpet. (Courtesy of Michel Van de Wiele NV)

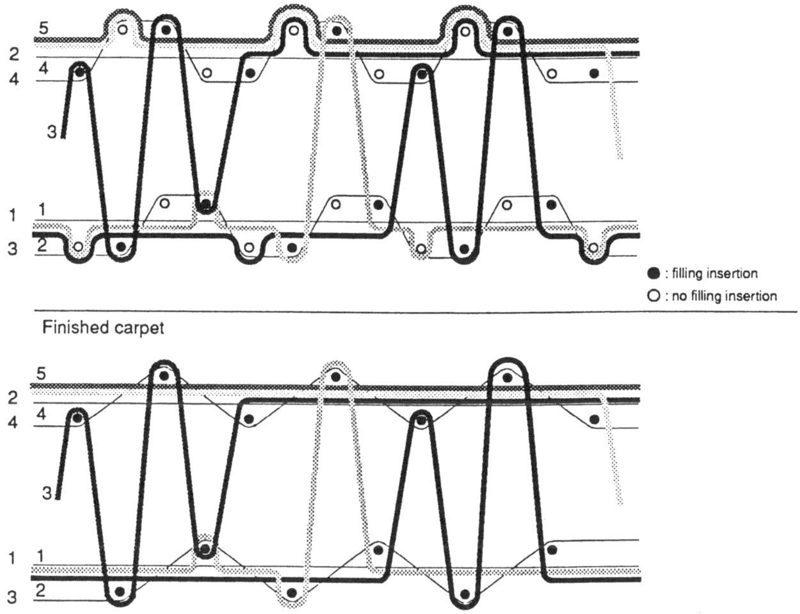

Figure 11.17. Effect of the rapier switch-off system on weave structure: 1/2 V giving 1/1 V with incorporated dead pile; ground weave 2/2 finally 1/1. (Courtesy of Michel Van de Wiele NV)

The clearest advantage of triple-rapier weaving is apparent in three-shot constructions; there is a 50% increase in production rate compared with double-rapier weaving. However, improvements in quality or versatility are often regarded as more important advantages.

With the triple rapier system it is possible to insert two weft shots simultaneously in the same fabric without having to move a rapier up or down, so that handover of the weft becomes more reliable. During weft insertion, the incorporated pile does not move, contributing to the smooth operation of the loom cycle and economising on dead pile, which lies straighter. The middle rapier can be fed with a finer weft than the outer two: this weft is always on the inside of the carpet sandwich, improving the ratio of the surface pile to total pile in the carpet.

By using the rapiers in different sequences (alternately, simultaneously or not at all) a wide range of structures can be woven:

- *Single-rapier constructions:* the middle rapier is not used and the bottom and top rapiers are inserted alternately.

- *Double-rapier constructions:* all double-rapier weave structures can be produced. By continuously inserting the middle rapier and using the top and bottom rapiers alternately, a 2/2 V-weave with two weft shots for each carpet, design through to the back, can be woven.
- *Triple-rapier weaving:* a 2/3 V-weave three-shot structure can be woven in two revolutions of the machine.

The following illustrations show how some of these constructions can be achieved. Figure 11.18 shows the first weaving cycle in the production of a 2/3 V-weave using all three rapiers simultaneously. Two shots of weft are inserted into the double shed of the top carpet and one weft is shot across the shed for the bottom carpet. In the second cycle, shown in Figure 11.19, the selected pile has moved from top to bottom, two shots of weft are inserted in the double shed of the bottom carpet and one shot of weft into the top carpet. A row of pile is woven in each carpet every two revolutions of the machine, and six shots are inserted, averaging three weft shots over the top and bottom carpets.

In the production of a 2/2 V-weave using the rapier switch-off system, in the first weaving cycle two shots are inserted in the top carpet and none in the bottom one. In the second cycle two weft shots are inserted in the bottom carpet, but none in the top. Thus, a row of pile is woven every two picks.

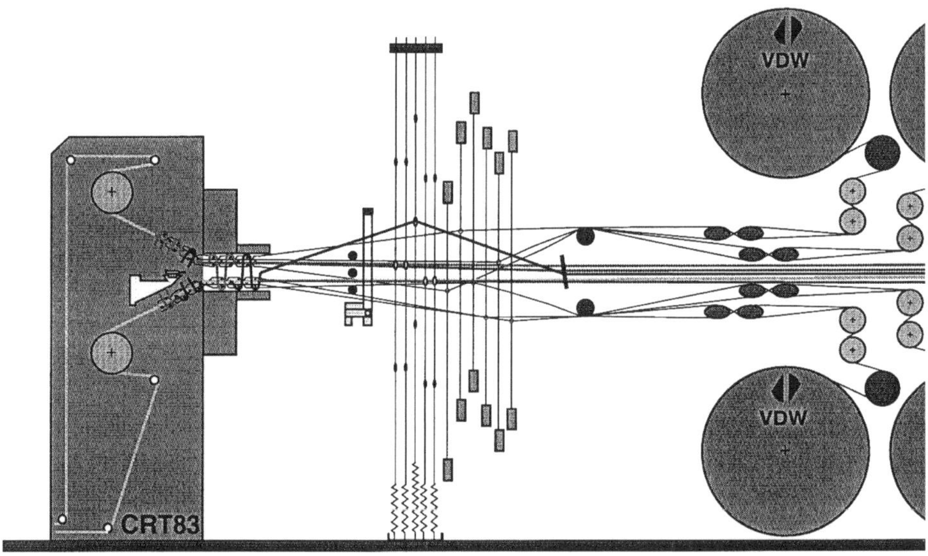

Figure 11.18. First cycle in triple-rapier weaving, inserting two weft shots in the top carpet and one in the bottom carpet. (Courtesy of Michel Van de Wiele NV)

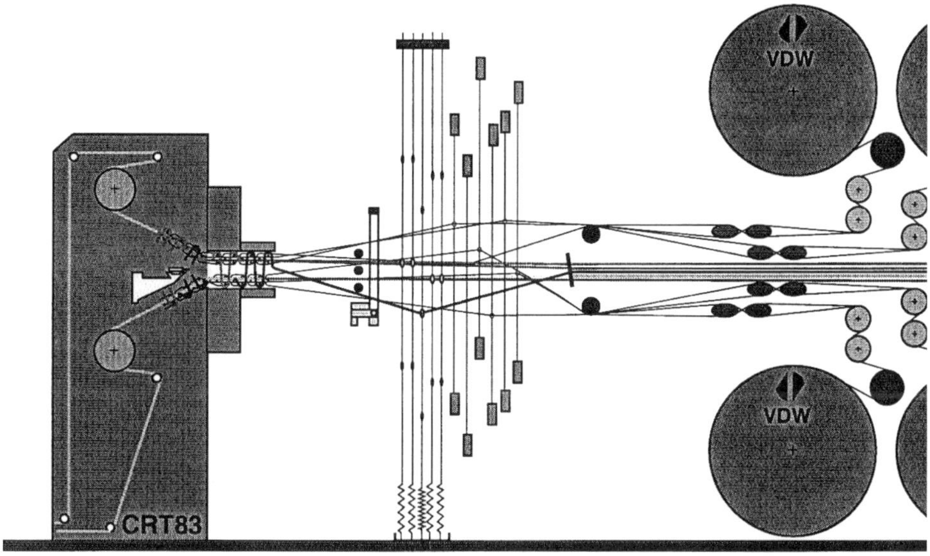

Figure 11.19. Second cycle in triple-rapier weaving, inserting one weft shot in the top carpet and two in the bottom. (Courtesy of Michel Van de Wiele NV)

11.4.4 Triple-rapier Weaving of Loop Pile

Van de Wiele's carpet weaving machine SLC83 is designed to produce the sisal-look carpets that became popular towards the end of the twentieth century (Plate 22). Sisal-look carpets have a structured pile surface and are usually woven in a small number of colours. Using the special loom, the structured surface is formed by looping the pile material over one or more thick wefts (Figure 11.20).

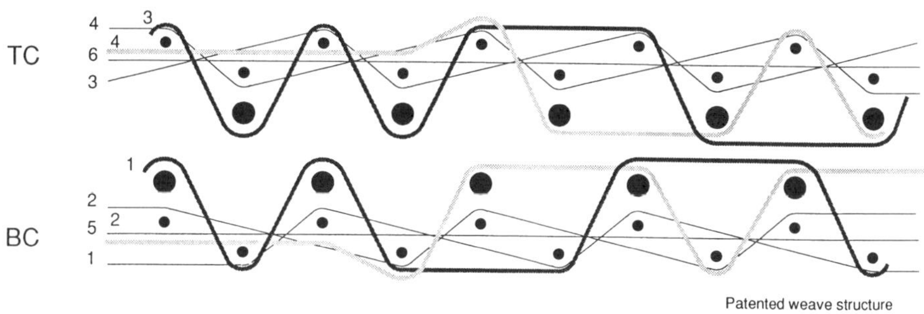

Figure 11.20. Sisal-look weave structure produced with three rapiers on the SLC83 face-to-face loom. (Courtesy of Michel Van de Wiele NV)

During weaving, three weft shots are inserted simultaneously. The top and bottom rapiers insert shots for the respective backings and the middle rapier inserts a thick weft that is woven alternately into the top and bottom carpets. The pile yarns are selected by the Piletronic PTX electronic jacquard and are raised only to a height above the thick weft: there is no pile linking the two backings to be cut, and no cutting mechanism is required.

The special electronic jacquard mechanism allows wall-to-wall or rug designs to be produced one after the other. Moreover, the designs and colours of the top and bottom carpets can be different (eg, it is quite possible to weave simultaneously 4.2 m wall-to-wall carpet and two lines of 2.1 m rugs in a different colour combination). Figure 11.21 shows two lines of various loop-pile rugs emerging from a SLC83 loom.

Note that patterns of texture known as false bouclé can be produced on the later models of more conventional double- and triple-rapier looms by floating pile over wefts of standard thickness. Designs incorporating carved effects or areas of very low loops may be produced in this way.

Figure 11.21. Different patterned loop-pile squares emerging from the SLC83 face-to-face loom. (Courtesy of Michel Van de Wiele NV)

11.5 POPULAR WEAVE STRUCTURES

Table 11.1, supplied by Van de Wiele, exemplifies the carpets that can be produced from single-rapier looms. Similar data are available for double- and triple-rapier looms. Note the wide variety of pile fibres that are used. Although some carpets in the lower price brackets are produced with cotton, jute or rayon pile, it should be appreciated that the majority of the mass produced carpets have polypropylene pile, while the higher qualities are generally made from semi-worsted or worsted wool yarns.

Table 11.1. Examples of materials and constructions for carpets woven on single-rapier looms.

Pile	Reed (m)	PPcm	Ground weave	Yarn count - Pile -	Yarn count - Filling	Type
BCF PP	280	4	2/2	290 tex, 3,5/1 Nm, 2600 denier	Jute 2x1175 tex, 0,85/1 Nm, 1,4/2 NeL	Floating
	280	5,5	2/2	290 tex, 3,5/1 Nm, 2600 denier	Jute 2x675 tex, 1,5/2 Nm, 2,5/2 NeL	Floating
Heat set PP	320	8	2/2	225 tex, 4,5/1 Nm, 2100 denier	Jute 2x280 tex, 3,6/2 Nm, 6/2 NeL	Floating
	320	7,2	1/1	260 tex, 3,9/1 Nm, 2300 denier	Jute 2x480 tex, 2,1/2 Nm, 3,5/2 NeL	Floating
	350	11,3	1/1	135 tex, 7,4/1 Nm, 1200 denier	Jute 2x240 tex, 4,2/2 Nm, 7/2 NeL	Floating
	350	6,8	1/1	225 tex, 4,5/1 Nm, 2100 denier	Jute 2x420 tex, 2,4/2 Nm, 4/2 NeL	Incorporated
PAC	320	5,8	2/2	200, 5/1 Nm (70-30)	Jute 2x675 tex, 1,5/2 Nm, 2,5/2 NeL	Floating
Jute	280	6	2/2	2x330 tex, 3/2 Nm, 5/2 NeL	Jute 2x480 tex, 2,1/2 Nm, 3,5/2 NeL	Floating
	350	5	2/2	2x210 tex, 4,8/2 Nm, 8/2 NeL	Jute 2x720 tex, 1,4/2 Nm, 2,3/2 NeL	Floating
	350	6	2/2	2x210 tex, 4,8/2 Nm, 8/2 NeL	Jute 2x420 tex, 2,4/2 Nm, 4/2 NeL	Floating
Viscose Staple	500	10	1/1	2x(2x37) tex, (27/2)2 Nm, (16/2)/2 Nec	Cotton 4x100 tex, 10/4 Nm, 6/4 Nec	Incorporated
Cotton	300	5,5	2/2	2x100 tex, 10/2 Nm, 6/2 Nec	Jute 2x560 tex, 1,8/2 Nm, 3/2 NeL	Floating
Acrylic	350	7,5	1/1	2x95 tex, 10,5/2 Nm	Jute 2x240 tex, 4,2/2 Nm, 7/2 NeL	Floating
	350	9	2/2	2x95 tex, 10,5/2 Nm	Jute 2x280 tex, 3,6/2 Nm, 6/2 NeL	Floating
	500	10	1/1	2x95 tex, 10,5/2 Nm	Jute 2x210 tex, 4,8/2 Nm, 8/2 NeL	Floating
Wool	350	8	2/2	2x84 tex, 12/2 Nm	Jute 2x420 tex, 2,4/2 Nm, 4/2 NeL	Floating
	350	10	1/1	3x66 tex, 15/3 Nm	Jute 2x210 tex, 4,8/2 Nm, 8/2 NeL	Floating
	380	6,2	1/1	3x59 tex, 17/3 Nm	Jute 2x420 tex, 2,4/2 Nm, 4/2 NeL	Incorporated
	420	12	1/1	3x59 tex, 17/3 Nm	Jute 2x210 tex, 4,8/2 Nm, 8/2 NeL	Floating
	462	11	2/2+1/1	3x59 tex, 17/3 Nm	Jute 2x210 tex, 4,8/2 Nm, 8/2 NeL	Incorporated
	500	10	1/1	3x59 tex, 17/3 Nm	Jute 2x210 tex, 4,8/2 Nm, 8/2 NeL	Floating
	500	12	1/1	3x59 tex, 17/3 Nm	Jute 2x165 tex, 6/2 Nm, 10/2 NeL	Floating

11.5.1 Plain Carpets

The simplest construction possible for a plain broadloom carpet involves the pile warp and ground warp operating in a 1/1 liftplan, with pile over each pick. Using a fairly heavy pile yarn (eg, R620 tex/2 woollen spun), carpet with an adequate pile mass is produced from this weave. Alternatively, two lighter yarns (possibly semi-worsted) can be introduced in the same dent to give a finer texture in a similar, or heavier, pile mass.

Such a simple construction tends to have low tuft bind. Improved tuft bind can be achieved with a W-weave (having the tuft interlaced over three weft shots) but at the expense of tuft density, so other techniques tend to be used. One of the pair of chain warps may be held under high tension with the other slack so that one warp remains straight in the carpet while the other is sinusoidal. This helps to bind the tufts, thicken the backing construction and increase the pile density. Plain carpets are always latexed on the back to improve tuft bind.

11.5.2 Wall-to-Wall Patterned Carpets

Whereas Axminster weaving has tended to focus on the higher quality carpets for the hospitality markets, face-to-face weaving of wall-to-wall carpets is tending to take over the mass production of patterned residential carpets. At the time of writing, there is a dynamic product development situation and no firm description of products can be given. However, the principal pile fibre is polypropylene and there is a trend towards the use of heavier yarn linear densities (sometimes exceeding 4000 dtex) than are used in squares. Most yarns are heat set BCF, but some cabled yarns and spun yarns are also used as pile.

Although it is not important for the design to show through to the back in wall-to-wall carpets, weaves with pile through to the back are sometimes used to obtain structures that can be beaten up, with the weft shots for the top and bottom carpets almost superimposed, to increase the pile density (rows/dm) and to give good tuft bind. Figures 11.22 and 11.23 show structures suitable for high and medium density wall-to-wall carpets, respectively.

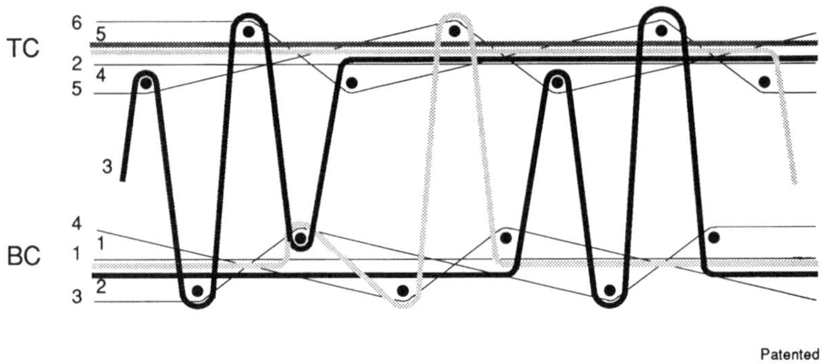

Figure 11.22. Double-rapier 1/2 V-weave with rapier switch-off giving 1/1 V with incorporated pile floating, ground weave 4/4 + 2/2 (finally 2/2 + 1/1), suitable for wall-to-wall carpet 70–140 pile rows/dm. (Courtesy of Michel Van de Wiele NV)

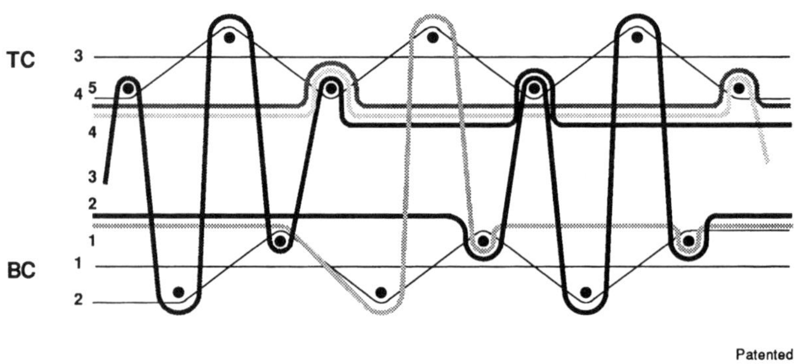

Figure 11.23. Double-rapier 1/2 V-weave with rapier switch-off giving 1/1 V with incorporated pile floating, ground weave 2/2 (finally 1/1), suitable for wall-to-wall carpet 50–80 pile rows/dm. (Courtesy of Michel Van de Wiele NV)

11.5.3 Squares

Three-shot 1/3 V-weaves

Figure 11.24 shows a weave structure with incorporated dead pile and with the design showing through to the back. With all the weft shown, this is a 1/3 V-structure (one row of pile per three shots) and has a number of advantages:

- Very good design definition, every pile tuft is able to have a different colour.
- Every tuft is clearly visible on the back, as in hand-knotted carpets, and incorporated dead pile is well covered.
- Good tuft bind because tufts pass through to the back and are locked by wefts.
- Good weavability.
- Uniform product structure.

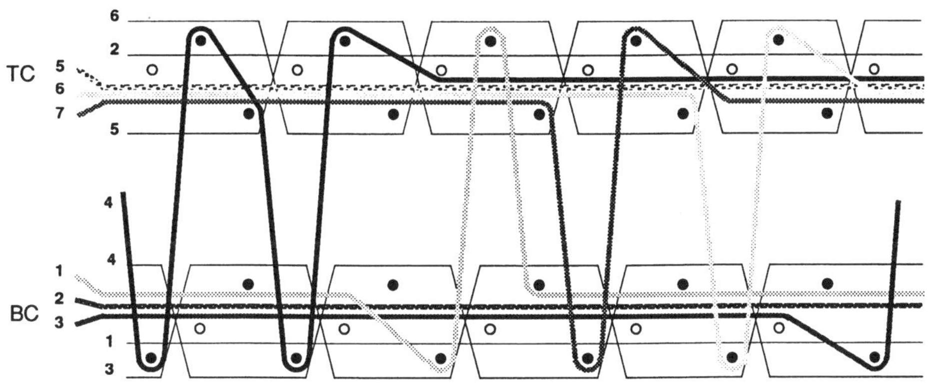

Figure 11.24. Double-rapier 1/3 V-weave with incorporated dead pile, giving 1/2 V-weave with rapier switch-off. (Courtesy of Michel Van de Wiele NV)

This structure is suitable for carpets in the range 100 000–400 000 pile tufts/m^2. If more than 270 picks/dm are required, however, too light a weft must be used and weaving becomes inefficient.

Note that in the diagram one in three of the weft shots is only outlined to indicate that these shots could be missed with a rapier switch-off system, effectively converting the structure to a 1/2 V-weave and allowing the pile density to be increased after beating up.

Two-shot 1/2 V-weaves

Although carpets of impeccable quality are produced, the principal disadvantage of the 1/3 weave is low output, since three weft shots (three cycles of the loom) are required for each row of pile. For this reason, weaving of rugs has become focused on 1/2 V-constructions, which have virtually the same quality as 1/3 V-constructions. The design can also be visible on the back and tuft bind is excellent.

Additionally, the dead pile lies very straight, minimising pile consumption relative to a three-shot weave, and the ground weave is thinner, also minimising consumption of pile yarn. Figure 11.25 shows a suitable weave, with six frames of pile yarn: the same basic structure can accommodate up to 14 frames of pile yarn.

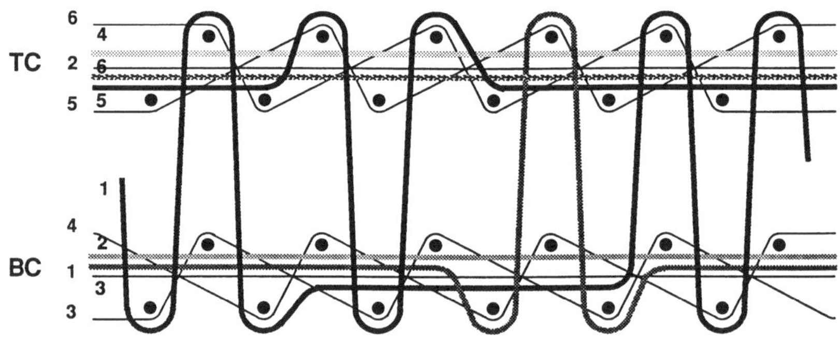

Figure 11.25. A 6-frame 1/2 V-weave with incorporated dead pile, produced on double-rapier looms. (Courtesy of Michel Van de Wiele NV)

This structure has proved so satisfactory that it has taken over much of the high quality rug market that was the province of 1/3 V-weaves. It can also be used to produce qualities as low as 80 000 tufts/m^2. With bulky BCF polypropylene pile, acceptable cover of the backing can be achieved with only 25 tuft rows/dm. The highest qualities that are practicable have up to 400 000 tufts/m^2. For higher pile densities it is better to use a 1/1 V-structure.

A disadvantage of the 1/2 V-weave was the problem of mixed colours in cases when a transition occurred between a colour to be incorporated in the bottom carpet and a colour that had been incorporated in the top carpet. Since the introduction of electronic jacquards, software has been developed to eliminate such defects.

Single-shot 1/1 V-weaves

Single-shot structures are favoured when very high pile densities are required, since a row of tufts is created on every pick. Densities in the range 200 000–700 000 tufts/m^2 can be produced.

Single-rapier looms normally produce 1/1 V-weaves, but Table 11.1 shows that most of their products have yarn floating on the back. However, double-rapier looms having the rapier switch-off facility can also produce 1/1 V-weaves with most of the advantages of double-rapier weaving outlined earlier, especially the incorporation of dead pile in the backings of both carpets. For example, see Figure 11.17 which shows how the rapier switch-off system can convert a 1/2 V-weave with a 2/2 ground weave to a 1/1 V-weave with 1/1 ground weave.

Constructions from triple-rapier looms

Figure 11.26 shows a 6-frame 2/3 V-weave with incorporated dead pile woven using all three rapiers. This construction can be produced at a remarkably high rate and at high weaving efficiency, for the reasons given in Section 11.4.3, and with the same advantages as indicated above for the 1/3 V-structure.

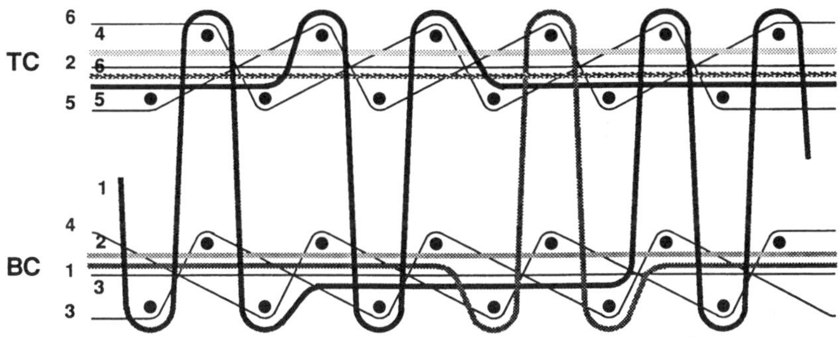

Figure 11.26. A 6-frame 2/3 V-weave with incorporated dead pile, produced on triple-rapier looms. (Courtesy of Michel Van de Wiele NV)

Figure 11.27 illustrates the production of a 2/2 V-weave (one row of pile on every pair of weft shots) produced with the rapier switch-off facility. The loom is set up to produce the 2/3 V-weave shown in the upper part of the diagram, but the missed picks result in the denser 2/2 V-weave shown in the lower illustration. The technique takes advantage of the smoother weaving associated with triple rapiers.

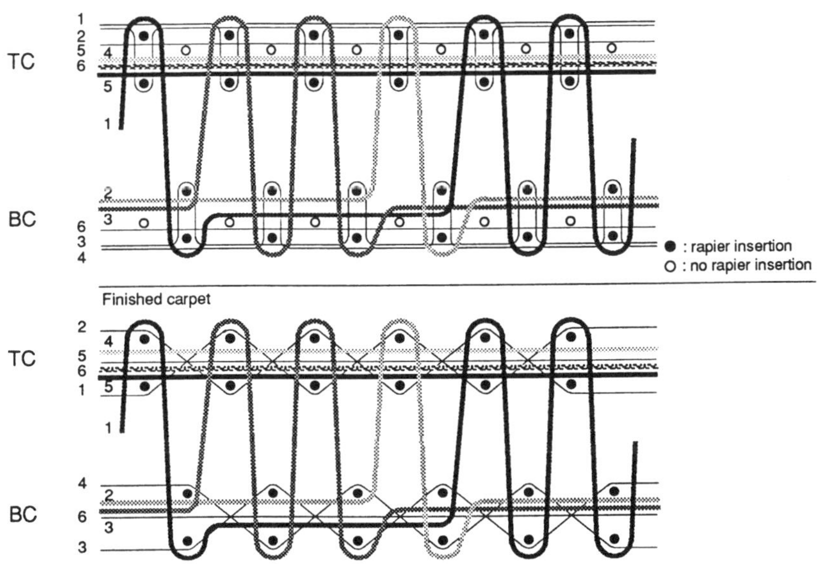

Figure 11.27. A 6-frame V-weave with incorporated dead pile, produced on triple-rapier looms with rapier switch-off: (above) 2/3 V-weave with missed picks; (below) final 2/2 construction. (Courtesy of Michel Van de Wiele NV)

11.6 RATES OF PRODUCTION

The output of a face-to-face loom depends on its speed and width, carpet construction and weaving efficiency. As explained earlier, the practicable speed of the loom depends on the weave structure, type of loom being used (single-, double- or triple-rapier) and whether or not a rapier switch-off system is used (in the case of double- or triple-rapier looms). Additionally, the feasible speed depends strongly on the width of the loom: a wide loom cannot be run as fast as a narrow loom, although the greater width more than compensates for the slower picking rate. If attempts are made to operate the loom too fast for the construction being woven, poor weaving efficiencies will be achieved. Finally, the output in square metres is heavily dependent on the density of the backing construction.

To exemplify some of the foregoing qualitative statements: a double-rapier loom in 2.5 m width may be expected to run at 170–200 ppm, depending on the carpet construction, whereas a 4.2 m wide loom could only operate at 140–150 ppm. However, when weaving a construction having 50 pile tufts/dm, the narrower loom would produce about 50 m^2/h, whereas the 4.2 m loom would produce about 65 m^2/h. Opening the construction to 25 pile tufts/dm would not significantly affect the feasible loom speeds, but would double the areas produced per hour. Weaving a carpet of modest quality, a standard double-rapier loom can produce as much as 130 m^2/h.

Van de Wiele's triple-rapier Carpet and Rug Tronic CRT83 is characterised by a high rate of production, especially when the three rapiers are inserted at the same time (3 x 150 ppm). To cope with the increased evolution of jute dust resulting from the high speed, the weft accumulators and weft tensioners are installed in boxes connected to a central dust extraction system; and programmable pneumatic brakes on the weft optimise the take-over. The knife motion is designed for optimal cutting at high speed. Higher weaving efficiency results from the fact that the incorporated pile yarns are stationary during weaving: only the selected pile yarns move. Depending on the weave structure, density of construction and yarns used, rates of production in the region 50–130 m^2/h are achieved.

Although the CRT83 can weave a variety of structures (see Section 11.4.3), a special weave structure, the 1 + 2/3 (Figure 11.28 and Plate 23) has been devised to take advantage of its capabilities. The 1 + 2/3 structure is well suited to high tuft densities (1 million tufts/m^2) and is characterised by better tuft bind and a clean back, as the incorporated pile is at a different level to the stuffer.

Associated with the CRT83 is the Piletronic PTX8400 electronic jacquard having a capacity of 6 Gb, operating from ZIP disk, CD-ROM or network. Up to 12 colours per reed dent can be controlled. A microprocessor machine control unit incorporates aspects such as machine speed, lubrication interval, stop motion, autodiagnosis and production analysis.

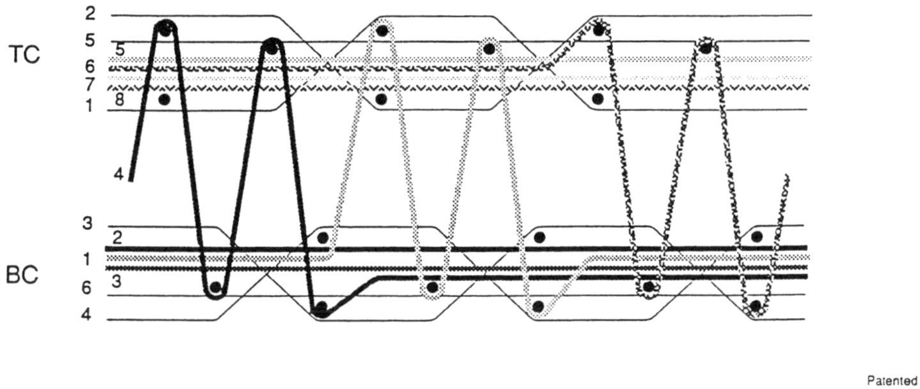

Figure 11.28. 1 + 2/3 V-weave developed for the CRT83 triple-rapier loom. (Courtesy of Michel Van de Wiele NV)

REFERENCES

1. N Gheysen. Face-to-face carpet weaving for creativity and productivity, *Textile Technology International*, 1994, 137–145.

12 *Flat-Woven Carpets*

Although they are not produced in comparable volumes to pile carpets, a variety of flat-woven textile floorcoverings are manufactured by hand and machine, some in significant quantities. Many of them clearly do not have a pile surface, but some of them are engineered so that there are components that may be distinguished as backing and use-surface.

Flat-woven rugs are commonly produced by hand weaving, and range from lightweight kelims in traditional designs to very heavy materials in extremely simple designs. Machine woven rugs that attempt to simulate both types are also available.

There has been a major revival of mattings made from plant fibres, mostly intended for wall-to-wall applications, and some of these are upgraded by backcoating. More recently, wall-to-wall products, usually woven from wool or wool blend yarns, have been developed using laminating techniques; and some quite sophisticated products having simulated loop piles have been engineered.

12.1 KELIMS

The longest established style of flat-woven rug is the kelim. It has no pile but the weaving technique is such that the weft, usually wool, forms the surface. Yarn forming a particular colour in the design is interlaced with the warp, once to the left and once to the right, so that both sides of the warp are covered. The next colour is similarly inserted, but there is no overlap between colours, so there is a small hole in the kelim at each colour change. Some kelims have concealed weft threads across the full width, improving the strength of the rug and preventing the holes from opening.

Kelims form reversible rugs, since both sides of the warp are covered with the surface yarn that provides the design. Clearly, kelims are produced more rapidly and, therefore, are cheaper than knotted rugs. Kelims are also used as blankets, saddlebags, wrappers and wall hangings.

12.2 INGRAIN CARPETS

The historic origins of ingrain carpets and their development are described in Section 1.2.2. Attempts to revive their manufacture are occasionally made. One objective is to manufacture environmentally friendly carpets composed entirely of natural fibres and without the use of latex. Such products are mostly reversible rugs in two colours, with the colours of the design inverted on the reverse side.

Figure 12.1 illustrates three alternative traditional structures for ingrain carpets. The upper structure (Figure 12.1(a)) was commonly made with a worsted warp and a woollen weft, and gives a two-colour design. The jacquard mechanism is used to bring to the surface warp and weft dyed to similar colours. The design on both surfaces is the same, but the positions of the colours are reversed.

Figure 12.1(b) illustrates the use of a stuffer warp to add to the weight and improve the binding.

In the construction illustrated in Figure 12.1(c) a greater variety of colour can be introduced, and the fabric is thicker and stronger.

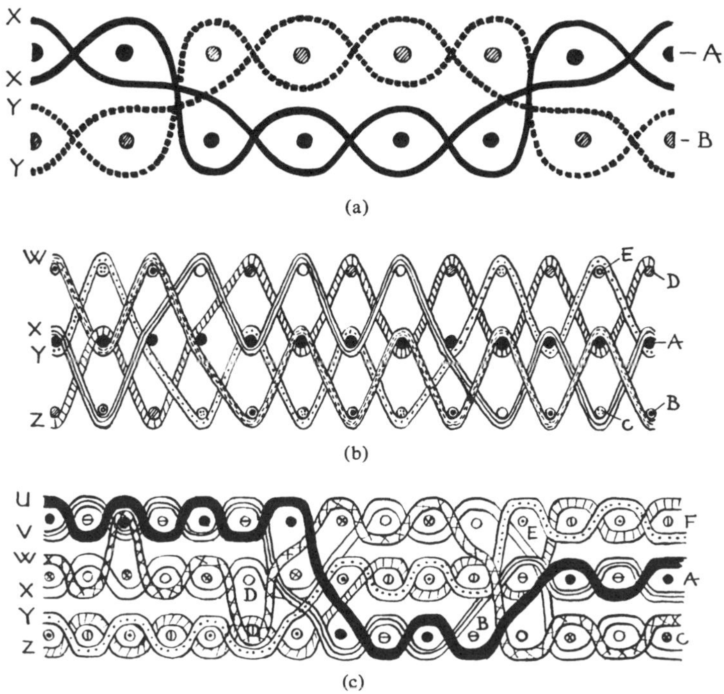

Figure 12.1. Ingrain carpet weave structures: (a) 2-ply warp and 2-ply weft – A and B warp yarns of different colours, X and Y weft yarns in corresponding colours; (b) 3-ply warp and 2-ply weft – A black stuffer, BCDE coloured warp yarns, WXYZ coloured weft yarn; (c) 3-ply warp and 3-ply weft – ABCDEF coloured warp, UVWXYZ corresponding coloured weft yarns [1].

12.3 PLAIN WEAVE WITH WEFT SURFACE

Heavy rugs composed of cotton warp and extremely thick wool weft are made in a number of countries, and are usually woven on hand looms. The cotton warp becomes embedded in the thick wool weft so that the weft is the most prominent feature of the product.

Yarns are often hand spun when the rugs are produced in developing countries, or are core spun (see Section 3.2.4) in the more developed areas of the world. Because the very thick yarns have few turns of twist per metre, they may advantageously be felted (see Section 4.3.2) to improve fibre bind.

Aesthetic interest is often provided by means of colour mixtures or slub effects in the weft yarn. For example, in South Africa, pigmented karakul wool, lightly mixed, is often used as the major (or only) component in a yarn produced simply by attaching an armful of fibre to a rotating hook and drafting the wool into the twisted fibre assembly while walking away from the rotating hook.

Simple patterns, for example a central motif in a single colour coupled with a border of the same colour and a background of a different colour, can be produced by changing the colours manually on the hand loom.

12.4 LAMINATES OF MACHINE WOVEN FABRICS

Carpets that are quite durable in use may be made simply by laminating a secondary backing fabric and a fabric woven in both warp and weft from yarns normally used for carpet pile, using a secondary backing line of the type available in the tufted carpet industry (see Section 18.4). Because of the smaller volumes produced relative to tufted carpets, heated drums rather than stenter ovens are often used for curing the binding latex.

Products range from rather thin materials that lack the impact absorbing properties of pile carpets to heavier products woven from yarns of high linear density. The latter are commonly produced in the UK from wool blend yarns woven into simple dobby designs.

12.5 LAMINATES OF WARP-FACED FABRICS

Using special weaving techniques, a 'pile' warp may be woven with 'backing' yarns in such a manner as to induce it to form the surface of the fabric. The techniques depend on the choice of yarn linear density and selective application of tension.

Figure 12.2 illustrates the principles used to achieve a 'loop-pile' fabric without the use of wires. If the tight warp in Figure 12.2(a) is a typical cotton (or cotton/polyester) chain warp and the slack warp is a typical pile yarn (wool or sisal) the pile is brought to the surface. Using two ends of slack pile warp to one of tight cotton warp emphasises the pile even more and the cotton virtually

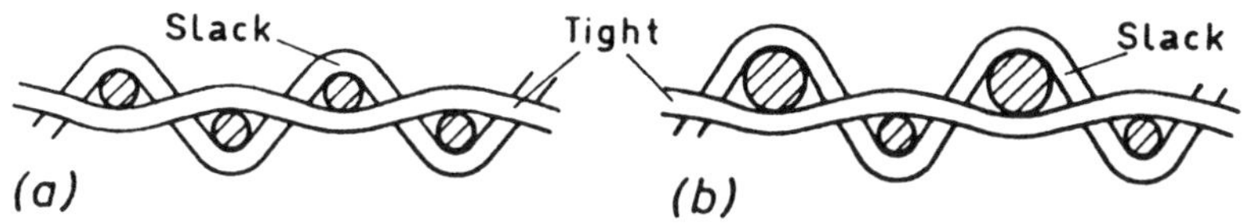

Figure 12.2. Use of tight and slack warps to produce a carpet with warp face [2].

disappears from sight. Add the technique shown in Figure 12.2(b) of using alternately thick and thin jute wefts (or double and single weft shots of the same yarn) and the loops of pile material become even more prominent.

Simple ribbed fabrics made in this way are used in textile floorcoverings that are a little thicker and more resilient than the lighter materials described in Section 12.4. They may be laminated to secondary backing fabrics of the type used for backing tufted carpets: some are sufficiently rigid (dimensionally stable) as to be converted into 'carpets' by applying a latex foam.

Warp-faced fabrics with simple patterns are made on dobby looms. The basic construction has the surface loops alternately over two weft shots or one shot, to give alternate thick and thin ribs. By interchanging the positions of the long and short loops, this effect is reversed. Interesting surface patterns can be made in this way.

12.6 DURALITE™ CARPETS

A special loom has been invented by Brian Carlson of the Duralite Carpet Corporation, New Zealand, to produce structures of the type shown in Figure 12.3.

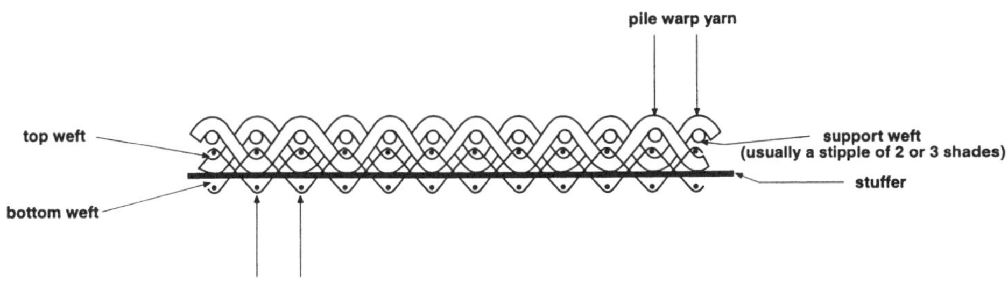

Figure 12.3. Structure of the DuraLite™ carpet. (Courtesy of DuraLite Carpet Corporation)

Typical backing yarns predominate on one face of the fabric and the other face is composed of typical pile yarns. The structure is somewhat similar to that of a single-shot Brussels carpet, except that there is additional pile material in the support weft. Details of the Carlson loom have not been released but it is said to have double the production rate of a wireloom at 1.4 m width, and offers greater gains at wider widths.

In practice, DuraLite™ carpets have a low, but extremely dense, pile and this is their principal selling feature. They are being promoted as aircraft carpets on the basis of their low total weight (1550 g/m^2, with a surface pile mass of 835 g/m^2) and their high durability. Their high density favours performance in aircraft flammability tests (particularly important when installed on vertical surfaces) and contributes to ease of cleaning and low fibre loss (reducing costs for maintaining air conditioning systems).

12.7 PAPER RUGS

The Finnish company, Woodnotes Oy, in the early 1990s established a market for high quality rugs, around 2200 g/m^2, made with a surface of paper dyed or printed to dark shades. Production details have not been released but basic paper-making techniques (grinding and pulping timber, laying a mat and calendering it) are used to create the paper, which is then slit and twisted. At weaving, the paper twine is used as weft across a cotton warp. In addition to unique styling, ease of recycling is quoted as a sales feature.

12.8 SISAL, COIR AND ALOE PRODUCTS

Probably the most numerically significant flat-woven floorcoverings are composed of sisal or coir. 'Sea grass' matting is made in similar structures.

The principal products are warp-faced fabrics from heavy warp yarns and much lighter weft yarns. Weave structures are simple, such as broken ribs or herringbone. A typical construction has 24 ends/dm and 7 picks/dm.

Sisal and coir products are woven on robust looms, common widths being 2 and 4 m. Coir matting is often woven in India by hand. The simplest looms are operated entirely by hand. There are also treadle looms, but having the shuttle thrown by hand. Fully automatic weaving is performed on heavy duty shuttle looms.

A unitary latex backing or light foam is often applied to improve dimensional stability and prevent fraying. Sisal and coir products should be conditioned before laying and the floor, if concrete, should be thoroughly dried: changes in moisture content are associated with dimensional changes so that installations may ruck or bubble if these precautions are not taken.

Synthetic products in structures similar to those of sisal and coir matting are woven from heavy filaments that may be regarded as plastics rather than fibres.

REFERENCES

1. G Robinson. *Carpets*, 2nd edition, Textile Book Service, UK, 1972.
2. A T C Robinson and R Marks. *Woven Cloth Construction*, The Textile Institute, Manchester, 1973.

13 Production of Needlefelts

Needlefelts constitute the second most important form of textile floorcovering in terms of square metres. They provide the basis of the major proportion of carpet tiles (see Chapter 19), although most needled floorcoverings are marketed in the form of roll goods. In many countries, they are sold primarily for contract applications. Product areas include outdoor applications such as textile sports surfaces (see Section 21.7.3). In some countries the residential market is very significant. In France and Germany, for example, flat forms of needlefelt are used as a non-directional base over which carpet squares (hand-knotted or machine woven) are laid. For this purpose, the absence of pile, which can cause squares to migrate in the direction of the pile lay, is an advantage.

Originally, needled floorcoverings were made only with a level surface, and aesthetic interest was provided by colour and fibre blends of mixed colour and diameter. Later developments allowed surface textures to be produced, simulating the loop and cut piles of tufted or woven carpets. By localising the textured effects, surface patterns can be produced.

Because spinning is obviated, the manufacturing cost of needled products is lower than that of tufted carpets and so the price of the average needlefelt is lower than that of the average tufted product. Needled floorcoverings, therefore, tend to occupy their particular level in the market.

13.1 PRINCIPLES OF NEEDLING

Figure 13.1 shows the principle of manufacturing needlefelts, as developed in Britain in the second half of the nineteenth century. In a way, the process simulates the behaviour of wool fibres in felting. Whereas wool fibres migrate and become entangled under the influence of the mechanical action of a machine due to their ratchet-like scale structure and unique wet rheological properties, needlefelting relies on the positive mechanical displacement of fibres within a web. Any fibre may be used: indeed fibres ranging from fine cotton to coarse glass are used in commercial products of various kinds, although a narrower range of fibres is used for floorcoverings.

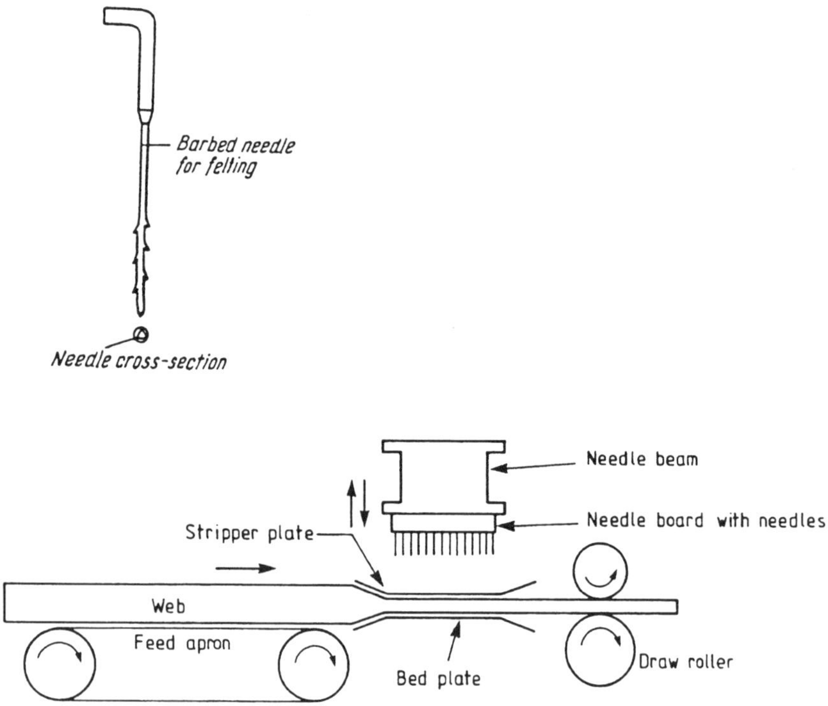

Figure 13.1. The principle of manufacture of needlefelts. (Courtesy of The Textile Institute)

The webs (or batts) for needling are made either on woollen-type cards or garnetts and are cross-lapped to obtain the required mass per unit area and to change the fibre orientation. A web prepared in this way is fed directly to the needleloom. A scrim fabric may be incorporated at this stage to improve the mechanical properties of lightweight products, but most needled floorcoverings are made without scrim. As the web moves under the needleboard it is subjected to multiple actions of barbed needles that entangle the fibres and consolidate the web. A perforated stripper plate prevents the web rising during the withdrawal stroke.

The needles are usually convex triangular in cross-section and have barbs that move the fibres preferentially in one direction (Figure 13.1 inset). Needles are available in various gauges and designs to take into account the fibres to be used and the products to be manufactured.

Some products require more than one needling machine, and a texturing machine using forked needles may also be incorporated in the line. After needling, some products are reinforced in the web with resin (usually synthetic latex) and others are backfinished. If the needlefelt is to be converted into tiles, the backfinishing process may be quite sophisticated.

13.2 FIBRES

Polypropylene is the most widely used fibre in needled floorcoverings and, as with tufteds, pigmentation is the standard method of coloration. However, colour mixtures are commonly used to provide aesthetic interest. Fibre types or blends can cover a wide range. Most products fall in the range 15–100 dtex, the most favoured value being 30 dtex, and fibres may be round, trilobal or delta cross-section. Particularly coarse fibres, 150–250 dtex, may be used in entry mats or to provide visual effects.

Nylon is the second most popular fibre in needled floorcoverings. It is used principally on account of its ease of coloration, allowing needled products to be printed. Because nylon blends that contain a proportion of conductive fibres are readily available, nylon products can easily be used as a basis of conductive floorcoverings. Polypropylene/nylon blends (eg, 85/15) incorporating conductive nylon and a conductive backcoating to create antistatic products are important in the contract market. Nylon is also valued for its resilience.

Nylon fibres are commonly round, crimped only for processability, and are around 17 dtex. Blends with coarse fibres (70–80 dtex) are used for some products, and very coarse fibres (125–135 dtex) are sometimes incorporated to provide visual effects.

Polyester fibres are principally used to meet stringent flammability specifications. For general use, they are regarded as less resilient than nylon and more expensive than polypropylene. Needlefelts have been used as an outlet for polyester fibres made from recycled polyester bottles.

Animal fibres constitute a significant minority fibre in the needlefelt sector of the industry. Wool has only occasionally been used, sometimes as uncarded staples for visual effect, but coarse animal hairs are regularly used in specific products. Pig bristle was used in the first carpet tiles and remains the most common animal fibre: its medullation means that it appears lightly dyed in a blend, contributing to its use as an effect fibre.

Waste fibre, which may be local mill waste, edge waste, garnetted tufting or weaving yarn, or spot lots of fibre, is an important raw material for the lower priced products.

13.3 BLENDING

The proportions of solid colours to create new mixture colours can be predicted from colour measurements of the individual components, using software specially developed for the purpose. A larger viewing area is required for measuring the colour of mixtures than for general textile colorimetry. Blending software is particularly useful whenever a blend is to be made from waste fibre where the colours of the individual components cannot be specified accurately: the software enables mixture colours to be predicted surprisingly accurately. Visual assessment of the mixture effect, as distinct from the average colour which is measured by a colorimeter, is also necessary. If two different fibres are to be used in a blend, the proportions of solid colours in a mixture are calculated separately for each fibre.

Blends are typically prepared in 3 tonne lots using the system of layering in bins described in Section 3.2.1. A blending plant may consist of a bale breaker, opener and an arrangement of bins, etc, to permit three blending operations.

13.4 CARDING

Preparation of the fibre web is usually on a single-swift woollen type card (see Section 3.2.1) with a pre-carding section consisting of one or two cylinders. Two-swift cards are used by some manufacturers, but these are less common. An example of an overall layout of a card for a nonwoven line is illustrated in Figure 13.2. Such cards are available up to 3.5 m wide: 2.5 m cards are commonly used.

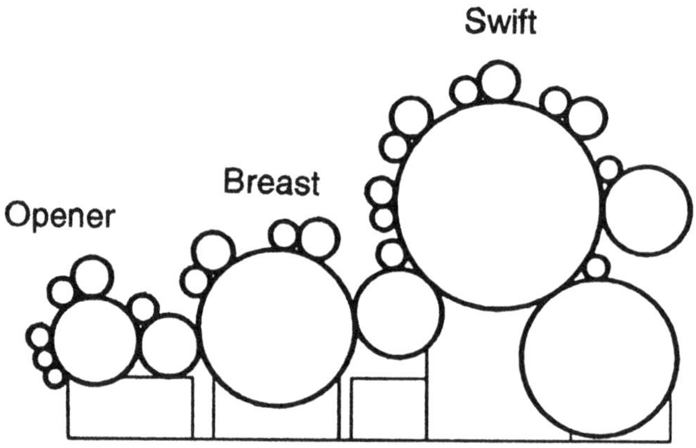

Figure 13.2. Configuration of the CA10-225 PP card for use in needlefelt manufacture. (Courtesy of Thibeau, NSC Group)

Feeding of the cards may be by continuous volumetric hopper or a weigh-pan system. Control of input weight is of critical importance, since this controls the long-term regularity of the finished product and is used for fine adjustment of the weight per square metre of the product. Weigh-pan systems are controlled by devices such as Garnett Controls' Microweigh XL2 (see Section 3.2.1).

The input of a volumetric feed unit may be measured by the product of flow rate and density (as measured by a beam radiation emitter) as in NSC's Servo-X autoleveller system. Alternatively, the mass of the fibre feed may be measured by means of a lightweight roller assembly on strain gauges, as in Garnett Controls' Rollaweigh system (see Section 3.2.1).

Alternative doffing systems are available to suit the fibres to be used and the products to be made. These include comb doffing, rotary comb doffing and roller doffing.

Output modules are available to give the required characteristics of the web, including fibre orientation and isotropic strength. Figure 13.3 illustrates the principal alternative possibilities.

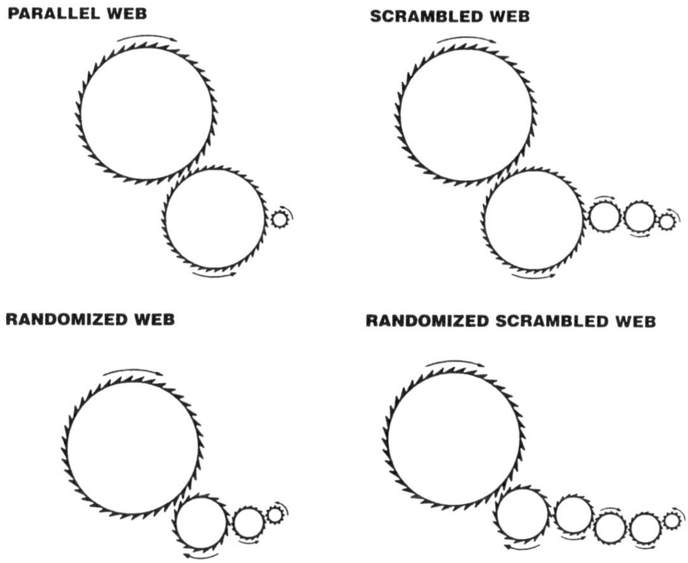

Figure 13.3. Systems for generating the various types of web. (Courtesy of Thibeau, NSC Group)

Modern cards for nonwoven lines are controlled from a central panel with mimic effect displaying input weights, roller speeds, output speeds, etc. Card control is just one menu in the central control panel for the whole needlefelt production line.

13.5 CROSS-LAPPING

Cross-lapping of the carded web has three principal objectives:

- To create the required product mass per unit area by layering the appropriate number of webs.
- To reconstruct the web into the required width.
- To change the preferential fibre orientation away from the machine direction.

The production line turns through 90° at this point, which is often regarded as a desirable layout.

Inevitably, the fibre orientation after cross-lapping is preferentially transverse to the machine direction so that a machine known as a batt drafter is sometimes used to improve the isotropic strength (ie, to improve the ratio of tensile strength and elongation in the machine direction and cross-direction). Such machines may be located immediately after the cross-lapper or after a pre-needling machine.

Simple cross-lappers would create a nonwoven fabric having a concave profile, as illustrated in Figure 13.4(a). Three factors contribute towards this defect:

- tension in the card web causes widthways shrinkage,
- overfeeding of fibres during carriage reversal, when lay down speed is briefly zero,
- shrinkage or stretching of the material during the needling process.

Control devices are available that over-correct for the first two problems to give a web of convex cross-section, of the type shown in Figure 13.4(b), anticipating the effect of consolidation in needling and leading to a final cross-section similar to that shown in Figure 13.4(c). A sophisticated cross-lapper producing a result of this quality operates by relaxing the web tension and minimising distortion at lay down; accumulating the excess of card web available during reversal of the carriages and distributing it uniformly across the width of the web; and counteracting the subsequent distortion peculiar to the consolidation technique by underfeeding the edges of the web during the acceleration phase of the carriages.

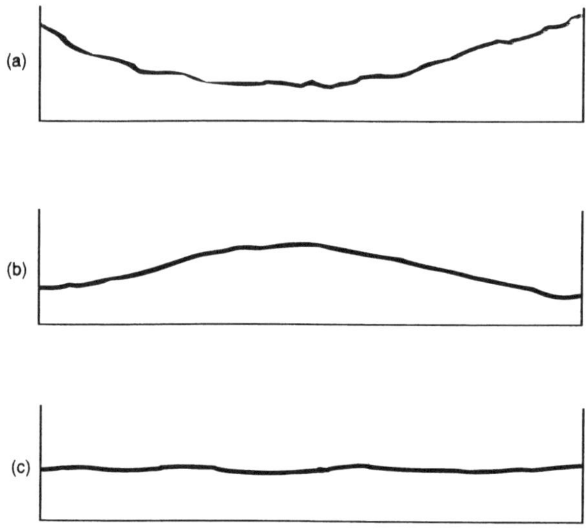

Figure 13.4. (a) Cross-section of a needlefelt fabric, conventionally cross-lapped, then bonded; (b) Cross-section of a batt produced with the aid of a sophisticated correction device (Profile Generation Technology); (c) Cross-section of a consolidated needlefelt from a profiled batt as shown in (b). (Courtesy of Asselin, NSC Group)

The mass of the batt is continuously monitored at particular points across its width and can be displayed on the batt profiling screen of the master control panel to check if the required profile is being achieved.

13.6 NEEDLING

Needling may be accomplished in one operation but often a pre-needling machine having a low needle density is used to consolidate and stabilise the batt while needling without draft. This allows high-speed needlelooms to be used for the second step, while preserving the uniformity of the fibre assembly. If the product is to be textured, a speciality needleloom is required, so a production line may include two or three needlelooms and possibly one or two batt drafters.

Devices that feed the batt from the cross-lapper into the needling zone are also required to compress it. A well designed batt feeder may obviate the need for pre-needling in some circumstances. Figure 13.5 illustrates a double-apron guidance system that transports the batt, without distortion, to within 14 mm of the first row of needles.

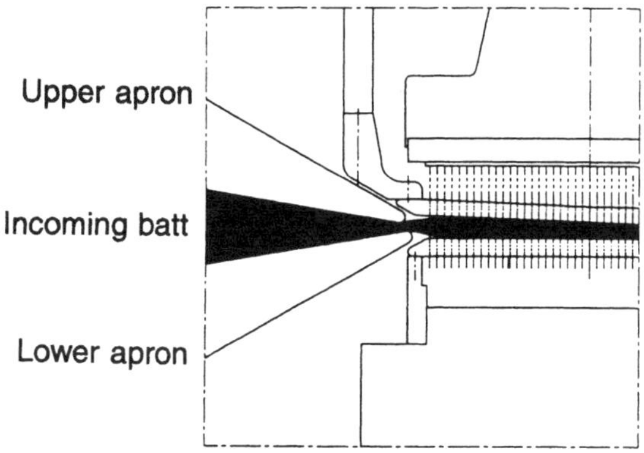

Figure 13.5. The double compression introductory nip (DCIN) batt feeding system, comprising double aprons, adjustable for the gap between the bed plate and stripper plate. (Courtesy of Asselin, NSC Group)

Although needlelooms having working widths up to 16 m are available for other purposes, machines for producing floorcoverings fall in the range 2.6 to 5.5 m. Needled floorcoverings are commonly manufactured 2 or 5 m wide.

The design of a needleboard is determined mathematically. There may be 3000–5000 needles/m width in each needleboard: these must not be in a regimented assembly, or stripes in the machine direction will be created in the product. Semi-random arrays of needles were first determined empirically, but the configurations of needles on modern machines are the result of computerised design programs.

The needles themselves are selected according to the fibre to be processed and the construction of the product. Needles are specified according to the gauges of the shaft and the operative part, and the length and spacing of the barb. An example of a needle used for 30 dtex fibres is 15×18×32×3RB (ie, 15 gauge shank; 18 gauge reduced part; 32 gauge operative section; 3 in. length; Regular Barb spacing (1/4 in. between barbs).

Improvements in the design of eccentric mechanisms, lubricating/cooling systems and low reciprocating masses have enabled speeds around 3000 strokes/min. to be achieved. Depending on the design of the machine and its purpose, practicable stroke rates used in the floorcoverings industry are in the range 1200–2500 strokes/min. Needling densities are typically 120–150 stitches/cm^2 of product. The depth of penetration of the needles, usually up to 16 mm, is controlled by the setting of the bed plate.

Fibrous dust may be generated by needling, and some machines incorporate a dust extraction system. Fibre damage and therefore dust can be minimised by optimum lubrication of the fibre.

Machines may be simple downstroke machines, as shown in Figure 13.1, upstroke, double-board, combined upstroke/downstroke or tandem downstroke/upstroke machines. The configuration of a tandem downstroke/upstroke machine is shown in Figure 13.6.

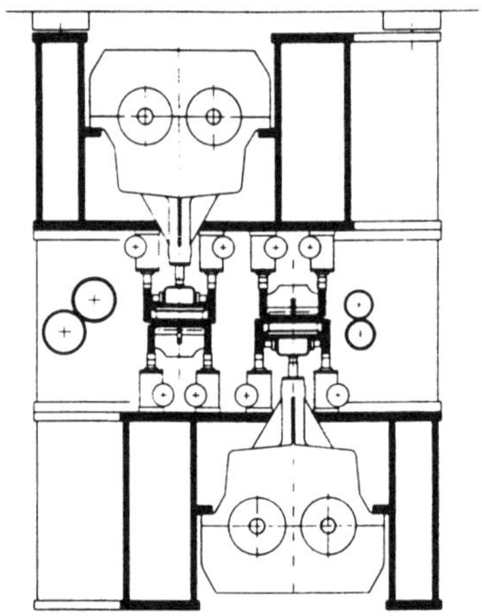

Figure 13.6. Configuration of the NL 2000 / SRS Supertandem needleloom. (Courtesy of Textilmaschinenfabrik Dr Ernst Fehrer AG)

Output systems are required to withdraw the needled product without distortion and to provide a degree of compression. A common system is a double-roller press configured so that the needled material is surface driven over a contact angle of about 45°. Output presses with greater contact angles are available for transporting lightweight or pre-needled materials (Figure 13.7).

After the final needling step, the edges are trimmed and the product is rolled up into batches on transportable A-frames. Conveniently, the batching frames stand on load cells so that the mass of a batch may be input to the central control panel. As the length of the batch is also monitored, the average mass/m^2 is directly available from the control panel.

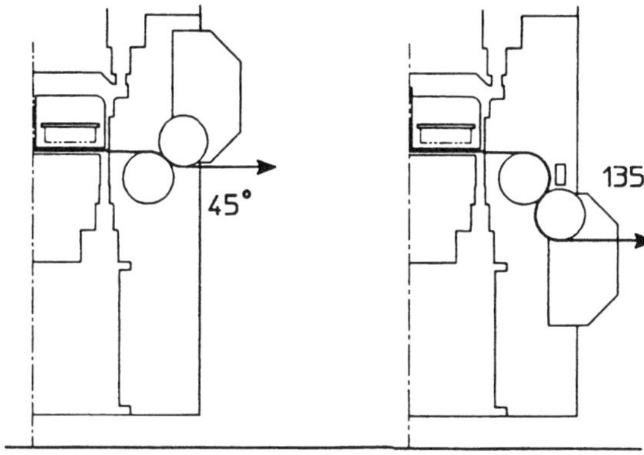

Figure 13.7. Alternative output presses for withdrawing material from the needling zone. (Courtesy of Asselin, NSC Group)

13.7 STRUCTURING AND PATTERNING

Whereas a conventional needlefelt is very dense and has a rather hard and uniform surface, surface structures may be created by a further needling process using forked needles [1]. Ribs, velours and patterns can be produced, resembling the structures of tufted or woven carpets, but the effects are composed of fibres rather than yarns. As well as changing the appearance of the product, texturing makes the surface softer and more compressible. Double-layered floorcoverings can be made to give enhanced colour effects without special textures.

Note that while the basic needling process increases the density of the fibre assembly, the creation of structured effects decreases the density of the assembly. It is, therefore, important that the base needlefelt has a sound structure, since strength will be lost during the texturing step.

Double-layered structures

For producing a two colour effect with less intimate mixing than the effects achieved by blending the colours before carding, two layers of needlefelt of different colours are needled together. The upper layer is prepared in the same way as a single-layer flat product, needled to a high stitch density (120–150 stitches/cm^2), except that the edges are not trimmed. The colour effect material (the bottom layer) is separately processed (pre-needled) but to a lower density (70–80 stitches/cm^2).

Finally, the two layers are fed together (with the upper layer inverted) to an upstroke machine, optionally with a scrim between them. Coloured material is transferred by the barbed needles to become a component of the upper layer, creating fine point patterns. Trimming takes place when the layered product is batched.

Loop effects

To produce a looped structure, forked needles of the type illustrated in the inset in Figure 13.8 are used. These are known as D-needles. When needling from top to bottom with forked needles mounted parallel to the needleboard (at

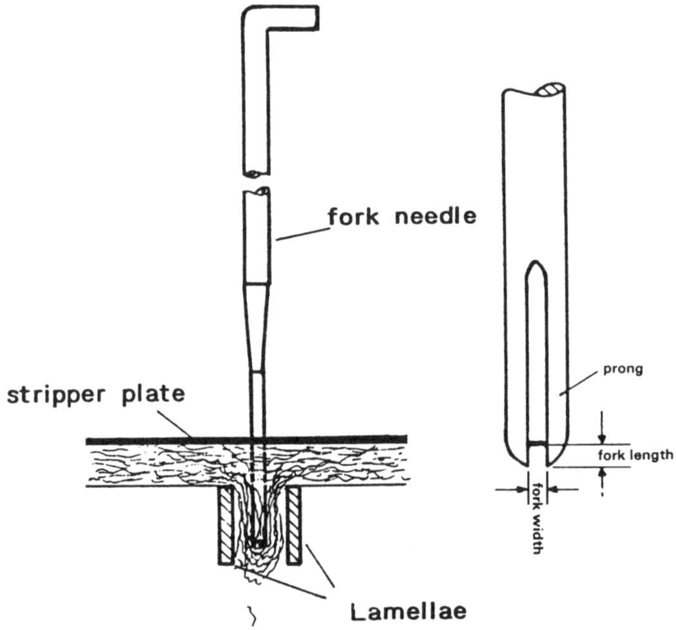

Figure 13.8. Formation of a loop using a forked needle. (Courtesy of Oskar Dilo Maschinenfabrik KG)

right angles to the direction of manufacture), loops are formed on the underside of the fabric. The normal perforated bed plate is replaced by a grid table consisting of parallel lamellae, and loops of fibre are formed between the lamellae, as shown in Figure 13.8. Feeding of the base fabric and take-up of the product are intermittent so that the loops are always formed in rows and a ribbed effect is created. The spacing of the ribs is equal to the spacing of the lamellae and is regarded as the gauge of the machine. Loops cannot be formed at random. After the first half of the passage under the needleboard, fibre transfer is essentially complete: the remainder of the needling action improves the uniformity of the looped structure.

As the base fabric supplies fibre for the loops, a compromise has to be achieved between a uniformly dense surface and satisfactory residual properties of the base material. An adequate proportion of fibres should remain horizontally oriented. Stitch densities for loop formation should be restricted to 100–150 stitches/cm^2. Depths of penetration are 8–10 mm.

Velour effects
Velour effects can be produced on a similar machine to that producing the loop effect, but the forked needles are turned so that they are parallel to the machine direction (and at right angles to the needleboard), rather than in the cross-direction (V-needles). Fine-fork needles have also been developed for producing velour styles. However, special velour machines that give a random distribution of dense surface pile are now favoured.

Figures 13.9 and 13.10 illustrate the principle of the Di-Lour machine. The main feature of this machine is the brush apron (belt), which serves as a stitching base instead of a lamellae table, and allows a dense random array of needles to be used. The brush apron also transports the material and serves as temporary storage for the fine loops of fibre.

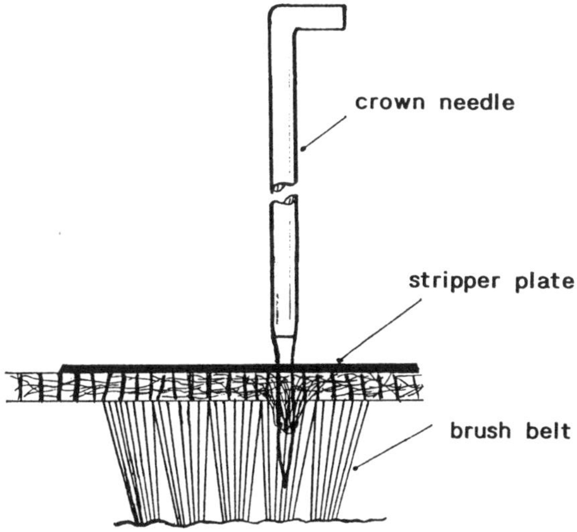

Figure 13.9. Principle of the Di-Lour machine using a brush belt and crown needles. (Courtesy of Oskar Dilo Maschinenfabrik KG)

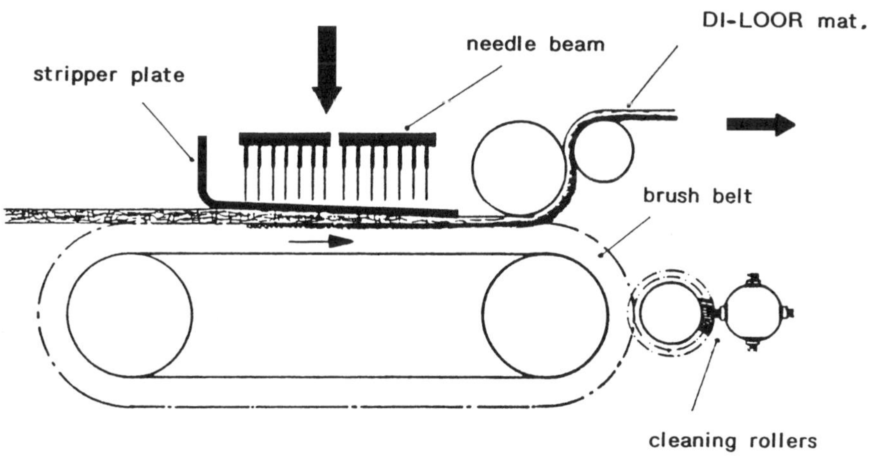

Figure 13.10. General configuration of the Di-Lour machine. (Courtesy of Oskar Dilo Maschinenfabrik KG)

Although fine-fork needles can be used on the Di-Lour machine, special needles having three or four barbs located at about the same height are favoured – so-called crown needles (Figure 13.11). The needle density used depends on fibre fineness and product weight, but needle densities are high at 8000–15 000 needles/m working width. Stitch densities are also high at 350–500 stitches/cm^2. Penetration depths are 10–14 mm.

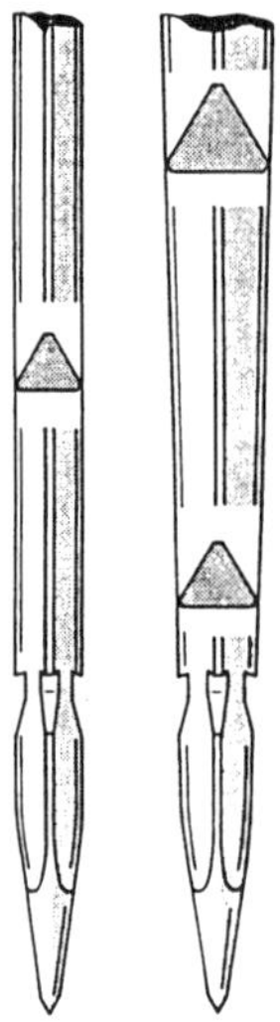

Figure 13.11. Crown needles as used in the Di-Lour machine. (Courtesy of Singer Spezialnadelfabrik GmbH)

The Di-Lour needleloom is capable of handling a wide range of product qualities (100–1800 g/m^2). The lower weights are used outside the floorcoverings industry (footwear, headgear, furnishings) but automobile products include boot linings, parcel shelf coverings and general linings that are similar to carpets.

Using a scrim (eg, a lightweight spunbonded polyester fabric) fed to the structuring loom below the base fabric, a very high proportion of the base material can be converted into pile. The needles drive the fibres through the scrim, giving excellent cover, although it is advisable to use a dark coloured scrim to avoid grinning of the scrim through the pile.

Special visual effects can be obtained by variations on the velour structuring technique.

The double-layering technique described earlier may be applied in velour manufacture. The Di-Lour DS (DS = double structuring and design) machine is designed for this purpose and has two needleboards operating on the same brush apron. In the first board, the principal base fabric is converted into a velour structure. A second needlefelt of a different colour, lightly needled, is fed into the second needling unit. Fine point patterns are achieved by raising and lowering the brush under the second needling unit and by using special needling arrangements. Alternatively, the machine may simply be used to increase pile density and resilience as a result of supplying additional fibre of the same colour from the second needlefelt (or pre-needled batt).

The Yarn Tacker GST is a machine that feeds yarns of different colours and colour sequences between the two needling zones of the Di-Lour DS machine. Fibres from the yarns are needled from the reverse side and appear as stripes in the pile.

By omitting rows of needles in the machine direction of the Di-Lour DS machine, longitudinal texture stripes can be created.

Surface designs
Patterns of loops can be produced using the type of machine utilising forked needles and a lamellae table. Typically, the needles in the board are arranged in a geometric design. Needling takes place with the base needlefelt almost stationary. After the desired number of strokes, the lamellae table is lowered and the material is advanced by a distance equal to the design repeat. An alternative to lowering the lamellae table used in some machines is to interrupt the drive as the material is advanced. Further sequences of needling and advancing take place. Three values can be programmed into the patterning system: rib length, repeat length and length of the high-speed draw-off. Otherwise, the machine operates on a standard cycle.

Bold designs are a necessary feature of the products from such pile sculpturing systems, since fine designs would quickly become diffuse during wear. Distinct from providing aesthetic interest, sculptured piles can have a practical purpose, increasing the cleaning effectiveness of entry mats made from very coarse fibres.

13.8 INTEGRATED MANUFACTURING

A modern production line for textured needlefelts consists of a fibre feed unit taking a blend from a bin to a card, a woollen-type card, a cross-lapper, a batt feeding system, one or more needlelooms, a texturing machine and a roll-up unit (Figure 13.12). There may be one or more batt drafting machines in the line. The card would be located behind the cross-lapper.

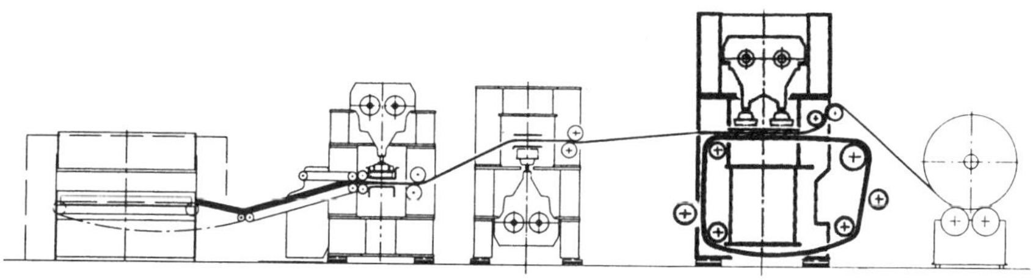

Figure 13.12. Example of an integrated production line consisting of (left to right): cross-lapper, batt feeder, downstroke needleloom, upstroke needleloom, texturing loom, and roll-up unit. (Courtesy of Textilmaschinenfabrik Dr Ernst Fehrer AG)

All these machines are fitted with DC motors controlled from a central panel so that they can be operated as one unit. The speeds are coordinated so that cumulative draft is applied to the material down the line, and on start-up all the machines ramp up to speed together. A central computer covers the whole system and monitors all the different stages of production: reference has been made in other sections of this chapter to screens displaying processing conditions, including mimic diagrams, for the individual machines.

One of the screens provides an overview of the whole line. The fibre input mass can be selected on the panel and the output mass can be calculated by the computer from the various machine settings. During processing, β-gauges monitor the material (eg, the profile of the web), and from the various displays the quality of the product can be predicted.

13.9 FINISHING

The principal finishing process for the roll goods in non-textured styles is resin bonding to improve durability and prevent pilling. Resin bonding also imparts firmness to the product. Most commonly, carboxylated styrene-butadiene (X-SBR) latex is used as the bonding agent. For flame-resistant products, X-SBR is unsatisfactory and ethylene vinyl acetate (EVA) is the preferred product. Acrylic resins and styrene/acrylic copolymers are favoured for a few speciality products.

Application of the bonding agent is in the form of foam: this technique economises on energy at the drying stage and there is virtually no effluent from residue in the applicator. The foamed latex or other formulation is metered over the surface of the product at the required application level and the material is mangled to assist penetration and break the unstable foam. Policies on resin bonding vary from company to company and with the quality of the product. Some densely needled products are not bonded at all; others are fully impregnated with bonding agent; and some qualities are treated so that only the upper layers of the pile are penetrated. After application, the latex or other resin is dried and cured (see Section 17.4).

Generally, textured styles are not latexed from the face, since this would negate one of the objectives of texturing, which is to soften the touch. Depending on the construction of the product and its technique of manufacture, a textured product may not be resin bonded at all. Often, textured products have latex applied to the back via foam applicator and mangle. The objective is to impregnate the base layer to reinforce the structure without affecting the texture of the surface.

Some needlefelt products incorporate melt-bonding fibres (see Section 2.5), usually bicomponent polyester or single component polypropylene/polyethylene fibres, which provide supplementary bonding, making the application of resin bonding agents unnecessary. Activation of the melt-bonding fibres is by calendering the material through hot rollers. It is important that the pressure and temperature be uniform across the width of the material. The problem of roller distortion that occurs when pressure is applied from the ends can be overcome by using the Swimming Roller principle of Küsters (ie, uniform hydraulic pressure is applied across the width of the roller). The Küsters Hot-S roller also incorporates a countercurrent hot oil system that can provide the surface temperatures up to 250°C that may be required for polyester. Figure 13.13 shows how oil introduced at 260°C may cool to 256°C by the time it leaves the roller but, because it travels first through an annular space right to left and returns through another annular space left to right, the average oil temperature at any distance across the roller is 258°C.

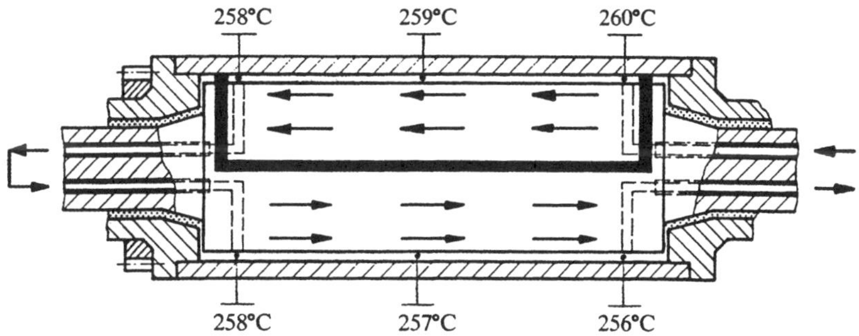

Figure 13.13. General design of a calender for thermal bonding showing the counterflow hot oil heating system. (Courtesy of: Eduard Küsters Maschinenfabrik GmbH)

Backcoating of material to be converted into tiles is a specialist operation and is described in Section 19.1. Some roll goods are not backcoated at all: they are commonly installed by a direct-stick technique and it is not considered necessary to improve their dimensional stability. When an antistatic product is to be produced – a high proportion of contract qualities are antistatic – the conductive fibre in the product may be augmented by the application of a conductive unitary backing containing an antistatic additive. Secondary backing fabrics may be applied to upgrade selected products. The techniques used are the same as those described in Section 18.4 for the secondary backing of tufted carpets.

13.10 NEEDLED PRODUCTS

Fibre masses for needled floorcoverings range from 400 to 2000 g/m^2. The lower weights are for plain domestic qualities and are fully reinforced with resin. Good quality products for contract applications range from 750 to 900 g/m^2. Ribbed entrance mats are the heaviest product type, in the range 1500–2000 g/m^2. Textured products mostly need to be relatively heavy because of the requirement to retain good physical properties in the fibrous base as well as provide a pile. However, velour textures having a spunbonded scrim can be as light as 400 g/m^2. For general products, scrims are avoided for reasons of cost and because recycling of edge waste is more feasible without scrim material.

Product areas include the following:

- domestic roll goods,
- roll goods for contract applications such as offices and educational establishments,
- textile sports surfaces for indoor and outdoor sports,
- textured products for the automobile industry,
- entry mats, stiff or soft,
- stick-down tiles for general contract floorcoverings,
- loose-laying tiles for access to sub-floor installations,
- patterned tiles for interior design work.

There are also related product areas that are not strictly floorcoverings (eg, wallcoverings, acoustic panels, lining materials for automobiles and boats, and thermal insulating material).

REFERENCES

1. V P Gupta. New generation of needlelooms for structured nonwovens. In: *Proceedings of Hofer Nonwovens Seminar*, November 1989.

14 Other Methods of Carpet Manufacture

Apart from the major methods of carpet manufacture – tufting, weaving and needling – there are techniques that are operated in insufficient volume to appear in statistical information, but which may be very important to individual companies and in particular product areas. This chapter is not an exhaustive survey of the relatively minor methods of carpet manufacture, but describes some of the special technologies that are operated by the industry.

14.1 ADHESIVE BONDING

A wide range of products, including high quality wool velours, bristle entry mats and simple ribbed structures made from fibrous webs, can be made by implanting yarn or fibre in a layer of adhesive carried by a backing material [1]. In some cases, the patent rights and/or know-how have, in the past, been licensed on a basis that has restricted the spread of the technology, but in many such cases the machinery is now available on general sale [2].

14.1.1 Fusion Bonding, U-tuft

The principle of a face-to-face process originating around the Belgian/French border is illustrated in Figure 14.1. Two backing fabrics (usually jute) are spread with PVC plastisol (a paste of PVC powder with plasticiser) and fed towards each other and to the point of carpet formation.

Pre-dyed pile yarn (usually woollen spun) is let off from the beam at a specified rate and is led towards the point of fabric formation via a reed that may be reciprocated laterally to break up any linearity in the carpet texture, if this is considered necessary. The curtain of pile yarn is folded and pressed into the adhesive by two reciprocating blades. The carpet construction is determined by the relative speeds of the pile material and backing fabrics, the spacing between the backing fabrics and, particularly, the setting of the pile folding or pleating mechanism.

Following this process of assembly, the sandwich passes through an infrared oven which causes the plasticiser and PVC to fuse into a stiff gel, binding the yarn to the backing material. After a very short period of cooling, the gel sets

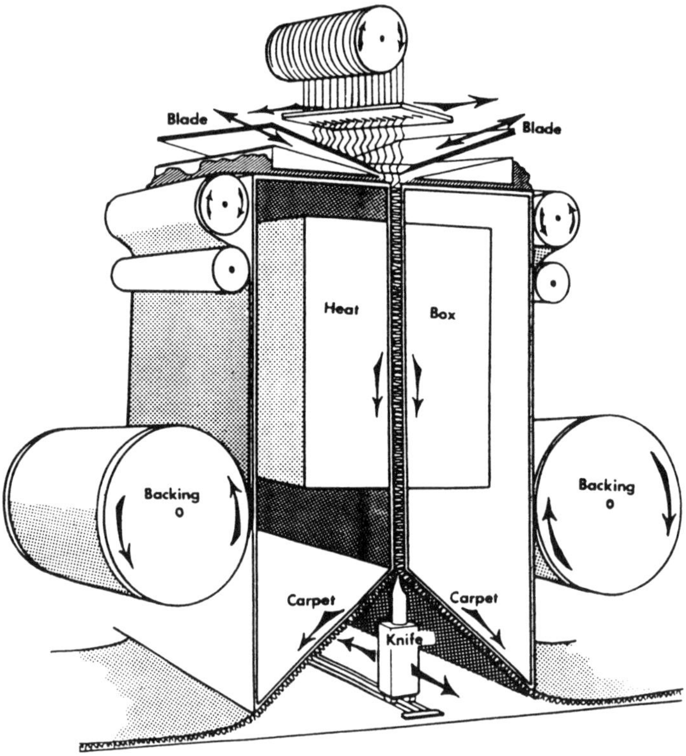

Figure 14.1. The major elements of the Stoddard Fusion Bonded machine. (Courtesy of David Crabtree and Son Ltd)

and the sandwich is cut with a reciprocating knife. The two resulting carpets are rolled up.

Such machines are available in widths of 4 or 5 m. Production rates depend on the carpet construction but are typically in the region of 70–150 m^2/h.

In most existing machines, the pile pleating mechanism is driven by excenter cams on four shafts, and setting up the machine for a new construction is rather slow. The Belgian manufacturer of fusion bonding machinery, Cotinfi NV, has designed an electrical position control system, in which high-torque servo-motors drive the pleating blades. Dedicated software controls this system, improving the flexibility.

The same machine may be used to manufacture loop-pile carpets, simply by omitting one of the coated backing fabrics. Cotinfi's servo pleating system includes the option of a sequence of 'software cams', allowing combinations of implantation movements (ie, patterns of texture). High/low loops and, obviously, tip-shear effects could be produced by this machine design.

The chief skill in operating the basic face-to-face process is to match the heat input to the mass throughput of material and the detailed constitution of the PVC plastisol. If the gel is insufficiently fluid, it does not penetrate the base of the tufts sufficiently well to achieve fibre and tuft bind. If it becomes too fluid, it can wick up the pile, reducing the effective pile above the backing.

Fusion bonding is capable of producing high-density velour carpets having excellent surface finish in a one-step process from a vertical machine that occupies remarkably little floor area. Pile yarn is processed under low tension, so that requirements for its mechanical properties are not stringent. The products have good dimensional stability, and only a small proportion of the pile material (15% is claimed) is buried in the adhesive. Tuft bind is excellent. Plain carpets of this kind are used either as broadloom carpets or as a basis for tiles.

Interface Flooring Systems operates a version of the face-to-face fusion bonding system, Vertron™, dedicated to the production of tiles. The Vertron process incorporates a fibreglass layer for enhanced dimensional stability. It is said to be particularly suitable for processing high-twist yarns, and friezé textures can be produced on the machine.

The Debron™ process (Figure 14.2) is an alternative face-to-face fusion bonding process that was developed by Interface specifically to produce carpet tiles. The sandwich is formed in two stages and processing is horizontal rather than vertical. A fibreglass layer is incorporated in the product. Apart from the different configuration of the machine, there are two notable differences from the Vertron machine. The yarn pleating system is said to be more precise and is particularly suitable for fine yarns: about 88% of the yarn is exposed above the polymer layer. Splitting of the sandwich is achieved with a band knife, which is claimed to give a particularly good pile surface requiring no shearing.

Fusion bonded carpets may be patterned by printing. This is more common for tiles than for broadloom.

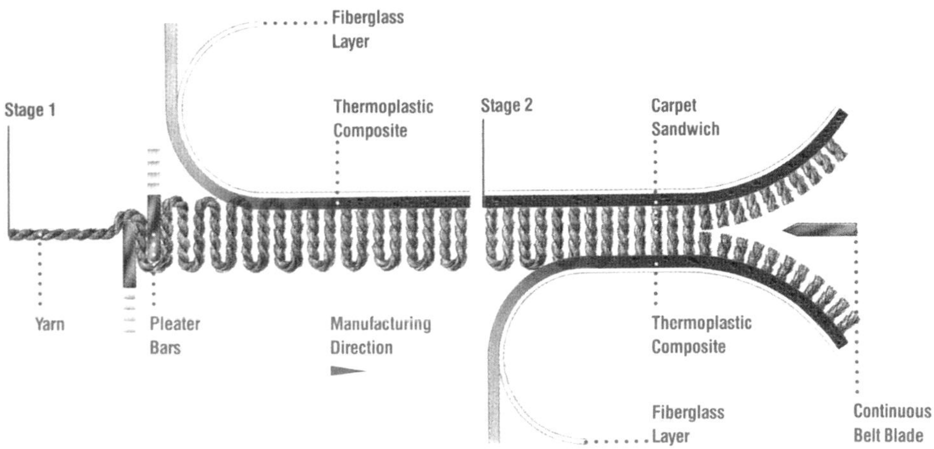

Figure 14.2. Principle of the Debron™ face-to-face bonding process for producing U-tuft carpets. (Courtesy of Interface Flooring Systems Inc)

14.1.2 Fusion Bonding, I-tuft

Tufts formed from a single leg of yarn are theoretically the most easily packed into a uniform pile structure and allow the yarn buried in the adhesive to be reduced to a minimum. However, the fusion process must be particularly well controlled if good fibre and tuft bind are to be achieved.

Interface's Quadron™ process (Figure 14.3) is a face-to-face fusion bonding technique that utilises an assembly of I-tufts. A pile yarn curtain is cut with a guillotine and the material is packed into vertical columns having bi-directional symmetry. A thermoplastic layer and associated fibreglass scrim are bonded first to the upper surface of the pile assembly and, in a second stage, to the lower surface, to form a sandwich. Finally, the sandwich is sliced into two carpets using a continuous belt blade.

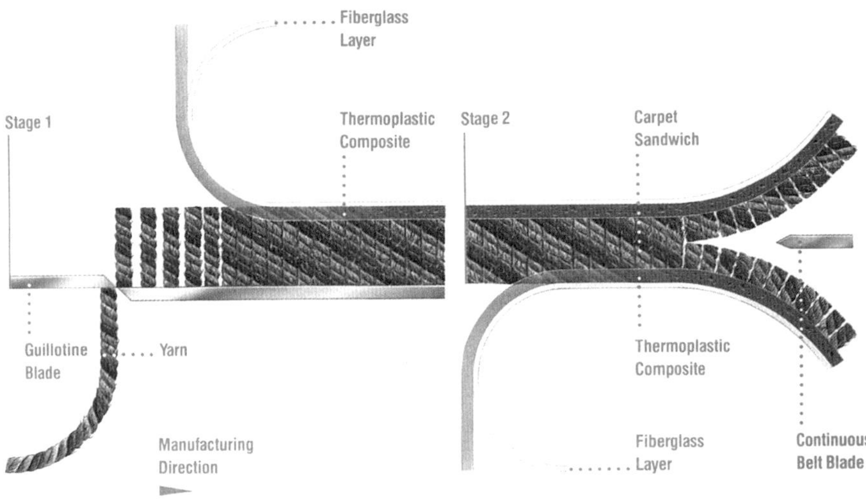

Figure 14.3. Principle of the Quadron™ bonding process for producing I-tuft carpets. (Courtesy of Interface Flooring Systems Inc)

A feature of the Quadron™ process is that it can achieve very high pile densities (MaxPac is Interface's trademark for the construction). Around 88% of the pile is present in the use-surface. A further advantage is the ability to handle multi-ply and cabled yarns of high linear density, including mixed yarns, allowing special textures to be produced.

14.1.3 Fusion Bonding Patterning Systems, U-tuft

There have been many attempts to develop a carpet manufacturing system with full patterning scope, but with the high rate of production associated with adhesive bonding. At the time of writing, the only full systems in commercial production are U-tuft systems dedicated to the production of carpet tiles.

The first machine having full patterning scope was the Bondax™ system of Bond Worth, UK. Pattern generation is by a spool system, as used for spool-gripper Axminster weaving described in Section 9.2. Backing material is coated with hot-melt adhesive and rows of tufts cut from the spools are pressed into the adhesive with a steel blade. This simple process of pressing tufts into the adhesive is quicker than transferring them to the weaving position with grippers and locking them in place with weft, as in Axminster production.

The Axtile™ process of Nissin Co, Japan, uses an electronic jacquard to present the pile to the adhesive system in the form of a design. The associated Design-Aid software is used to prepare the design, either by the Scanner-input-soft or Retouch-soft technique, in the required production size and construction.

Four production steps are required to produce the finished tiles: application of adhesive, implantation of pile, fusion of adhesive and shearing. In the first step, narrow width backing fabric is coated with adhesive and cut into tile sizes. The bonding machine itself, called Comeu-1, has a central jacquard-gripper system that prepares rows of pile as required by the design and feeds two conveyor belts, one on either side, each carrying two rows of tiles. A slider driven by a servo-motor places the pile accurately in position for the bonding. A double-bar gripper holds the row of cut tufts over a slot and a press knife forms them into U-tufts in the slot and presses them into the adhesive. Fixation is completed in a conveyor oven and, finally, the pile surface is finished by shearing under a special narrow width shearing head.

The pitch of the machine is equivalent to 1/8 gauge, and 4–10 tuft rows/in. can be inserted. Eight colours are possible on the standard machine, and machines for 12 colours can be supplied.

The tiles can be laid on the floor to create the effect of traditional repeating Axminster designs, or coordinating tiles can be assembled to give the appearance of rugs, perhaps as the centrepiece of a repeating design.

14.1.4 Fusion Bonding Patterning Systems, I-tuft

The Westbond system, Denmark, operates in 1 m width and is dedicated to the production of tiles. It is a face-to-face system that can produce either plain carpet for screen printing or random patterns that can be reproduced.

The Westbond Flex system utilises ropes of pile yarn in up to 12 colours as the starting material. To create the tufts, the ropes are cut into lengths of twice the required pile height and the groups of tufts are gripped between tapes to convey them to the implantation position. Pile is built up across the direction of manufacture between two fibreglass scrims coated with PVC adhesive. After fusing, the sandwich is sliced and further backing material is applied before the sheet material is cut into tiles.

Westbond Flex tiles typically display coloured flecks on a plain coloured or tweedy background. The tiles have the advantage of the high proportion of effective pile above the adhesive associated with I-tuft systems. However, the outstanding advantage of the Westbond system is flexibility, since lots as small as four tiles can be produced.

An interesting piece of history is the production of fully patterned wool squares (rugs) by Karastan, using the Italian SICI or Bondi system. Assembly of the pile yarns into the design had nuances of the chenille Axminster process (see Section 1.2.5) but the temporary fabric was produced on an electronic loom and carpet production was entirely different. Pile yarns in the required sequence were shot across a light nylon warp and the resulting fabric was folded precisely into a block representing half a rug. The edge of the block was bonded to a jute backing fabric coated with adhesive and a section of pile was sliced off to create half a rug. The sequence was repeated on other sections of backing fabric until half the block was consumed. The remainder of the block was used to build up the second half of the rugs with the mirror image of the design.

14.1.5 Latex Bonding of Fibrous Webs

An apparently simple process, but one which requires detailed know-how for successful operation, creates ribbed floorcoverings that occupy roughly the same market level as needlefelts.

The essential component of the machinery (Figure 14.4) is a drum bearing ribs into which a carded web, prepared as described in Chapter 13, is folded. In the illustration, folding is accomplished by means of a reciprocating blade, while in other versions of the technique the blades for folding are mounted on an associated drum, and mesh with those of the main drum. The ribs of the principal drum have a high-friction surface, whereas the folding blades are smooth, so that the web remains in the intended place. Latex is sprayed on to the web. Backing fabric pre-coated with latex is married to the latexed web on the surface of the drum, pressure being applied by heated presses supplemented by tensioning the fabric over the curved surface. The version that is illustrated requires a pre-coated backing fabric. Pre-coating may be avoided by drying and curing the latex in stages: partial drying is effected before laminating; the drum is heated; and the process is completed in a continuous oven.

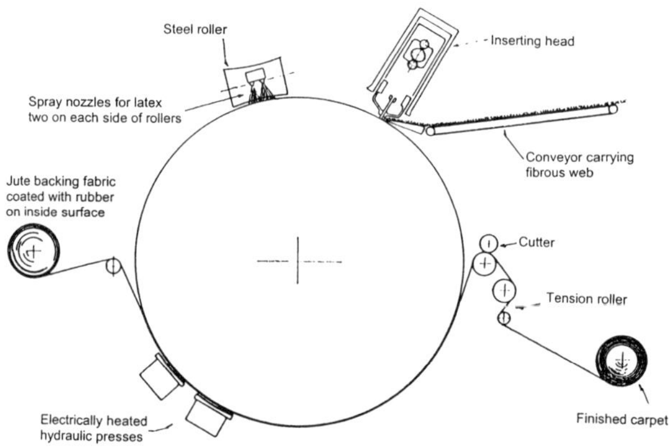

Figure 14.4. System for providing ribbed floorcoverings direct from a fibrous web [3].

Loop-pile products can be made in a similar way using yarns instead of card sliver. The Vernier Satellite machine utilises a disc system, rather than straight blades, for pressing the yarns into place.

14.2 ELECTROSTATIC FLOCKING

Low dense pile products with a velvet texture can be made from chopped fibre (flock) without the need for spinning, by bonding it to a backing fabric, using an electrostatic process developed in France around 1960 to parallelise and orient the fibres. Because of the low pile, absence of tuft definition, use of synthetic fibre (usually nylon) and the presence of a waterproof layer, the products have a unique ability to be wet cleaned by rotary scrubbing machines in a similar way to hard or resilient floorcoverings (vinyl, etc). This makes them particularly suitable for healthcare establishments.

Flocked floorcoverings are widely used in schools, colleges, shops and supermarkets, as well as in hospitality and commercial locations. They are also used for indoor sports surfaces. They are made up to 2 m wide and may be used as sheet material or converted into tiles. Patterning is by screen printing, die cutting, waterjet cutting or laser cutting.

The principle of electrostatic flocking is shown in Figure 14.5. Chopped nylon fibre 2–3 mm long is fed to a perforated metal box which is charged to 20 000–150 000 V, using an electrostatic generator, and the flock is brushed through a metallic sieve. A backing fabric coated with adhesive (eg, PVC plastisol), which passes under the applicator, is earthed. The fibres develop a dipole so that one end is preferentially attracted to the backing material, resulting in the formation of a vertical pile. Bonding of the flock to the backing is accomplished by fusing the adhesive in an oven. After cooling, flock not bonded to the adhesive is extracted by suction.

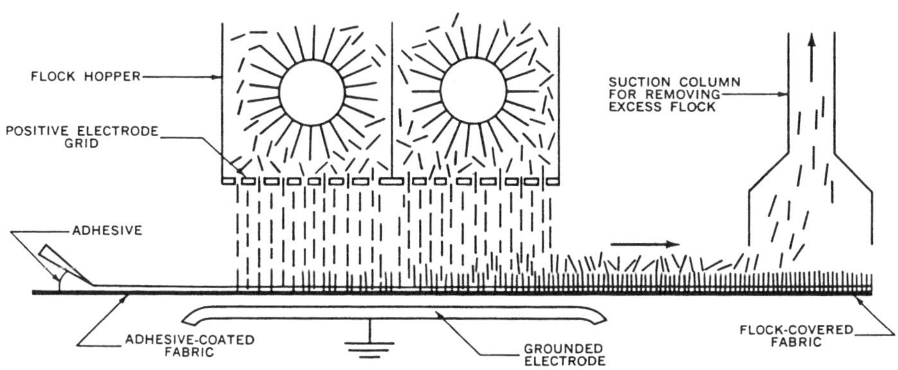

Figure 14.5. Principle of electrostatic flocking. (Courtesy of Bonar Floors Ltd)

Flotex™, by Bonar Floors, uses dyed nylon 6.6 as the pile fibre and the backing contains a layer of woven glass fibre to provide dimensional stability. Pile heights are 2 mm and the total thickness of the product is up to 4.3 mm. There are about 80 million fibres/m^2 of pile. A proportion of conductive fibre may be optionally included in the flock, enabling the product to meet specifications for body voltage generation and electrical resistance (see Section 21.2). If the product is to be cut into tiles, a special base is prepared that includes two scrims and three layers of PVC, two of which are gelled before flocking. The product is available in solid colours and as prints.

14.3 KNITTING

Warp knitting has been used extensively in Eastern Europe for producing carpets, and a few firms in Western Europe have specialised in the process. In the USA, warp knitting is used for producing textile sports surfaces (see Section 21.7.3).

The basic principle of warp knitting as applied to the production of loop-pile carpets is shown in Figure 14.6. Essentially, a backing fabric is knitted from flat filament yarn and the pile is laid in. The fact that the pile yarn does not have to pass through critical needle eyes and is not subjected to significant forces during the cycle of stitch formation has been seized on as an advantage by some carpet manufacturers: they use exceptionally bulky and irregular fancy yarns as pile material. Loop-pile products range from extremely heavy 100% wool carpets to lightweight BCF polypropylene.

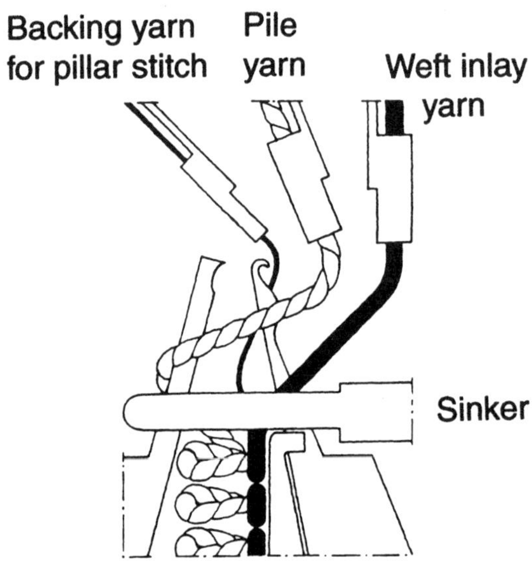

Figure 14.6. Stitching mechanism for producing loop-pile knitted carpets.

To a lesser extent, warp knitting has been applied to the production of cut-pile and face-to-face carpets. The cutting mechanism for the single-face carpets involves disc knives, one for each needle, impinging on the face of the looper. This arrangement has the advantage of producing pile tufts of uniform length, with no tendency to J-tufting (see Section 6.5.2). The disadvantage is that the knives need resharpening frequently.

Because of vibration, the sandwich resulting from face-to-face knitting cannot be cut on the knitting machine itself. A separate machine is used for slicing the sandwich into two carpets.

The products from the knitting machine are even less stable than those from tufting, so that an essential part of the manufacturing procedure is to laminate a secondary backing fabric to the basic cloth.

In the past, most knitted carpets have been in plain colours or have been produced from multicoloured effect yarns. The relatively new Kändler Carpet Knit system [4] produces loop-pile carpets in five colour jacquard designs. This warp knitting machine has three yarn feed systems:

- Warp yarns that are knitted into closely constructed pillar stitches, providing longitudinal stability and connecting with the weft.
- Weft inlays underlapped across five needles, providing lateral stability.
- Pile yarns in five colours per needle that may be selected for the surface loop or underlapped by pile sinkers into adjacent needles and knitted into the weft inlay.

The four colours of pile yarns not required on the surface support the pile loop, as shown in Figure 14.7, and contribute positively to the pile density.

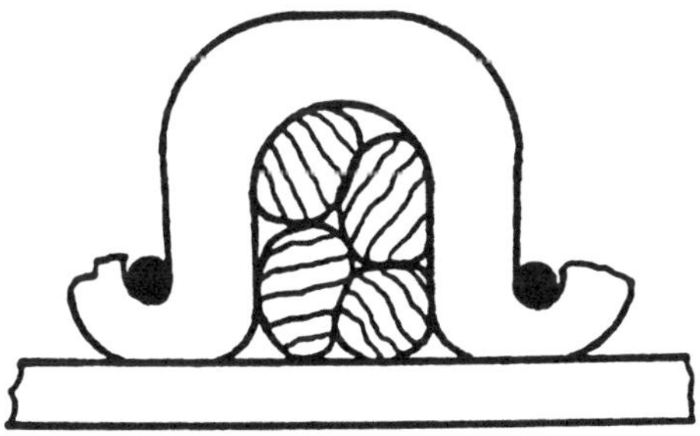

Figure 14.7. Configuration of the pile loop in the Kändler Carpet Knit system [4].

The machine is supplied in 5 or 8 gauge and in widths up to 4 m. Patterning is controlled from a Grosse EJP-2 electronic jacquard.

An example of a carpet produced on the Kändler machine follows.

Machine gauge	8
Pile yarn	nylon 1350 dtex x 2
Knitting yarn	polypropylene 100 dtex
Weft	polypropylene 850 dtex
Stitch length	1.75 mm (1.66 mm after finishing)
Pile sinker height	6 mm
Finished weight	1540 g/m^2

14.4 STITCH BONDING

A whole variety of stitch bonding processes, trade names include Malipol and Arachne, originating in Eastern Europe, have been used for the economical production of fabrics. These include knitting of a proportion of the fibres in a cross-laid web; overstitching a cross-laid web with yarn in the warp direction; laminating a compacted web by stitching to an existing stitched web; stitching together warp and weft yarn sheets; and combinations of the foregoing [5]. Stitch bonding has been used to a minor extent for producing floorcoverings that compete with needlefelts.

REFERENCES

1. J Hospied. Bonded carpets. In: *Proceedings of Tifcon '78*, The Textile Institute, Manchester, 1978.
2. P Billiet. Fusion bonded carpet manufacturing. In: *Proceedings of Tifcon '89*, The Textile Institute, Manchester, 1989.
3. G Robinson. *Carpets*, 2nd edition, Textile Book Service, UK, 1972.
4. B Klemm. Kändler Carpet Knit – a novel knitting technique for carpet production. *Melliand Textilberichte*, 1996, 77(3), 128–129, (E32).
5. J Tröger. Stitch-bonding technology Malimo for the production of automotive textiles. *Melliand International*, 1997, 1, 29–30.

15 Coloration and Chemical Treatments

The development of carpet coloration technology, especially piece dyeing and printing, paralleled the development of the tufting process itself and equally influenced the rapid growth of the industry in the 1960s and 1970s. Quick response to market demand through late-stage coloration processes gave tufting an essential advantage over weaving, although it should be recorded that a continuous exhaust process was used by Firth Carpets, UK, to dye wool Wilton automobile carpets in the 1930s. For woven carpets, stock dyeing and yarn dyeing are now the accepted methods. Needlefelts are most commonly produced from pigmented or dyed fibre, although some are printed.

Despite the outstanding commercial advantages of piece coloration, all coloration routes possible are used by the carpet industry, namely:

- polymer pigmentation,
- stock dyeing,
- yarn dyeing by exhaustion (hank and package dyeing),
- continuous yarn dyeing or printing (space dyeing),
- piece dyeing by exhaustion, including differential dyeing (mostly winch/beck dyeing),
- continuous piece dyeing,
- carpet printing.

In different countries and in different areas of the trade, the various methods are used to different extents, for good reasons, including the relative popularity of different pile fibres, trends in carpet styling, the volumes of carpet produced per company, energy costs, environmental issues and the mood for investment in new technology[1, 2, 3].

The principal fastness requirements for dyeings on carpet pile material are to: light, shampoo, rubbing (people sit on carpets), water spillages, salt (in coastal locations) and further mill processing.

15.1 POLYMER PIGMENTATION

Producer coloration for the carpet industry is essential for standard polypropylene, the second most important carpet pile fibre. Polypropylene fibre producers have largely accepted the fact that attempts to make dyeable polypropylene would detract from the fundamental cost advantage of the fibre and have concentrated their efforts on refining pigmentation techniques.

If a whole batch of polymer is pigmented before extrusion, a colour change is expensive and large quantities of fibre per colour must be produced. Modern extruders are more flexible. Typically, the extruder is fed with bulk polymer chips and heavily pigmented chips are metered into the melt in the barrel of the extruder in the proportion required by colour physics to achieve the required colour (most colours can be obtained using three primary pigments plus black). Rapid dispersion of the pigment concentrates is achieved through the use of a proportion (20–30%) of polyolefin wax plus a special dispersing agent.

High standards of reproducibility of colour from batch to batch are achieved. Similar systems are used for staple fibre and filament.

Pigmentation at fibre production is cheaper than a separate dyeing process would be and there is no aqueous effluent, but the outstanding technical advantage of pigmentation is the high level of colour fastness that can be achieved. Exceptionally high fastness to light of pigmented fibre has led to polypropylene dominating the market for outdoor carpets. Automobile carpets require high standards of light fastness that are difficult to achieve in all shades by dyeing, but are entirely feasible with pigments. Pigmented colours (and polypropylene itself) are resistant to bleach, providing an advantage when carpets are to be installed in institutions.

Although nylon is eminently dyeable, the foregoing advantages have led to substantial production of pigmented nylon. Pigmentation is the most satisfactory way to produce black nylon yarn. Polymer pigmentation is the most important coloration technique for carpet materials in the USA.

Pigmented filament yarns in different colours can be intermingled to create special effects, and pigmented staple can be lightly mixed to create tweed styles.

15.2 STOCK DYEING

Although stock dyeing of wool, nylon and acrylic fibre is least advantageous from the point of view of quick response to orders, it is associated with a number of advantages.

- Stock dyeing is essential if multicoloured yarns such as tweeds, mixtures and flames are to be spun.
- Large weights of yarn in a single uniform colour can be produced by blending a number of dye lots, minimising the risk of streaks and stripes in plain carpet.

- Colour matching by blending different dye lots, dyed to converge on the standard of colour, can be extremely accurate.
- Dye cycles and liquor ratios are short, giving savings in energy and water relative to yarn or winch (beck) dyeing.
- Stock dyeing of small lots of disperse-dyeable polypropylene has been recommended to compensate for shortfalls in quantities of pigmented polypropylene.

15.2.1 Stock Dyeing Machinery

Although continuous dyeing machinery, mainly based on the pad/steam principle (in one machine, dielectric heating was used to fix the dye), has been used in the carpet industry, technical developments have been focused more strongly on piece coloration, and the industry mostly uses classical batch dyeing equipment. Totally enclosed liquor circulating machinery (as used for dyeing yarn packages or tops) having the fibre pressed into a cage is ideal, but older designs, such as the Longclose conical pan and pear-shaped machines, are satisfactory for the purpose.

Modern installations concentrate on efficient materials handling, for example a machine has been designed to dye bales of synthetic fibre without the need for breaking the bale and uniformly repacking the material in a perforated cage. Figure 15.1 shows a line comprising loading and pressing station, dyeing machine, centrifuge and drying machine that is based on a cage lining common to the first three stages. A continuous centrifuge developed by Rousselet can be used as the basis of an alternative line – dyed stock is fed to the centrifuge at a steady rate and hydroextracted stock flows from it direct to the dryer.

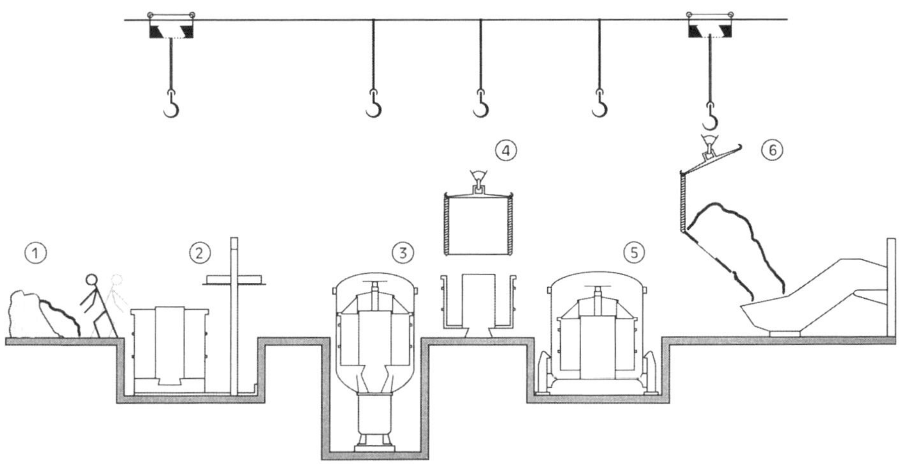

Figure 15.1. Materials handling system for stock dyeing: 1 raw material; 2 loading and pressing; 3 dyeing; 4 unloading from carrier; 5 hydroextracting; 6 material unloading. (Courtesy of Argelich, Termes Y Cia SA)

15.2.2 Dyestuff Selection

Dyestuff selection for stock dyeing, which is almost always carried out on individual fibre types, not fibre blends, is primarily for colour fastness and environmental requirements rather than for level dyeing or partition qualities. Chrome dyes have been abandoned in most countries because of the risk of soluble chromium compounds polluting dyehouse effluent. Metal complex dyes, which have excellent fastness properties, are widely used.

Recommended dyestuff classes

Nylon

- Selected acid milling dyes for bright and subdued colours.
- Selected metal complex dyes for subdued colours.

Recipes developed for one brand of nylon are likely to be invalid if another manufacturer's nylon is substituted. A change from nylon 6 to nylon 6.6 or vice versa, often requires re-formulation of the recipe.

Wool

- Selected acid milling dyes.
- Selected metal complex dyes.
- Reactive dyes for special cases (eg, heavy reds where milling dyes would give inadequate fastness to shampooing).

Note that the most appropriate individual dyes for wool and nylon may not be the same.

Wool / nylon blends

If logistics permit, the two fibres are dyed separately, and blended subsequently. When dyeing wool/nylon blends, the partition characteristics of the dyestuffs between the fibres must be considered, so that a limited selection of trichromatic combinations is available. Partition tends to favour the nylon in pale and medium colours, and the wool in dark colours. In pale and medium colours, nylon blocking agents are necessary if similar colours are to be obtained on the two fibres.

Acrylic

- Selected cationic dyes.

15.2.3 Continuous Stock Dyeing

Accuracy of colour matching achieved by blending different dye lots is an important advantage of batchwise dyeing of loose stock and this may account for the fact that continuous dyeing has not achieved widespread popularity.

Continuous dyeing machinery based on the pad/steam principle has been produced by several machinery manufacturers and is used by a few carpet companies in different parts of the world. Dye fixation by radiofrequency radiation is applied in alternative continuous systems developed by Fastran Engineering and Smith Whitworth. The impregnated fibre is transferred to the dye fixation unit by an hydraulic ram. It is subjected to the radiofrequency field for about 15 minutes, becoming uniformly hot, and dwells for a further 10–15 minutes. Afterwashing and radiofrequency drying complete the process.

15.2.4 Drying Machinery

Mechanical drying is usually by centrifugal hydroextractor. Batch operation is the most common, but continuous hydroextracting is a practical alternative. The installation shown in Figure 15.2 requires a regular feed from dyed batches: it can then deliver directly to a thermal dryer.

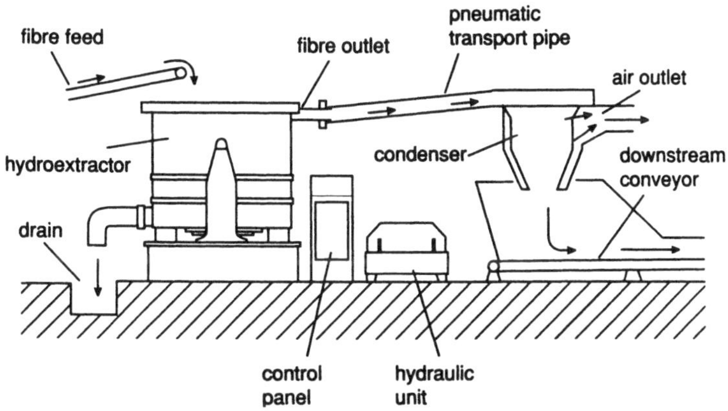

Figure 15.2. Rousselet continuous centrifuge for loose stock [2].

Thermal dryers operate continuously and are either of the suction drum type or carry the fibre through the oven on brattices. Control equipment is available for optimising the thermal efficiency. The Streat Instruments' moisture meter uses the conductivity principle and is capable of measuring materials with highly variable moisture levels. It can be used to control a dryer by regulating either the speed of the machine or the energy of the dryer.

An increasingly popular form of drying for loose stock depends on radiofrequency energy. A feed hopper is necessary to control the mass and packing density of the fibre, which is then fed through the drying zone on a polypropylene belt. A combination of radiofrequency energy and airflow provides high thermal efficiencies and ensures that the temperature of the fibre does not rise above 60°C. The principle advantage of radiofrequency drying, however, is the fact that energy is directed towards the wetter parts of the fibre, so that the final regain is uniform and can be controlled to a desired value.

15.3 HANK DYEING

Most pile material for woven carpets is dyed in yarn form, for a number of reasons:

- Yarn dyeing allows rapid response to orders. Dyeing machines covering a wide range of lot sizes are usually available in a carpet yarn dyehouse.
- Spinning is more efficient than when spinning dyed fibre because ecru fibre is usually spun into very large lots of standard yarns and undyed

waste can easily be recycled into the blend, especially on the woollen system.

- Optimum yarn bulk is developed as the material is relaxed during dyeing.
- In the case of wool, the twist in the yarn is set by the action of boiling water, giving good tuft definition in cut-pile carpets and allowing special textures, such as friezé carpets, to be produced economically.

For similar reasons, nylon, wool and wool/nylon yarns for patterned tufted carpets are commonly dyed in yarn form. As piece dyeing of wool carpets is not quite as straightforward as it is for nylon, tufted wool and wool/nylon carpets in plain colours are often produced by the yarn dyeing route.

It is not easy to achieve completely level results and highly accurate shade matching on yarn. However, improved chemical technology and various improvements centred around colour physics have made yarn dyeing much more controllable.

15.3.1 Hank Dyeing Machinery

Single-stick Hussong-type machines, as shown in Figure 15.3, have been used by the carpet industry for many decades. Although the rate of flow of liquor is rather low and by no means uniform, incremental improvements in the design of impeller, baffles and supplementary impellers, location of heating pipes and design of sticks have enabled improved results to be obtained. It is usual to reverse the direction of flow at intervals during the dyeing cycle.

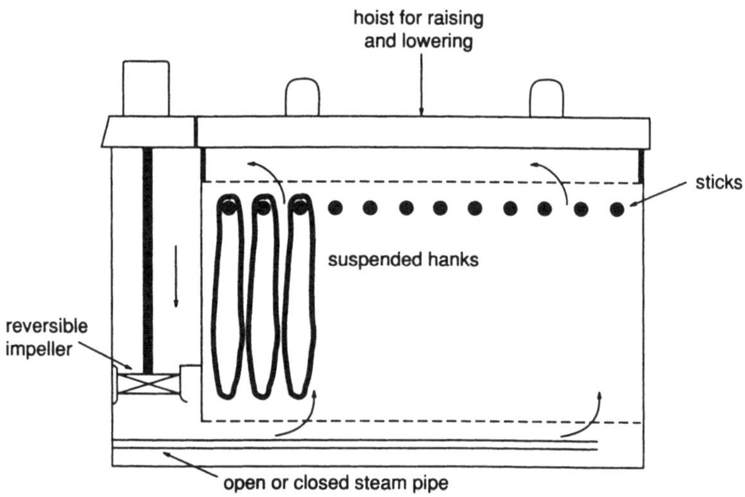

Figure 15.3. Hussong-type hank dyeing machine [2].

Quite small hank dyeing machines are available to provide colours used in small areas of a design or for sampling. Conversely, two or four large machines may be linked to provide dye lots of up to 4000 kg.

Perforated-cage machinery of the type used for dyeing loose stock is used for dyeing hard fibres, such as coir and jute, in hank form. Such yarns are not bulky and can be packed in cages without detriment.

15.3.2 Dyestuff Selection

Wool

Advances in dyestuff technology have been more rapid than those in machine technology, so the possibilities for achieving level results to high standards of fastness improved significantly in the last two decades of the twentieth century. The following classes of dyes are available for dyeing wool carpet yarns.

Equalising acid dyes were once the most widely used. Their use has declined because of the difficulty in achieving reproducible results and poor wet fastness in deeper colours. Hard-twist (friezé) yarns are commonly dyed with equalising acid dyes because they diffuse well into tight structures.

Half-milling (weakly acid dyeing) dyes are now used for a wide range of colours. Better exhaustion than equalising acid dyes and the availability of highly compatible trichromatic combinations assist reproducibility of results.

Sulphonated 2:1 metal complex dyes are used for subdued medium to dark colours. In heavy shades, wet fastness standards may limit their use to plain carpets.

1:1 metal complex dyes, once only applicable in the presence of damaging sulphuric acid at low pH, can now be applied at pH 3.5 to 4 by using special auxiliary products. Their good levelling characteristics are retained and they give good fastness properties in pastels to heavy shades.

Milling dyes are difficult to apply level in single-stick machines but are used in bright medium colours for patterned carpets, where minor unlevelness can be tolerated.

Reactive dyes are used when it is difficult to achieve the required result (brightness coupled with fastness) from other classes of dyes.

Wool/nylon blends

Because the fibres are well blended, it is important to include samples of the separate wool and nylon in the dyebath so that the partition of colours can easily be observed. Failure to achieve a reasonable match between the two fibres will result in patchy colours developing in the installed carpet owing to differential wear creating a nylon-rich surface.

Half-milling dyes are particularly suitable for dyeing wool/nylon blends. Metal complex dyes give higher standards of wet fastness in the deeper colours. The various dyestuff manufacturers recommend selected dye combinations for dyeing wool/nylon blends.

Nylon

Special ranges of acid dyes selected for high light and wet fastness, compatibility and reproducibility have been developed for carpet applications. Ranges of metal complex and sulphonated metal complex dyes are available for carpet applications.

Acrylic

Cationic dyes selected for high standards of fastness are widely used.

15.3.3 Colour Control

An important target in a carpet yarn dyehouse is to be able to match colours with such accuracy that different dye lots are indistinguishable to the human eye and can be used in the same carpet. Large reductions in waste yarn can be made if this target can be achieved, and production planning can be simplified. An alternative approach is to dye only the exact length of yarn required for a particular carpet. Quite apart from savings that can result from improved stock control, accurate colour matching is an important element of service in the contract carpet trade. By no means every carpet company is able to achieve the goal of indistinguishable dye lots, but it is practicable to achieve such high standards by paying attention to many different procedures. The use of instrumental colour measuring equipment that is more discriminating than the human eye is essential.

Sample viewing is critical. Clearly, the same dyed yarn has a very different appearance when viewed from the side, as in loop-pile carpets, than when viewed from the end, as in cut-pile carpets. For accurate matching of cut-pile materials, the state of relaxation of the sample must be exactly the same as in the bulk. This applies equally when viewing visually and when presenting a sample for objective colour measurement. Figure 15.4 indicates the principle of a reflectance spectrophotometer meter: this type of instrument is more commonly used in the carpet industry than a tristimulus colorimeter.

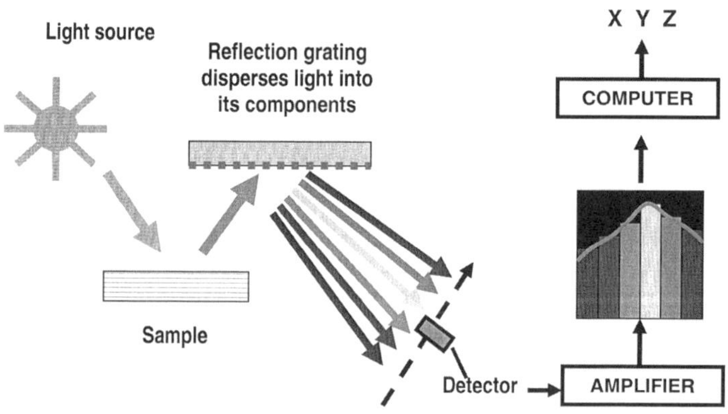

Figure 15.4. Principle of a reflectance spectrophotometer.

Fibre blends must be consistent. Although wool may vary from farm to farm, it is an important task for the wool buyer to produce consistent blends from the materials available.

Laboratory dyeings are required to forecast the results of bulk dyeings, to give 'right-first-time' results. This is achievable using modern equipment and skilled staff.

Accurate weighing and dispensing of dyes and chemicals for bulk dyeings is essential. Computer-controlled access to only the materials specified in the recipe, coupled with automatic weighing and dispensing, are features of modern dyehouse kitchens. Dyestuffs must be accurately standardised.

Accurate weighing of the goods to be dyed and the use of the same liquor ratio each time are just as important as the weighing of the dyes. In the case of wool, it is the mass of wool base that is important (ie, allowance must be made for the moisture regain at the time of weighing).

Dye cycle controllers are universally used to control the dyeing procedure. Thought must be given to the physical conditions required to achieve uniform flows and temperatures throughout the dye pack.

Software for calculating shading additions to achieve the final colour is available, although shading additions add considerably to the cost of dyeing, and 'right-first-time' dyeings should be the target.

15.3.4 Drying Machinery

Mechanical drying is carried out in centrifugal hydroextractors.

Thermal drying of jumbo hanks using the traditional system of poles carrying the hanks through hot-air ovens is not entirely satisfactory. A special dryer designed for jumbo hanks has a very sophisticated air circulation system and yarn carriers that turn alternate rows of hanks in opposite directions, so that yarns do not rub against each other. An alternative system takes hanks individually on a series of hooks and conveys them through the dryer. A fully robotic system provides mechanical drying by squeezing the hanks, transferring them to a tunnel dryer and removing the dry hanks from the carriers.

Radiofrequency drying, combined with air circulation, is also applied to the drying of hanks. As with loose stock dyeing, the principal advantage is that drying is uniform and the yarn can be dried to a preselected moisture content.

15.4 PACKAGE DYEING

Package dyeing machinery provides better hydrodynamics than hank dyeing machinery, operates at lower liquor ratios to give savings in water and energy, and is more controllable. Packages are more easily and cheaply wound and handled than hanks. Despite these advantages, package dyeing is used only to a minor extent in the carpet industry, although its use is expanding slowly.

A serious disadvantage is that yarn bulk is by no means as well developed as in hank dyeing. The difference in yarn handling costs between cross-wound packages and hanks diminishes as counts become heavier, and many of the criticisms of processing yarn in hank form have been answered by the development of 'jumbo' hanks. Perhaps the most promising application of package dyeing in the carpet industry is to dye singles yarns for use as components of multicoloured plied yarns.

15.5 CONTINUOUS YARN DYEING/SPACE DYEING

Machinery for dyeing carpet yarn continuously has been produced by a few machinery manufacturers. The Accu-Strand process [4] is recommended for producing small lots of dyed yarn. The material is processed in the form of tubular knitted fabric (sock) and is later deknitted. Dye liquor is applied by padding and the material is steamed for at least 11 minutes. A countercurrent afterwashing system doubles as a seal for the steamer. Chemicals, such as antistats or stainblockers, may be applied from a further bath following the afterwash.

Random colour application (space dyeing) in a continuous manner is very fashionable from time to time. Use of 100% space-dyed yarn in a carpet gives a somewhat crude effect, but inclusion of space-dyed yarn as one component of a design can usefully enhance the number of colours in the design in an interesting way.

One commonly used technique is the knit-deknit process. The temporary fabric is dyed to a ground colour, bars of colour in various widths are overprinted, and the material is steamed and dried. The socks are deknitted directly at the backwinding machine.

The Belmont Textile Machinery Company developed a continuous dyeing machine capable of dyeing 3.6 tonnes of yarn per 24-hour day. The dye applicator has an eight colour capability which can produce space-dyed yarns with a spacing of 10–250 mm. Up to 36 ends of yarn (48 in some versions of the machine) are pre-steamed to develop bulk; dye is applied; post-steaming gives a light set; the yarns are coiled on a perforated stainless steel belt which takes them through a fixation steamer; spray washing completes the wet process; excess water is removed pneumatically; after thermal drying, the yarns are cooled and accumulated for winding on a 36 or 48 spindle auto-doff take-up unit. Solid colours are applied either by overprinting the same colour at multiple stations, or by locking the print rollers in the print position.

The Ecodye/MI system of Superba SA is a low investment unit having a steamer in-line with the colour applicator. It can process up to 36 ends of 800–5000 dtex nylon yarn at rates up to 150 kg/h. Colours can be applied by a flooding spray to create short (5–50 mm) spots or long (50–500 mm) streaks on a background colour. Using associated software, the user can simulate patterned carpet on a computer screen and design the corresponding space-dyed yarn. The Ecodye machine can alternatively dye yarn in solid colours.

Nickel-modified polypropylene may be space-dyed using Crompton and Knowles' Altcolene dyes which complex with the nickel. Steaming conditions of 10–15 minutes at 100–105°C are required to achieve complete metallisation of the dyes. Polypropylene yarn that has been space-dyed in this way may be cabled with nylon yarn and tufted into carpet for piece dyeing. The polypropylene is not stained by acid dyes, so that clear multicoloured effects are obtained.

Polypropylene modified with Chromatex polymer additive (Elf-Atochem, Belgium) to make it dyeable is recommended for space dyeing, as well as for winch/beck and continuous dyeing.

15.6 WINCH/BECK DYEING

Exhaust dyeing of carpets in piece form, along with continuous dyeing and printing, offers the advantages of quick response to the market, rapid tufting of plain carpets, simplified production planning and the low costs of stock keeping associated with the use of a limited range of ecru yarns. Additionally, patterned and semi-plain carpets can be produced through the use of differential dyeing techniques, most commonly using substrates containing standard nylon and variants having modified dyeing properties.

A further advantage of the technique that is sometimes quoted is that the mechanical action of the winch (or beck) most commonly used for exhaust dyeing develops cover and texture in loop-pile carpets. An extreme example of this is the use of tightly twisted (and lightly set) yarns tufted into fairly long loops that snarl to give an interesting texture (loop-pile friezé) during dyeing.

Exhaust dyeing in the piece is most commonly carried out in the winch (beck in American terminology). Winch/beck dyeing was once regarded as the most flexible form of piece dyeing, capable of much shorter runs than continuous dyeing or printing. Improvements in continuous machinery have since largely negated this advantage, but continuous machinery is still associated with high capital value and high total volumes of production. A piece dyehouse having five winch/becks may be regarded as a more viable production unit than a continuous range.

Installed capacity for winch/beck dyeing is particularly high in the USA and the technique is also the most popular one in Germany. The general trend is towards continuous techniques, however, because of potential savings in water, effluent and energy.

15.6.1 Winch/Beck Dyeing Machinery

The basic design of a carpet winch/beck is similar to that of a winch/beck for fabric dyeing, but the machinery is necessarily robust to handle the heavy wet carpet and there are special features mainly associated with the scale of the operation. Although dyeing in rope form is favoured for some styles of carpet, particularly in the USA, carpets are more often dyed in open width so that scroll openers and centring devices are fitted. Most carpet winches/becks are capable of handling the material either in open width or rope form.

Figure 15.5 illustrates Brückner's Supraflor winch/beck. The flat sloping floor reduces the volume of the bath and aids movement of the folded carpet. Note the circulation system taking liquor from the underside, and external heat exchanger on the left. Liquor from the base of the beck can be distributed over the carpet at two points: intermediate between the scroll openers and the reel, and over the reel itself, contributing towards the uniformity of the dyeing. Liquor

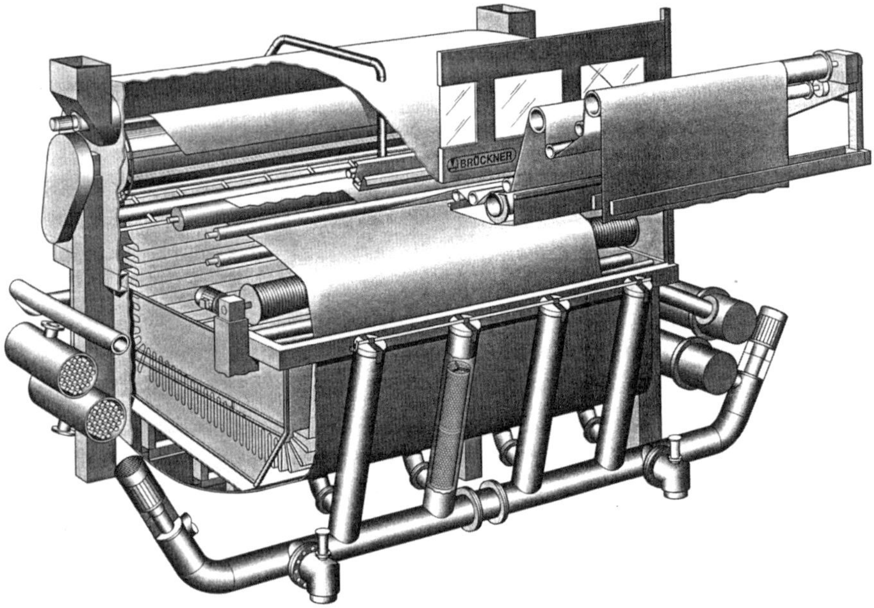

Figure 15.5. Supraflor carpet winch / beck. (Courtesy of Brückner Apparatebau GmbH)

filters at the front of the machine remove fibre flock that is commonly shed when cut-pile carpets are dyed.

When dyeing plain colours, considerable savings in water and energy may be made using a "standing bath" [5]. After completing a dyeing, the residual bath is reset for pH and temperature and a further batch of carpet is introduced. Using selected dyeing auxiliary products, three to five dyeings may be carried out from one bath. The traditional cooling cycle cannot be associated with the standing bath approach and an alternative procedure for rinsing and cooling must be used. Brückner's Frigoflor device, associated with the power reel for unloading the carpet, removes excess liquor with an air blast and the carpet is then sprayed with clean cold water. Another possibility is to remove excess water by vacuum extraction before the carpet is plaited into a truck.

Machines are available to accommodate carpets either 4 or 5 m wide and for batches up to 240, 300, 360 or 400 running metres. Two machines may be coupled to give double these batch sizes. Small machines are made for batches of 10–50, 10–80 or 20–120 running metres. Special sample dyeing machines made by Brückner hold a piece of carpet of fixed dimensions mounted on a perforated drum that rotates in the dye liquor. Such machines typically hold samples 1× 1 m or 4×4 m.

15.6.2 Application Procedures

During the processing cycle, reel speeds for nylon carpets are typically 60–100 m/min and rates of temperature rise 1–2°C/batch revolution. Delicate cut-pile wool carpets are reeled at lower speeds (eg, 20 m/min) to reduce mechanical action, and to ensure level dyeing, the temperature rise rate, which controls the rate of dyeing, is about 2°C/batch revolution (ie, 0.4°C/min for a 100 m batch). Other precautions are recommended for preserving the pile texture. Dyeing machines must be maintained in smooth internal condition to avoid the laddering of loop-pile filament constructions. Modern machines are fitted with microprocessor devices controlling time/temperature cycles, and many have pH controllers.

The dyeing machines are supplied from a dye dispensary, usually located on a mezzanine floor. Sophisticated control systems include automatic weighing and dispensing, favouring the use of dyes in liquid or granular form. Free-flowing granules are also favoured over powders for reasons of safety (non-dusting) and reproducibility (non-settling and less hygroscopic than powder).

15.6.3 Dyestuff Selection

Nylon carpets in plain colours are now rarely dyed with disperse dyes (originally the most popular for piece dyeing) because of poor reproducibility, low fastness to light and gas fumes, and risk of allergic reactions. Their principal merit was to cover irregularities in the dyeing characteristics of the fibre, but the modern approach is to control earlier processing sufficiently well as to ensure that variations in dyeability are not introduced. Equalising acid dyes are most commonly used, but metal complex and selected acid milling dyes are favoured for some colours. Application techniques are similar to those used for hank dyeing, controlling the rate of dyeing, and therefore levelness, by the rate of rise of temperature, or by pH swing allowing a more rapid rate of temperature rise to a top temperature of 95°C. For preserving tuft definition in cut-pile carpets and for energy saving, techniques are available for applying dyes of low molecular weight at temperatures as low as 65–70°C, using special levelling agents.

Wool and wool/nylon carpets are dyed with equalising acid dyes or half-milling dyes (eg, Sandolan MF) which provide an excellent compromise between fastness and levelling. When dyeing cut-pile carpets, an anti-set-release agent is incorporated in the recipe so as to preserve tuft definition.

Polyester carpets are dyed on the winch/beck with disperse dyes. Conventional polyester fibres require carriers for atmospheric dyeing, and carrier dyeing is now considered environmentally unacceptable. Modified polyester (eg, Trevira 210) can, however, be dyed on the winch/beck without carriers.

The dyeing properties of PTT (Corterra) are more favourable than those of conventional polyester (PET) because its glass transition temperature at 55°C is lower than that of PET (81°C) [6]. In practical terms, this means that dye uptake commences at 75°C and exhaustion is virtually complete at 95°C. Dyeing can be carried out at atmospheric pressure without the aid of carriers.

Nickel-modified polypropylene can be dyed with special disperse dyes (Altocene of Crompton and Knowles) that chelate with the nickel. The dyestuffs diffuse into the fibre at 60°C but metallise only slowly at this temperature, so a key aspect of the application procedure is to level them at 60°C and then raise the temperature to 100°C to complete metallisation.

Disperse-dyeable polypropylene is recommended for piece coloration, but little experience has been gained at the time of writing.

Differential dyeing of nylon carpets having latent patterns or intermingled yarns composed of fibre variants with different dyeing characteristics is a technique that is popular in the industry from time to time. Dyeability options may be cationic, light cationic, regular acid, light acid, deep acid and extra deep acid. Commonly one cationic dyeing fibre, regular fibre and one acid variant are incorporated in the carpet, and these are identified with sighting colours to ensure their correct placement at tufting.

Acid dyes that do not stain the cationic nylon are available for differential dyeing. Basic dyes always stain the acid dyeable yarns to some extent, but are selected for minimal staining. This situation means that the deeper colour in the design (or intermingled yarn) is dyed with acid dyes, so that staining with basic dyes is less noticeable. The recipe contains a dispersing agent to prevent interaction between the acid and basic dyes, and a cationic levelling agent that accentuates the contrast between the fibres. Dyeing is started cold (30°C) with the auxiliary products and phosphate buffer to pH 4.5–6.5 (depending on the nylon type). The acid dyes are added first and the goods are run for 10 minutes before the basic dyes are added. The temperature is raised to 95°C over 45–60 minutes and dyeing is continued for 45–60 minutes. The temptation to increase exhaustion by lowering the pH should be resisted, since this can affect the partition of dyes between the different fibres. In fact, dyeing to a standard procedure is likely to give a more reproducible result than that achieved by making shading additions and thereby varying the dyeing time.

15.7 OTHER BATCHWISE PIECE DYEING TECHNIQUES

Jet dyeing machines for dyeing carpets in rope form are produced by several machinery manufacturers. Transportation of the carpet through the dyebath is effected gently by a combination of a soft flow of liquor through a large nozzle and a winch. Such machines are suitable for nylon, wool and polyester carpets. High temperature versions are available that can dye carpets made from conventional polyester fibre. Liquor-to-goods ratios are in the range 8:1 to 16:1, giving significant savings in water and energy compared with the winch/beck. Despite these advantages, jet machines have not displaced winches/becks to a major extent.

In Japan, a star-frame machine was made specifically for dyeing carpets with acrylic pile. The carpet is spirally wound on a frame which is then suspended vertically in the circulating dyebath. The objective is to prevent contacts that could lead to the distortion of the acrylic pile, which is easily deformed under the conditions of dyeing.

15.8 COLD PAD/BATCH

The Carp-O-Roll system, developed by Brückner and the then Sandoz, is used in the industry only to a minor extent, but it is an elegant system for dyeing nylon carpets using a long exhaustion time at room temperature. Acid dyes of small molecular size and a special dyeing auxiliary product are key chemical features of the process.

In the Carp-O-Roll range, ecru carpet is straightened, steamed to develop pile bulk and cover, and dye liquor is metered on to it through a nozzle system to give a pick-up of around 300%. The carpet is batched, covered with plastic film to prevent surface drying, and the batch is moved to a dwell station where it is rotated slowly overnight or for 24 hours, depending on the depth of colour, to allow dyes to diffuse into the fibre.

15.9 CONTINUOUS PIECE DYEING

Continuous carpet dyeing ranges typically operate at 10 linear m/min, making them most appropriate for large carpet companies or for commission operation. However, the earlier systems that were only viable for production runs of several hours per colour have been superseded by more flexible systems which can economically produce as little as 500 m^2 (some claim 100 m^2) per colour [7], so that batch sizes are not so greatly different from winch/beck batches.

The successful commercial operation of a continuous dyeing range depends on flexibility, economy, reproducibility and ecology. To be allowed to run at all, the operation must satisfy environmental considerations; customers' needs must be satisfied via appropriate batch sizes and accurate colour matching; and the capital intensive range must be operated efficiently and economically. A further consideration is that the equipment has much in common with that required for printing, so that facilities for dyeing and printing are often combined in one range.

Consumption figures per kg of carpet [8] for a theoretically ideal range are 10–12 kg of water, 1–1.2 kWh of heating energy and 0.1–0.12 kWh/kg of electrical energy.

In 1994, water and heating costs were about equal but water (including effluent) costs were escalating faster than those for energy, pointing to the need to economise on water. Heating costs are highly dependent on the pick-up of liquor, but low levels of pick-up tend to be associated with lower dyeing quality, such as surface frosting, so that a compromise must be reached.

15.9.1 Continuous Piece Dyeing Machinery

A range for dyeing and printing comprises the following sections: entry unit comprising sew-on, accumulator, straightening and beating; pre-washing; pre-steaming; padding; printer; dye applicator; steamer for fixing dyes; afterwash; straightening; drying; and inspection and re-roll.

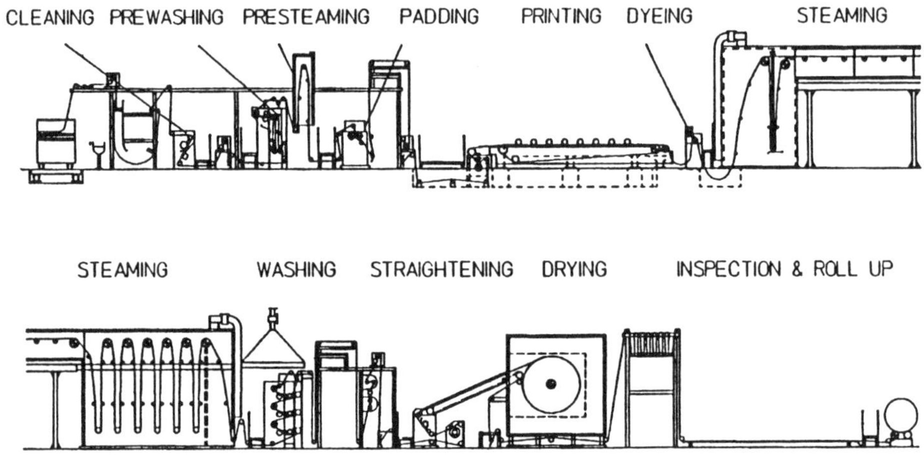

Figure 15.6. Components of a continuous dyeing and printing range for carpets. (Courtesy of Fleissner GmbH and Co)

An outline of one option is shown in Figure 15.6. Additional to the range itself is an effective colour kitchen.

Dye liquor preparation

The basic requirements are: weighing and measuring equipment (which may be automated), thickener mixing vessels, storage vessels for thickener concentrate, dissolving tanks for dyes and chemicals, and at least two dye liquor storage tanks per colour station. Use of liquid dyes can eliminate the need for dye dissolving tanks.

Küsters' Instacolor system produces dye liquor continuously from master solutions of primary colours (red, yellow, blue) and the associated Intermix system produces a colourless carrier liquid containing the dyeing assistants. The essential components are:

- buffer to the required pH value;
- coascervate, to prevent dye aggregation in concentrated solution and to achieve film formation over the fibre surface;
- steam activated foaming agent, to ensure dye remains on the pile surface despite condensation in the steamer;
- antifoam agent, to minimise foaming in the mixing/application system;
- thickener.

In the standard version of Instacolour there are six liquor control loops, two for each colour. While one loop is linked to the production system, the second loop can be set up to produce the next required colour. Rapid change of colour is achieved by switching to the alternative loop system. Additional master colour loops can be installed to cover the colours not satisfactorily achievable from the primary colours.

Selected acid dyes are used for residential carpets. For automobile carpets, requiring exceptionally high standards of light fastness, it is necessary to use metal complex dyes, which may present ecological problems in some countries. Differential dyeing of, for example, regular, deep and basic dyeing nylon, can be carried out continuously using basic and acid dyes selected for high contrast, with appropriate auxiliary products and a high level of applied liquor (300–500%), but the degree of contrast is inferior to that obtained by exhaustion.

Carpet entry system

The carpet must be presented for dyeing straight, clean, free from creases and without stoppages. Lengths of carpet from rolls or plaited on carts must be sewn together securely and a J-box accumulator or similar device must allow adequate time for this, without the need to stop the range. Removal of lint is achieved by means of a beater, brush and suction. The carpet is straightened with bowed and scrolled rollers before passing to the wet stages.

Pre-washing

Before dyeing, the carpet must be washed to remove fibre processing aids, fugitive tints, dirt and silicone. Early systems for pre-washing passed the carpet through hot water and detergent with agitators or circulation. Modern systems are more economical in the use of water.

Fleissner and Küsters offer competitive pre-washing units, differing considerably in detail, both of which work by impregnating the pile, allowing a dwell time for dissolving/emulsification, and removing the contaminated liquor via a suction slot. An alternative Fleissner system incorporates a foaming agent in the wash liquor, which is activated by steam. Dirty foam is suctioned from the pile surface of the carpet immediately after steaming.

Pre-steaming

A short steaming treatment is needed to bulk the pile and remove irregularities that could adversely affect the uniformity of the dyeing. Steaming is usually effected in a vertical loop steamer holding 7–10 m of carpet, or 12–21 m in a double-loop steamer. Relaxation of the carpet can be enhanced, if thought necessary, by backbeating at the steamer's exit.

Dye application [9]

Padding in the conventional way is unsatisfactory for most carpet applications because of the difficulty of achieving uniform liquor expression across the width of the broadloom carpet. However, the Küsters' Swimming Roll mangle, which applies hydraulic pressure uniformly across the roll, is used for non-critical applications, such as applying a ground colour before printing or for pre-wetting. The liquor pick-up (maximum 120%) is, in any case, considered too low for producing solid colours.

Dip/drain methods, despite their simplicity, are not widely favoured because of the risk of tailing and high liquor wastage. The carpet is impregnated and conducted vertically, giving an equilibrium pick-up of 400–800%, depending on liquor viscosity and carpet speed.

Curtain coating methods were originally based around a trough of liquor with a weir or doctor roller causing the dye liquor to flow on to the pre-wet carpet from an inclined plate, and such systems are now mostly used for overdyeing prints. Application levels are around 300%, giving a total liquor content at the steamer of 400%. To change the colour the trough must be drained, cleaned and refilled.

The Küsters' Quick-Change colour applicator (Figure 15.7) has a manifold system that sprays liquor over the width of an inclined plate that leads it on to the carpet pile. A duplicate liquor supply enables a rapid change of colour to be made within 1.5 seconds. There is no need to clean the plate: the new colour displaces the previous one in a carpet length of 150 mm.

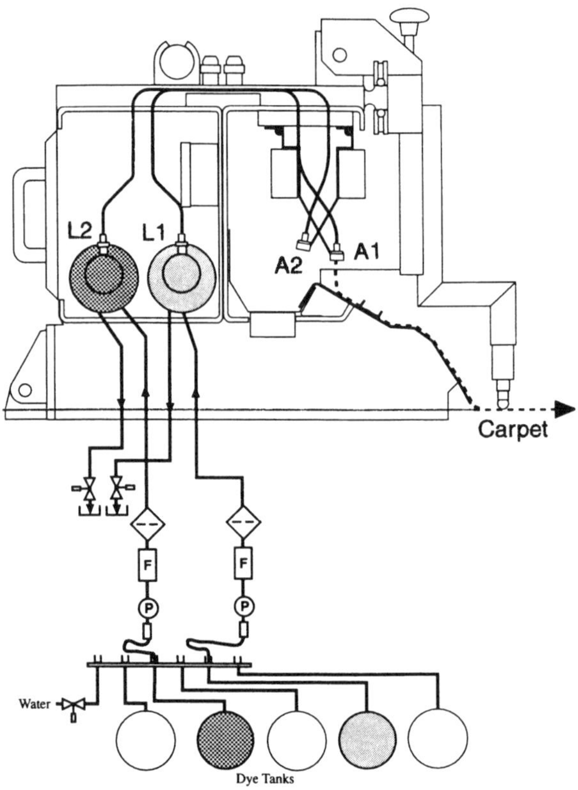

Figure 15.7. Küsters' Quick Change Applicator Type 271.26. L1 liquor being applied; L2 liquor for next colour; A1 applicator in use; A2 stand-by applicator. (Courtesy of Eduard Küsters Maschinenfabrik GmbH and Co)

Küsters' Fluidyer (Figure 15.8) also allows rapid changes of colour without the need to clean the system. The Fluidyer injects liquor directly into the carpet pile. A manifold system supplies dye liquor uniformly to a slot that distributes it across the pile. A pneumatic cushion below the carpet presses it against the slot and a gliding foil between the air cushion and the carpet backing reduces frictional drag. Pick-up is adjustable within the range 50–400%, to deal with ground colours for overprinting or for the production of plain carpet. The liquor pumps are synchronised with the machine speed so that the pre-selected application level is achieved. When the colour is changed, about 4.5 l of mixed liquor is applied to the carpet over a period of 2 seconds, corresponding to a very short length of waste carpet. The Quick Change Fluidyer applies liquor to the pile-down carpet (ie, from below) and has an even lower internal liquor capacity.

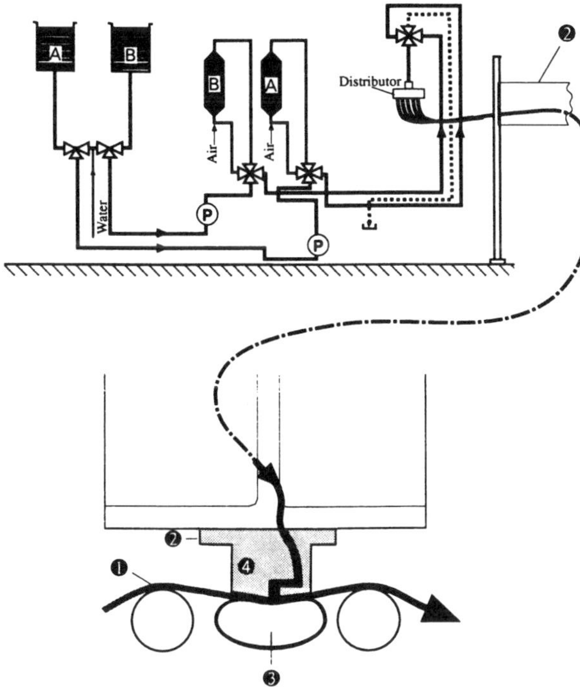

Figure 15.8. Küsters' Fluidyer dye applicator: 1. carpet; 2. dye head; 3. air cushion; 4. dye slot. (Courtesy of Eduard Küsters Maschinenfabrik GmbH and Co)

Hot dye liquor may be applied via the Fluidyer (and some other applicators), a procedure which allows shorter steaming times and gives more solid colours on coated nylon fibres.

Spray application methods may be used for plain colours or random dyeing. In the Otting system, oscillating sprays mounted half an inch apart deposit liquor on the pile surface. The spray can be varied from very coarse to a fine mist. Curtain coaters can be modified to break up the curtain of liquor into droplets of random size. Dropped on a carpet pre-coated with gum, the randomness of the effect can be enhanced. There was once a major fashion trend for 'shadow prints' and other random colour effects produced in this way.

Foam application through devices such as Autofoam (Datacolor International), Rotafoam (Mitter) and Fluicon (Küsters) reduces liquor pick-up and consequently energy costs and chemical requirements. Low pick-up levels tend to give frosty dyeings on nylon fibres having fluorosilicone finishes, however, so that foam applicators are not universally useful. Foam applicators are particularly useful for applying chemicals in line with a dyeing range.

Steaming

The impregnated carpet is steamed for 2–5 minutes for nylon or 4–10 minutes for wool, during which time dye diffuses into the fibre and becomes fixed. Too short a steaming time results in ring dyeing, which can lead to a colour change in the carpet during wear as the outer layers of fibre are abraded away. The longer steaming times are preferable for printing, where there is a risk of cross-staining from loose colour. Attention must be paid to the steam quality (temperature, air content, turbulence, saturation, presence of free water, etc) if reproducible results and quality dyeings are to be achieved [10].

For dyeing only, a relatively compact steamer having the carpet in vertical loops is preferable, whereas horizontal steamers are favoured for printing to prevent the paste from flowing before the dye is fixed. For ranges that combine dyeing and printing facilities, a satisfactory compromise is a steamer with a horizontal section to give preliminary dye fixation, followed by a compact vertical section to complete the process (Figure 15.9)

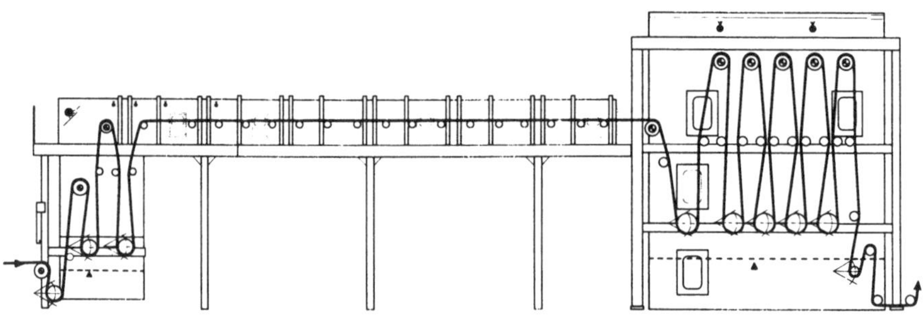

Figure 15.9. Steamer with horizontal and vertical sections, suitable for fixing prints as well as solid colours. (Courtesy of Eduard Küsters Maschinenfabrik GmbH and Co)

Dry saturated steam is the most effective fixation medium, so that air should be eliminated from the steamer. As air is heavier than steam, the most appropriate steamer configuration is for the carpet (with entrained air) to enter at the bottom and for steam to be supplied from an external saturator at the top. A cloud forms at the steam/air boundary. For satisfactory operation, the cloud should be under the carpet and a sensor mounted at this level can control the steam supply so as to achieve this condition.

Afterwash

Unfixed dyestuff, dyeing assistants and thickener must be removed before drying and finishing the carpet in order to avoid poor fastness to shampooing and rubbing, soiling problems and preserve the handle of the carpet. Countercurrent washers have largely been replaced by spray/suction systems that use less water.

15.10 PRINTING

Printed carpets originally had a down-market image because they could only be produced economically in long runs (leading to limited product variety) and because of low quality in terms of inadequate penetration of colour down the pile. Improved machinery and techniques have raised the quality, and designers of printed carpets are generating unique styles that are clearly not cheap imitations of Axminster designs. Computer controlled jet printing techniques allow short runs and quick response to customers' designs, so that even expensive wool carpets are now produced by this ultimate late-stage coloration route. Jet printing is the most popular method for printing carpet tiles, and is increasingly used for printing broadloom carpets, including those intended to be cut into rugs.

Fixation of dyestuff in carpet printing is achieved by steaming the wet carpet: there is no intermediate drying, as in fabric printing.

15.10.1 Relief Roller

The Stalwart system is based on rollers bearing the motifs in rubber tipped with foam, which are coated with print paste and pressed into the carpet pile. This simple system is used for producing shadow prints on a ground shade.

15.10.2 Flat Screen

Flat screen printing of carpets was developed in the late 1960s by the Bradford Dyers Association and P Zimmer. Flat screen machines take up a great deal of floor space and are slow and cumbersome by modern standards. The relatively long contact time and the possibility of multiple squeegee passes make the technique suitable for fairly thick piles, and a few machines survived in the late 1990s, possibly for this reason.

The machine illustrated in Figure 15.10 is a low production machine (average 2m/min), particularly suitable for printing rug designs. The carpet remains stationary while each colour component is applied in turn, taking the relevant screens from the stack.

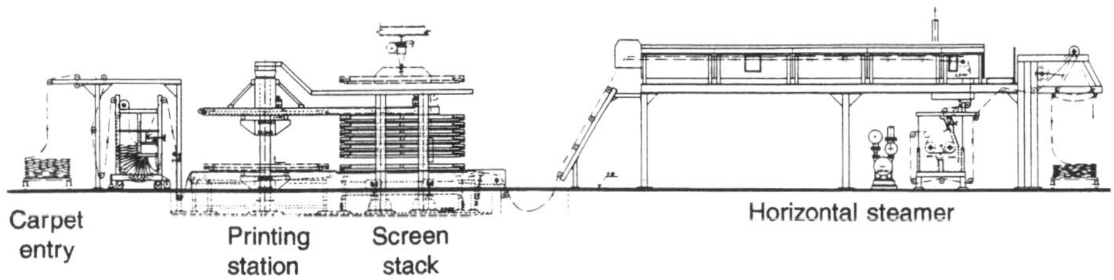

Figure 15.10. Low capacity flat screen carpet printing line for designs 3×4 m. (Courtesy of Mitter GmbH and Co Maschinenbau)

15.10.3 Rotary Screen

Rotary screen machines are highly suitable for extremely rapid (40 m/min) printing of low pile height carpets, although they can be set up to print piles up to 30 mm and can handle needlefelts as well as loop, cut and cut/loop tufted carpets (Plates 24 and 25).

A typical printing table is shown in Figure 15.11. When installing a carpet printing machine it is wise to leave room for future options, such as more colours, larger screens or combination systems. Zimmer's standard table, allowing 750 mm length for each colour, can accommodate screens with repeats of 923,1017, 1621 or 1998 mm. Very large screens with a repeat of 3000 mm require a table length of 1100 mm for each colour. Alternative squeegees for pressing print paste through the screen are a simple hollow roller squeegee, a magnetic squeegee activated by an electromagnet beneath the table or a hydroslot squeegee for applying large volumes of paste to thicker carpets. The screens can be adjusted to accommodate different carpet thicknesses and for lateral and skew registration: longitudinal registration of the different colours is motorised and is controlled from a central panel.

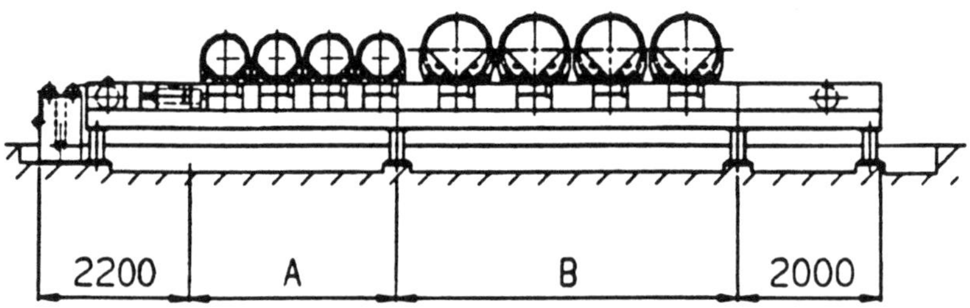

Figure 15.11. Rotary screen printing table with options for short or long repeats, emphasising the greater length (B) required for long repeats than for shorter repeats (A). (Courtesy of J Zimmer Maschinenbau GmbH)

Screens no longer have to be removed for cleaning to change the colourway; they are washed while still mounted on the machine. A colour economy pipe minimises waste of print paste during colour changes. A new pattern can be set up while printing is in progress, and open-end bearings on modern screens allow the squeegee systems to be removed easily so that changeover times for eight screens can be as short as 15 minutes.

An interesting alternative to the rotary screen is Mitter's triangular printing head (Figure 15.12) that is said to combine the advantages of flat screen and rotary screen printing.

Rotary screen machines are used for direct printing, displacement printing or resist printing.

Direct printing

Direct printing is usually carried out on a pre-padded pale colour to ensure that the base of the tufts and any gaps resulting from lack of register are coloured.

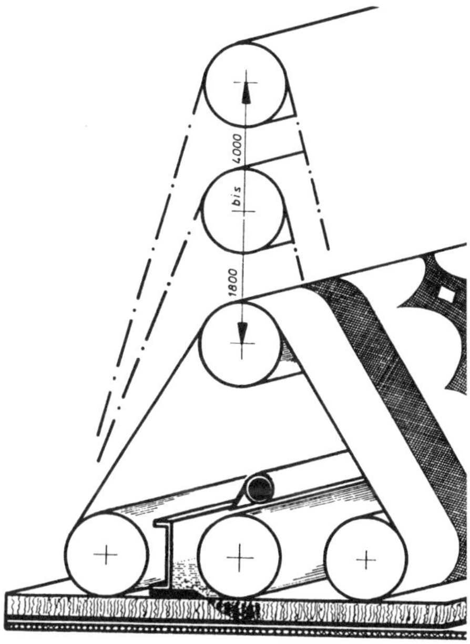

Figure 15.12. Continuously operating flat screen system offering long design repeats. (Courtesy of Mitter GmbH and Co Maschinenbau)

Positive printing of this kind was superseded in popularity in the 1990s by resist or displacement techniques (negative printing), to create effects that would be difficult to achieve by close registration of small areas of colour. The latter techniques are typified by styles that require large, possibly dark, areas of ground colour highlighted by small areas of bright colours (eg, floral motifs). Alternative possibilities include fine stripes, diagonals and small spots to create effects similar to those obtained by using multicoloured yarns.

Direct printing of nylon carpets uses the following general recipe:

- Equalising acid or metal complex dyes.
- Urea, which is occasionally used to aid dispersion and penetration of dyestuff.
- Buffer to pH 4-5.
- Thickener (12–15 g/l) to give the viscosity appropriate to the machine, screen mesh and carpet construction.
- Coascervate to aid solution of dyestuff and film forming over the fibre.
- Antifrosting agent to maintain dyestuff at the pile surface during steaming.
- Antifoam to prevent foaming of paste during processing.

Sometimes, two or more printing chemicals may be combined in a single auxiliary product.

Carpet is pre-wet to aid rapid penetration of the print paste, and pre-wetting is usually combined with the application of a pale colour. Pre-wetting is accomplished by padding to a pick-up of about 100%. The pre-padding liquor is lightly thickened (eg, 1–2 g/l thickener) and contains acid levelling dyes and dyeing assistants similar to those in the print paste.

Displacement and resist printing

As in the direct printing technique, a ground colour is applied by padding from a lightly thickened liquor, but this forms an important component of the design and may be quite a rich colour. The key to displacement printing (and resist printing) is the use for the ground colour of 1:2 metal complex dyes that can further complex with a cationic resist agent (Thiotan TR or TRN of Clariant are commonly used) in the printed areas, preventing fixation. The overprint (highlight colours) is with selected acid dyes that are not affected by the resist agent.

The procedure is:

- Pad to 100% pick-up with sulphonated metal complex dyes at pH 5.
- Overprint with acid dyes plus cationic resist agent at pH 3.5–5.
- Steam.

If white areas are required (ie, a discharge print), the cationic agent is printed at pH 8–10.

Should environmental considerations preclude the application of a few parts per million of metal into the carpet, acceptable results (but inferior in terms of contrast and fastness) can be obtained by padding at the higher pH value with selected acid milling dyes.

Resist printing is based on the same chemistry as displacement printing, but printing and dyeing are carried out in the reverse order:

- Print with equalising acid or milling dyes plus resist agent at pH 4–5.
- Flood to 300% pick-up with metal complex dyes at pH 5, slightly thickened.
- Steam.

A white resist can be obtained by printing a higher concentration of resist agent at pH 10.

As in the case of displacement printing, metal complex dyes can be avoided if absolutely necessary.

Selection of dyestuffs and auxiliary products is critical for the success of displacement or resist printing, and the recommendations of dyestuff suppliers must be followed closely.

15.10.4 Jet Printing

Jet printing is a non-contact system for applying designs to ecru carpet without the need to prepare screens [11]. Colour separation is achieved by means of CAD systems, and the patterning of colour jets on the machine is controlled by computer, so that designs can be changed instantaneously, and colourways very rapidly. There is virtually no limitation on design repeat, as for computer controlled weaving or tufting, and very short runs are economical. Preparation and storage of expensive screens is unnecessary. Jet printing uses lightly thickened liquor and is particularly suitable for relatively thick cut-pile carpets, since the print paste is wicked down the pile, although loop-pile carpets are also jet printed.

As for continuous dyeing, printed carpets are steamed to fix the dyestuffs and are washed to remove thickener, auxiliary products and any unfixed dye before drying.

Millitron
The first commercial jet printing machine for carpets was produced by Milliken [12]. Millitron machines operate according to a principle described by Textima [13]: continuous coherent jets of dye liquor are deflected selectively by air jets opposed at right angles to them (Figure 15.13). When not printing, the jets of liquor are continuously deflected and recirculated, but momentary interruption of an air jet allows the liquor to print the carpet. Bars of jets, one bar for each colour, run across the inclined bed of the transport belt for the carpet. The amount of liquor applied to the substrate can be regulated according to the 'firing time' of the jets on to the carpet, usually from 1 to 10 milliseconds.

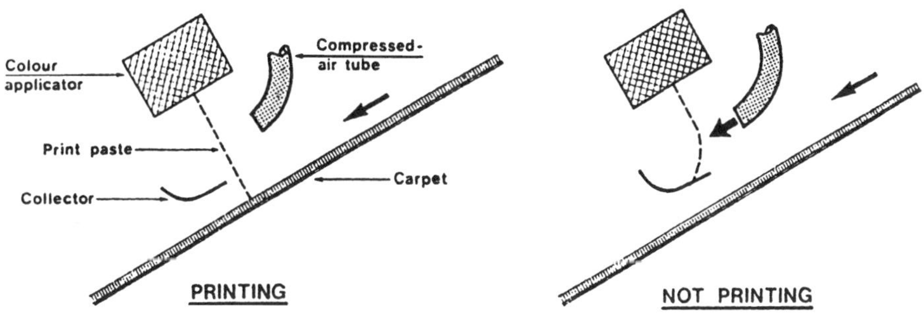

Figure 15.13. Principle of the Millitron computer injection dyeing machine [14].

Most of the broadloom machines that have been installed have 10 jets/in. (1/10 gauge), or 10 mesh in printing terms, but later machines are made in 1/16 gauge. Special machines are made for printing carpet tiles, mostly in 1/20 gauge (see Section 19.3.6).

Milliken developed the Millitron machine primarily for use in their own carpet factories and, indeed, their carpet operation has been largely centred around the capabilities of what they originally termed "computer injection dyeing" (Plate 26). Millitron machines have been made available to other companies on a restricted basis, with only one licensee per country.

Chromojet
Zimmer's Chromojet jet printing machine operates on the principle of 'drop-on-demand' [15]. Each jet is operated by a special electromagnetic needle valve that is controlled by a computer. Because of the greater bulk of the valves

compared with air valves, the jets of most versions are assembled into a packed array which is scanned across the carpet substrate while it is stationary (ie, the printing range operates intermittently). The quantity of dye liquor delivered per pulse is governed by pressure in the range 1–4 bar, and carpets with pile weights in the range 200–2000 g/m^2 can be printed. Clearly, high standards of liquor preparation using good quality water are necessary to achieve reliable operation of the numerous small valves.

Zimmer's Chromojet machine is freely available for sale and is in widespread use. It is used principally for printing nylon 6.6, nylon 6 and wool/nylon tiles, small goods and broadloom carpets.

Chromojet-LS is a sample printer for design and product development work. It operates with two jets per colour producing at 16 mesh and can have up to eight colours. A sample 1 m square is produced in 6 minutes.

Chromojet-LT is a small production machine for custom designs of mats, tiles and runners, operating at 30–60 m^2/h. It is a component of a continuous range consisting of printer, steamer, washer and dryer.

Chromojet-MS is a similar machine for printing tiles, mats and roll goods up to 2150 mm wide. It has 32 jets per colour, compared with eight for the LT machine, and produces at up to 600 m^2/h. The range is illustrated in Figure 15.14.

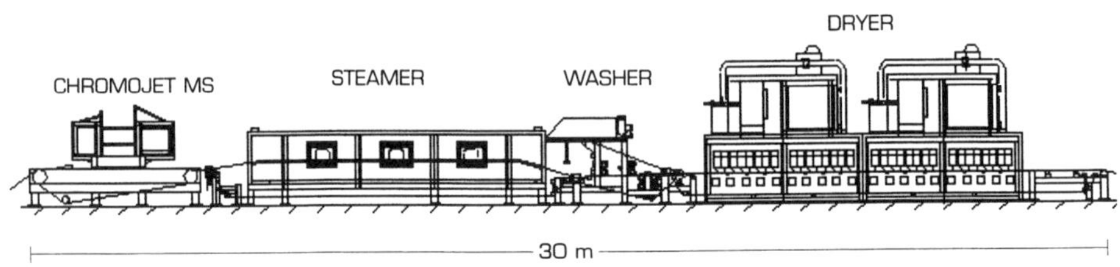

Figure 15.14. Line for printing tiles, mats and narrow carpets, based on the Chromojet-MS machine. (Courtesy of J Zimmer Maschinenbau GmbH)

Chromojet-MT prints broadloom carpets and is supplied in printing widths up to 5200 mm. Versions with 64 or 128 jets per colour are available and there are up to 12 colours. The largest machine produces at up to 1000 m^2/h.

Chromojet-F has a different configuration. The carpet moves continuously and there is a printing head for each colour, but the same system of direct operated valves is used. The printing table can accommodate up to eight heads, requiring 750 mm for each head. Alternative versions have print definition in the range 1/10, 1/12 and 1/16 gauge. Output is in the range 5–25 m/min, depending on the quality of the carpet and the capacity of the steamer. Designs can be changed without stopping the machine.

Patternjet

The Patternjet machine of Fine Jets Ltd is designed for the rapid production of either woven or tufted carpet samples from CAD data. It has a fixed print head that can apply four, six or eight colours at 18.8 dpi (mesh).

15.11 DRYING CARPETS

Mechanical drying at all required stages in a range is achieved by the use of suction slots serviced by high-power vacuum pumps. In some cases it is desirable to pre-wet the carpet to ensure a uniform final water content. To economise on energy in thermal drying, as much water as possible should be removed mechanically, and figures as low as 25% residual moisture have been claimed. After the slot, the carpet is backbeaten and possibly brushed to open the pile before stentering.

Thermal drying machines fall into three general categories: dual airflow systems, through-flow systems and drum dryers. A cross-section of Brückner's dual airflow stenter is shown in Figure 15.15. The main feature of such dryers is that the airflows over the face and back of the carpet can be controlled independently to accommodate different carpet qualities: different air jets or slots can be fitted to the upper or lower ducts if required. The carpet is conveyed through the dryer on heavy duty pins.

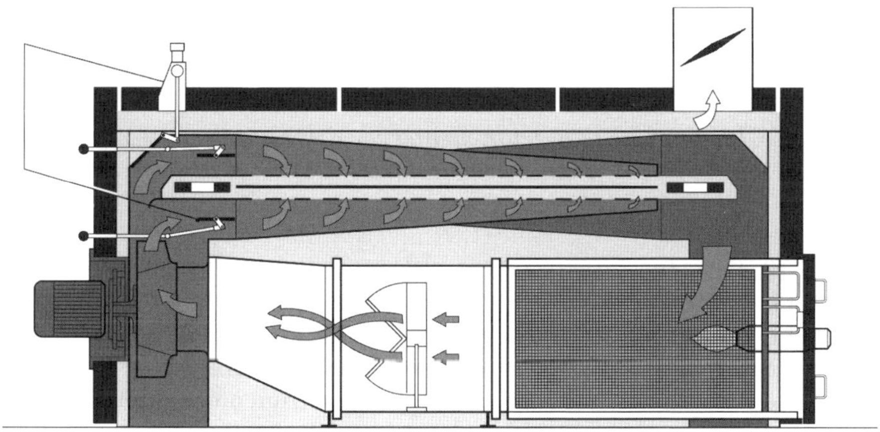

Figure 15.15. Cross-section through a dual airflow carpet dryer. (Courtesy of Brückner Apparatebau GmbH)

The Fleissner high-capacity circular stenter (Figure15.16) has airflow through the carpet from the pile to the backing, and is notable for its compact design. Its capacity is governed by the diameter of the perforated drum (up to 3.4 m) and the power of the fans.

In all cases, the efficiency of thermal drying is optimised by countercurrent airflow and an optimum balance between rate of airflow, temperature and moisture content of exhaust air, and temperature of air input [8].

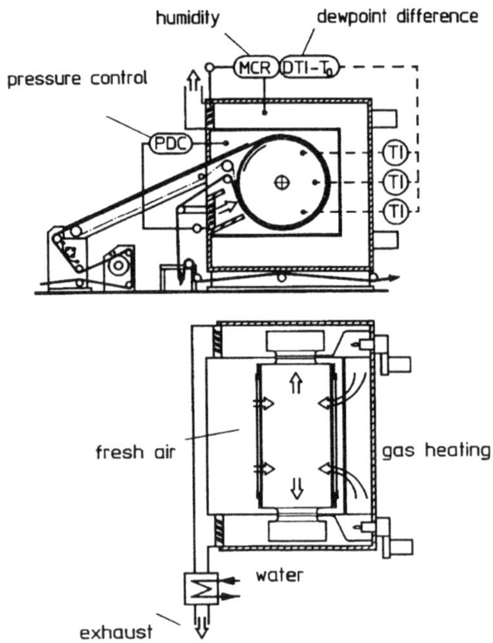

Figure 15.16. Outline of a drum-stenter for drying carpets. (Courtesy of Fleissner GmbH and Co)

Figure 15.17 shows the overall airflow path of Brückner's through-flow stenter. The carpet is cooled at the exit of the stenter by cold intake air. Pre-warmed air (from cooling the carpet) for the drying process is taken from the cooling zone, and is heated by direct gas burners protected by lint screens (or steam or oil heaters) and is conducted through the carpet and from one section to another, building its moisture content until it almost reaches wet bulb temperature. It is cooled by the entry of the wet carpet. Cool moist air is discharged at a rate controlled by a humidity measuring device.

Overdrying can be costly and can give a temporary colour change with some dyestuffs, confusing the colour matching operation.

15.12 CHEMICAL TREATMENTS

It is convenient to apply chemicals to enhance carpet properties at the stage of coloration. Huge quantities of nylon carpets are treated to improve their resistance to staining, and a wide variety of treatments are given to wool to upgrade various performance features.

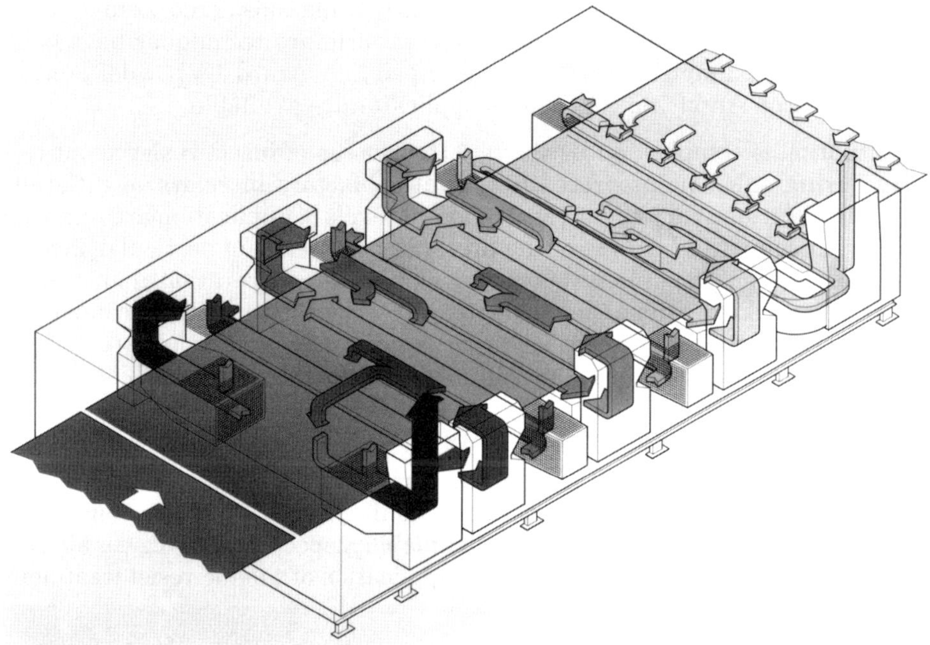

Figure 15.17. Countercurrent airflow in a through-flow dryer. (Courtesy of Brückner Apparatebau GmbH)

15.12.1 Insect-Resist Treatment of Wool

Of all wool textile materials, carpets are the most susceptible to attack by the larvae of certain moths, notably *Tineola bisseliella*, and a few species of beetle, because the hidden parts of some carpet installations provide the warm, dark, undisturbed environment that such insects prefer. Wool carpets have been treated against such insects since about 1960 and this has virtually eliminated the clothes moth and similar insects from the home: prior to this time, moth larvae tended to breed in carpets and migrate from there to damage expensive clothes.

The application of insect-resist agents has been most commonly at the stage of dyeing. The first product applied widely in the carpet industry was dieldrin, in the form of an emulsifiable dyebath additive. Although it was effective and cheap, in common with many other chlorinated hydrocarbons, it was not biodegradable and was quickly rejected as environmentally unacceptable. Synthetic pyrethroids, specifically permethrin, have proved effective over many years. Permethrin is effective against a wider range of insects than dieldrin, has low mammalian toxicity and is relatively easily absorbed from effluent by particulate matter. It is, however, toxic to the aquatic life that provides food for fish, so that great care must be taken in the generation and disposal of effluents containing permethrin. For this reason, other insecticides and alternative techniques to dyebath application are constantly being considered. Zero-effluent treatments are clearly desirable and a few proprietary techniques have been developed. For example, the WRONZ developed Lanaguard dry powder process and the Texicon Autofoam continuous applicator are available.

The situation is complicated by the fact that losses of insect-resist agent can occur during subsequent processing, dilution factors in rivers in different industrial locations vary widely, and different levels of application are required in different countries because of the presence or absence of particular insects there. Insect-resist treatment is a more sophisticated operation than simply adding a fixed proportion of pesticide to a dye formulation. Carpet companies need to take a responsible approach to the environment and should study their particular circumstances in collaboration with specialists in the subject.

15.12.2 Flame-Resist Treatments

Although wool carpets have good natural resistance to ignition and do not normally spread flame in a situation of a conflagration, they are sometimes required to meet particularly stringent flammability specifications (eg, for aircraft or high-rise buildings). For such cases, the application of a flame-resist treatment at the stage of dyeing is convenient.

The Zirpro™ processes, developed by the former IWS, are mostly designed to be carried out post-dyeing. An anionic zirconium complex is exhausted on to the wool or yarn, while circulating the bath in the conventional manner. As the liquor is strongly acidic, dye bleeding is prevented. Alternative recipes, differing in detail, are available for meeting different flammability specifications. Full technical details are provided to licensees by the Woolmark Company.

15.12.3 Stainblocking

Stainblocking treatments are largely applied in the USA and some other countries to nylon pile tufted carpets in plain colours to combat the problem of staining by acid colorants in spillages of food and drink. The processes were originally developed for application to tufted carpets at the stage of piece dyeing. They may sometimes be applied to nylon yarns destined for woven carpets, but are unlikely to be considered necessary for highly patterned carpets.

A stainblocking treatment has two components: a synthetic tanning agent (syntan) or dye-resist agent to block dye sites in the fibre; and a fluorocarbon to retard penetration of water or oil based stains into the fibre. The fluorocarbon is sometimes incorporated in the fibre by its manufacturer but is more commonly applied to the carpet by spray or foam applicator. The syntan is usually applied as a post-dyeing treatment.

As syntans tend to be susceptible to photoyellowing, the amount applied should be the minimum required to give the desired effect. The actual amount required depends on the particular nylon fibre and its heat history; nylon 6 requires more stainblocker than nylon 6.6, for example.

The various fibre manufacturers and 3M each provide their own detailed recommendations and set standards associated with their labelling schemes. A typical post-dyeing procedure is as follows.

Drop dyebath and rinse yarn.

Set fresh bath with:
x% stainblocker (on mass of yarn),
3% magnesium sulphate,
acid to pH 2.0–2.5.

Enter yarn.

Raise temperature at 2°C/min to 80°C.

Hold at 80°C for 20 minutes.

Drop bath and rinse.

The fluorocarbon is (except in the case of one brand of nylon) applied later to the carpet itself. When the pile has been cut, this procedure gives protection to the cut ends.

It is often convenient to apply both the stainblocker and the fluorocarbon in association with continuous carpet dyeing or printing. Küsters has developed a special foam applicator, Fluicon 2, for this purpose. Twin units are located between the afterwash and the dryer of the dyeing range (Figure15.18). The foamed stainblocker formulation is injected into the pile at a level of around 10% and is fixed by steaming in a unit typical of a pre-steamer. Vacuum extraction of the hot carpet as it emerges from the steamer reduces the moisture content before the fluorocarbon formulation is applied from a second foam applicator. The carpet then passes directly into the dryer.

The attitude of wool interests towards stainblocking is ambivalent. Many do not believe that any treatment is necessary; one process is based only on a combination of dye-resist and syntan chemicals, with no fluorocarbon [16]; and another process recommends co-application of fluorocarbon and syntan during hank dyeing [17].

15.12.4 Soil Resistance

The emphasis of the merits of topical fluorochemical treatments was for many years on their ability to repel water- and oil-based stains, thereby delaying the penetration of food colorants into the fibre. Du Pont's Duratech™ treatment, however, has the primary objective of improving resistance to soiling, and is claimed to be particularly durable.

Duratech is fluorochemically based and is applied to the pile material after dyeing. The carpet (or yarn) is then dried, but the product only adheres firmly to the fibre after it is exposed to higher temperatures at the stage of carpet backcoating/laminating.

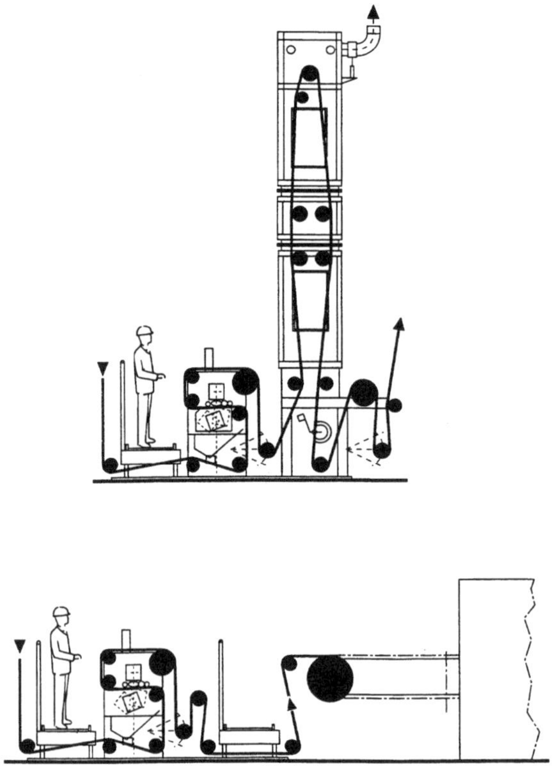

Figure 15.18. Equipment for the application of stain-resist chemicals in-line with continuous dyeing or printing: (above) stainblocker application; (below) fluorocarbon application. (Courtesy of Eduard Küsters Maschinenfabrik GmbH and Co)

The durability of Duratech is an important feature. Most fluorochemical treatments are applied to residential carpets only, because they wear off under heavy foot traffic and are removed by hot-water-extraction cleaning. Duratech is aimed at the contract carpet market, and is claimed to contribute to the long-term appearance retention of the carpet and reduces maintenance costs owing to a reduced need for wet cleaning. Initially, Duratech was available only for Antron Legacy™ and Antron Lumena™ solution-dyed nylon.

When rayon was an important carpet pile fibre, it was often treated with 'white soil' products, such as colloidal silica or titanium dioxide. The fine particles filled rugosities in the fibre, blocking these potential traps for particulate soil. Delustred wool (Fernplus EP), based on treatments with white pigments, is now available from New Zealand and may resist soiling better than untreated wool.

15.12.5 Antistatic Agents

Nylon and acrylic carpets are usually protected against electrostatic effects – principally uncomfortable shocks experienced when a person is earthed after walking over an insulating carpet – by incorporating conductive fibres and making the carpet backing conductive (see Section 21.2). Polypropylene carpets do not provide serious static problems owing to the favourable position of the fibre in the triboelectric series. An unusually durable antistat is available for wool and the most favoured point of application is in the dyeing machine as a post-dyeing treatment.

Disperstat W (Stephenson Thompson Textile Chemicals) is a cationic chemical that can be exhausted on to wool and is durable to wear and to carpet cleaning, including wet treatments. It reduces body voltage generation to levels below the threshold of sensitivity, but does not improve the conductivity of the carpet. Disperstat is, therefore, effective for most applications of carpets, but does not meet those specifications for carpets in computer rooms that are based on resistivity measurements.

A potential problem is complexing of anionic dyes with the cationic Disperstat W to give materials having unsatisfactory colour fastness. Such problems are largely eliminated by confining the cationic compound to the fibre surface. There are a few incompatible dyes, however. Lists of compatible dyes are available from the supplier of the antistat.

Disperstat W can be applied after either stock or hank dyeing. After dyeing, the dyebath is dropped and the dye pack is cooled, either by circulating cold water in the case of stock dyeing or by lifting the load of hanks into cool air. If anionic dyeing assistants have been used, the material should be rinsed. A dilute dispersion of Disperstat W (typically 0.8% on mass of wool) is prepared and added to the liquor in stages (for stock dyeing equipment with poor circulation characteristics) and exhaustion is completed in 30 minutes at 20°C and pH 5 7. For hank dyeing machines, which usually have good circulation characteristics, all of the Disperstat may be added before the yarn is entered.

A similar product, Disperstat A, is available for application during yarn scouring, twist setting or to carpet (in the mill or after installation). Mill application to the carpet is advantageously from foam, using typically an Autofoam unit or Küsters' Fluidyer. Other chemicals, such as fluorocarbons and insect-resist agents, can be applied concurrently.

Anionic antistats are available that are effective on wool or synthetic carpets, but these need to be re-applied at intervals during the life of the carpet. They are, therefore, usually applied to installed carpet rather than in the mill.

15.12.6 Protection of Wool Against Photobleaching

Wools in a finished carpet are yellow to some degree as a result of stains, process yellowing or the base colour of the wool itself. In dark shades, the yellow colour is masked and is irrelevant. All types of yellowness are susceptible to rapid photobleaching, however, so that in paler colours where the yellowness of the wool is a significant component, the carpet is likely to fade shortly after installation. The fading may take the form of a noticeable change in tone: greys turn reddish blue, for example. It may prove impossible to achieve certain colours (eg, light blues) when dyeing yellowish wool.

Clearly, it is important to use wools of good colour for dyeing pale or bright colours. As a guideline, dyeings below 1/12 standard depth require wools that are not discoloured and care to minimise process yellowing.

All wools are susceptible to yellowing in a boiling dyebath and it is necessary to combat this when dyeing pale colours. Protection can be achieved simply by incorporating a proprietary anti-yellowing agent in the dyebath. They are mostly based on hydroxylamine or mild chemical reducing agents, and tend to bleach the wool slightly. For protection against photobleaching effects arising from residual yellowness, an additional proprietary chemical, Lanalbin APB, is available from Clariant. The chemical applied yellows slightly under the action of light at the same rate as the wool photobleaches, so that the net colour is completely stable to light.

REFERENCES

1. J Park. Carpet dyeing in the 1990s. Part 1 – Fibre and yarn dyeing; Part 2 – Batchwise piece dyeing. *Journal of the Society of Dyers and Colourists*, 1993, 109(3), 92–95 and 109(4), 133–137.
2. F W Marriott. Wool-dyeing machinery. In: *Wool Dyeing*, D M Lewis (ed), Society of Dyers and Colourists, Bradford, 1992.
3. I Holme. Dyeing, printing and finishing machinery for the carpet industry. In: *Proceedings of Tifcon '97*, The Textile Institute, Manchester, 1997.
4. J Phipps. An unconventional approach to yarn dyeing. In: *Proceedings of Tifcon '98*, The Textile Institute, Manchester, 1998.
5. G Kolmer. Ecological and economic processing methods – carpet dyeing. *International Textile Bulletin, Dyeing / Printing / Finishing*, 1996, 2, 26.
6. H L Traub, P Hirt and H Herlinger. Dyeing properties of poly(trimethylene terephthalate) fibres. *Melliand Textilberichte*, 1995, 76(9), 702–706 (E179–180).
7. H Peak. Machinery for continuous carpet dyeing. *Review of Progress in Coloration*, 1988, 18, 12–15.
8. U Münstermann. Water and energy saving in continuous carpet dyeing. *Textil Praxis International*, 1994, 49(4), 255-259 (English VIII–XI).
9. A Keller. Developments in carpet dyeing and printing. In: *Proceedings of World Textile Congress on Floorcoverings*, University of Huddersfield, 12–13 July 1995, 82–104.
10. T L Dawson and K Dunkerley. The efficient use of steam and water in the

continuous dyeing and printing of carpets. *Journal of the Society of Dyers and Colourists*, 1978, 19(5), 177–189.

11. T L Dawson. Jet printing. *Review of Progress in Coloration*, 1992, 22, 22–31.
12. M V Locke. The Millitron machine: computer injection dyeing. In: *The Fabric Revolution*, Proceedings of the Annual Conference of The Textile Institute, Manchester, 1981.
13. Textima. BP 1 202 345, 1970.
14. G H Crawshaw and J Ince. Textile floorcoverings, *Textile Progress,* 1977, 9(2), p42.
15. T Naschberger. Chromojet – productivity and flexibility with carpet jet printing. In: *Proceedings of World Textile Congress on Floorcoverings*, University of Huddersfield, 12–13 July 1995, 210–214.
16. A D Fischer. Wool Shield™ – stain resistance on wool and wool-blend carpets. *Australasian Textiles*, 1991, 11(5), 16–21.
17. Scotchgard stain release for wool. *International Carpet Yearbook*, 1995, 16–17.

16 Designing

Carpet weaving has always provided the manufacturer with the outstanding advantage of being able to tempt the customer with the beauty of pattern as well as colour, and the design capabilities of tufting have progressively been extended. Carpet designers are much more than artists. They must be aware of trends in architecture and interior decoration in all the geographic market areas of interest to the company and must observe the public's changing interest in colour and pattern in a more general sense, allowing forecasts to be made of the acceptability or otherwise of carpet designs in future seasons. They need to consider trends in sales for the various qualities of carpet in order to be sensitive to the acceptability of designs in the different price brackets. In these respects the design studio may be considered to be a key component of the marketing department. Designers must also understand the capabilities and limitations of the available technologies and the cost consequences of stocking coloured yarns, the importance of 'quick response', and other commercial aspects of manufacturing. Designing links marketing and production. The purpose of this chapter is to provide the technologist with an insight into the designing process.

Any carpet design has two principal aspects: colour and pattern. Patterns of texture as an alternative or addition to colour are features of some designs. Designs are often produced with a choice of colourways: patterns may be modified significantly or perhaps drastically changed by changing the colours used.

16.1 COLOUR

Colour is a sensation produced when light reflected from a carpet (or other object) impinges on the human eye. Each of three components can affect the sensation: the quality of the light (eg, daylight or incandescent lamp); the dyestuffs/fibre combination; and the unique make-up of the observer (colour vision tends to change with age, and some people are defined as colour blind). Despite the potential for serious misunderstanding arising from these variables, it is possible to arrive at useful descriptions of colour and to measure colour by standardising some of the variables [1].

16.1.1 Light

Visible light is radiation falling within the wavelength range 400–700 nm, and it can be analysed into a spectrum of seven colours: red, orange, yellow, green, blue, indigo and violet – the colours of the rainbow. Red represents the longer wavelengths and violet the shorter ones.

Colourists and dyers like to view colours in indirect sunlight (north daylight in the northern hemisphere) but this is not always available. Colour matching lamps that simulate total daylight are, therefore, commonly used. These are fluorescent lamps that roughly match the spectrum of daylight, but they tend to have peak intensities at certain wavelengths and do not exactly match the smooth spectral curve of daylight. They can, however, be standardised, which daylight cannot.

Incandescent lamps, commonly used for residential lighting, generate radiation that is much richer at the red/yellow end of the spectrum than daylight, so carpets are likely to look redder (less blue) at night than during the day. Figure 16.1 illustrates the quite different spectral characteristics of incandescent (tungsten) and fluorescent lamps and daylight. However, it should be noted that the characteristics of daylight can vary significantly.

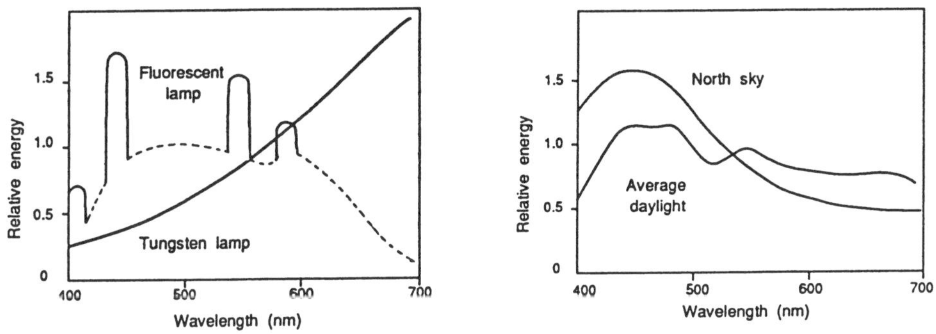

Figure 16.1. Spectral characteristics of incandescent and fluorescent lamps (left), compared to daylight (right).

16.1.2 The Eye

The retina of the human eye has two types of receptor: 'cones', which are highly responsive to colour, are concentrated near the centre of the retina and 'rods', which are more sensitive to the blue/green end of the spectrum and to subdued light, increase in density towards the outer parts of the retina. Colour is best viewed in intense light when the iris of the eye closes to concentrate vision on the cones. Under conditions of low illumination, the iris admits light to the whole retina and the less colour-responsive rods dominate the situation. In the extreme case, all objects appear grey at night. It is useless to evaluate the colour of a carpet when the level of light is low.

The colour response of cones is not uniform across the spectrum. There are three types of cones that are sensitive to the red, green and blue sectors of the spectrum, respectively. Consequently, any colour may be simulated by mixing red, green and blue lights in the appropriate intensities. For this reason, the colour programs of CAD systems are based on triads of colour.

About 8% of males and 0.5% of females have eyes with unbalanced sensitivity and are described as colour blind. Colour-blind consumers are usually aware of their problem and nothing can be done by the designer to alleviate their situation. Clearly, such persons should not be employed as designers or colourists. Persons with proven ability to work with colour do, however, vary slightly in their perception of colour.

16.1.3 Carpet Substrate

Loop- and cut-pile carpets having fibre dyed to the same colour differ markedly in apparent colour, owing to their different surface reflectivity, and designers must take account of this.

Because the trichromatic system of the eye responds selectively to particular regions of the spectrum, it is possible to match under one illuminant two colours with dyes having different spectral response curves which, however, may not match when the illuminant is changed. This phenomenon of change of colour match when the illuminant (or the observer) is changed is known as metamerism. It is the duty of the dyer to avoid dyeings that change markedly in apparent colour as the light changes: it is, for example, unacceptable for an olive green colour (in daylight) to turn brown when viewed under incandescent light. Designers should check the responses of their carpet trials under different illuminants.

16.1.4 Colour Matching

The designer makes use of both additive and subtractive colour matching and must be clear about the difference between the two.

When the primary colours of light – red, green and blue (the colours that stimulate the three types of cone in the eye) – are mixed in the appropriate proportions the result is white light. Such additive colour matching is used by the designer when operating CAD systems. In painting and dyeing (and in the printer of a CAD system), mixing the primaries proportionally and in increasing quantity subtracts colour and results in grey, and finally black.

The primaries in subtractive colour matching are different: yellow, cyan and magenta. These pigments or dyes absorb (subtract) certain wavelengths from white light impinging on the carpet or paper: for example, a carpet will appear red because the light reflected from it is deficient in the green end of the spectrum. The primary colours used in painting, printing or dyeing are similar to the colours obtained by subtracting the additive primaries from white light, as illustrated in Table 16.1.

Figure 16.2 illustrates the difference between additive and subtractive colour matching.

Table 16.1. Primaries in subtractive colour matching.

	Additive primary		Subtractive primary
From white light subtract	red	to give	cyan
From white light subtract	green	to give	magenta
From white light subtract	blue	to give	yellow

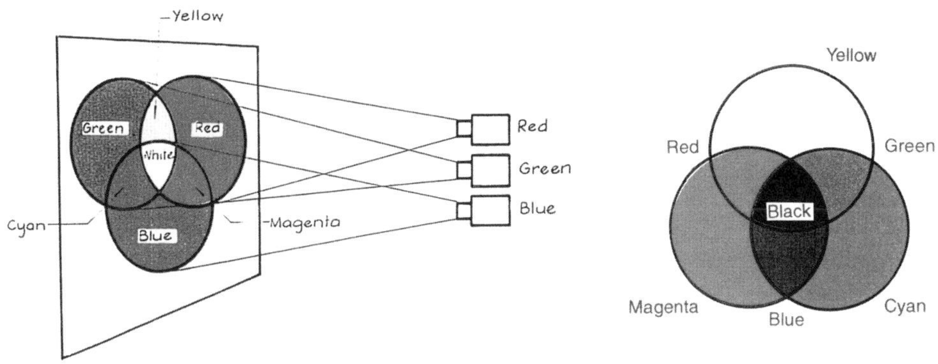

Figure 16.2. Additive (left) and subtractive (right) colour matching.

16.1.5 Colour Description

The artist's general description of colour is more meaningful than the dyer's description.

Colour is taken to mean the clear colour of the spectrum – the rim of the colour wheel – and black is the hub. A shade is a colour mixed with grey (the spokes of the colour wheel), while a tint is a colour mixed with white. A pastel is a tint mixed with grey. (Tints and pastels make up the tyre of the colour wheel.)

Specific descriptions of colour (eg, salmon, rouge, rose pink) are useful but are not standardised, which limits their value to the designer. There is a need for accurate means of communicating colour.

Colour reference systems [2]

The first comprehensive colour reference system, developed in 1905 and still widely used, was the Munsell System which consists of a collection of painted chips. Three numbers based on 'hue', 'chroma' (intensity or saturation) and 'value' (lightness or darkness) define each colour, so that designers in different geographic locations can communicate colour to each other.

The Munsell Colour System uses three mutually perpendicular axes (Figure 16.3) to construct a colour solid containing the whole gamut of colour that can be produced in paint. It consists of a sequence of colours (hue) around a central shaft (chroma) that is neutral in hue (value), ranging from black at the base to white at the apex. The intensity of the colour (chroma) increases from the centre to the outside.

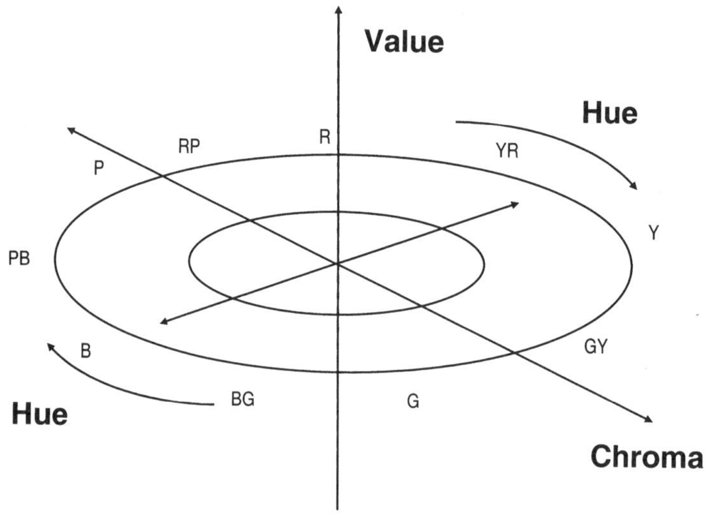

Figure 16.3. The three axes, hue, value and chroma, in the Munsell Color System. The 10 basic hues are arranged around a circular axis from red (R) through yellow (Y), green (G) and Blue (B) to purple (P).

For carpets, the Chromatone™ Carpet Colour Atlas [3] is more directly applicable and is widely used by designers to specify and communicate colours. The full system consists of a series of 24 racks, covering particular hue groups and each carrying dyed wool poms systematically divided into 45 shades, arranged as shown in Figure 16.4. The cut-pile of the poms is normally exposed, but poms may be removed from the rack to reveal the side of the yarn, and are used inverted when selecting colours for loop-pile carpets.

Chromatone™ colours are coded by six-figure numbers (eg, 00 70 50) in which the first two figures represent the hue (colour). The second two figures represent the relative lightness or darkness of the colour, with increasing numbers representing increasing lightness (value). The third pair represents increasing intensity of colour (chroma). Colours may also be interpolated between the physical standards, allowing the theoretical specification of over 472,000 colours.

Some colours do not fit conveniently into this Munsell-type classification and there are seven racks of supplementary colours (eg, Mineral Colours, Chalk Tones).

For architects and specifiers, abbreviated systems, the Chromatone™ "600" System (600 colours) and the "360" System are produced. The Chromatone™

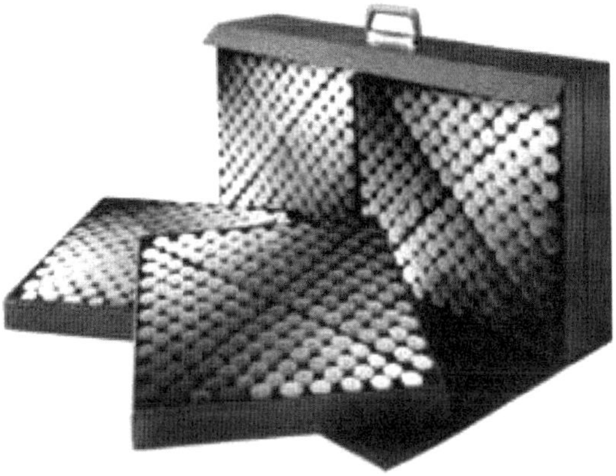

Figure 16.4. Arrangement of a hue group of wool poms in a tray of the Chromatone™ Carpet Colour Atlas.

Compact System is based on fringes of yarn in 24 hue trays, each holding 25 colours. It is favoured for fabrics, but is also widely used in the carpet industry.

The poms in the full system are designed so that they can be condensed to a standard density for colour measurement via the small aperture of a colorimeter. From such measurements, major dyestuff manufacturers can supply their clients with computer generated recipe predictions for nylon, as well as the original wool, against Chromatone notations.

16.1.6 Stock Colours

For reasons of cost, designers do not always have the full gamut of colour at their disposal. A colour library consisting of perhaps 100 colours may be taken as the colours normally stocked by a carpet company. A colour palette is a smaller number of colours (10–20) used to create a particular carpet or collection of carpets, and is sometimes regarded as the number of colours available on the creel of a weaving machine.

16.2 PATTERN

16.2.1 General Scope

A whole variety of styles may be used as the basis of a design. Oriental, traditional European, modernised traditional, floral and chintz, leaf and verdure, geometric, abstract, small repeats, and textured effects are normally found in the designer's library and may be used as starting points for new designs. Designers may also be inspired by outside influences, such as popular themes from films, archaeological discoveries, natural phenomena or space exploration. At the other extreme they may be required to incorporate a company's emblem and/or corporate colours into a carpet.

Figure 16.5 shows how a traditional motif may be used as a trigger for new work and how it may be elaborated or simplified in either a traditional or modern way to create the basis of an original design.

Figure 16.5. Variations on the traditional Paisley motif. (Illustrations provided by Jean Clements, head designer, Wools of New Zealand)

16.2.2 Pattern Repeats

In wall-to-wall carpets, with the exception of many custom contract carpets designed for particular locations, patterns often have fixed dimensions (eg, 36× 36 in. or 0.5×0.5 m) and repeat in both the length and width of the carpet. The repeats can be made to match in different ways. Some possibilities are as follows.

- Block or self match: this type gives a side match for each repeat of the design.
- Drop match: the matching point of the adjacent width is half way along the length of the design (ie, the pattern repeats diagonally).
- Brick match: the matching point of the adjacent length is half way along the width of the repeat.
- Reverse drop match: the matching point is the same as in the drop match but with the figure inverted.
- Half drop match: this method extends (doubles) a design widthwise. The same two figures only appear in line across two widths after two lengthwise repeats of the design.

Figure 16.6 gives examples of design repeats (see also Plates 16 and 27).

16.2.3 Traditional Designing of Woven Carpets

Usually, a design is originated in the form of a drawing or painting in which the designer's ideas can be expressed with as few restrictions as possible. Next, a sketch based on the original is prepared, but with a view to using the concept in a practical carpet. The possibilities and restrictions of the weaving technique are considered at this stage, together with the length of repeat, type of repeat

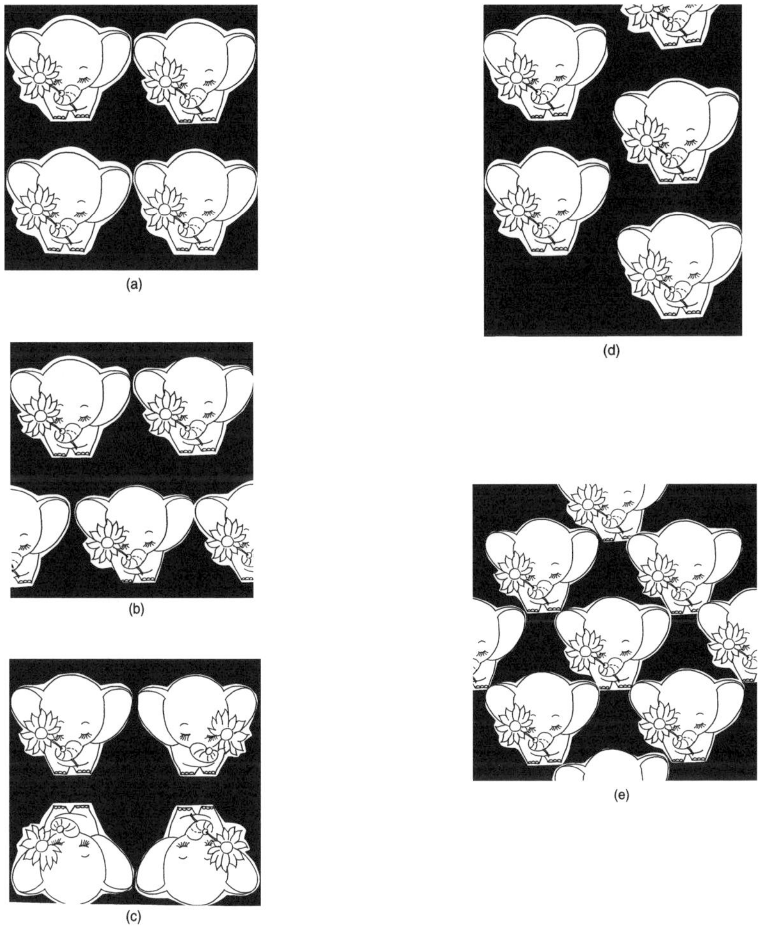

Figure 16.6. Examples of design repeats. (a) Self match; (b) brick match; (c) mirror repeat; (d) half drop match; (e) diamond match. (Courtesy of Jean Clements)

and number of colours permitted, including the possibility of 'planting' extra colours in a creel of yarn nominally of a single colour. The design is then adapted to point paper – a grid representing the tuft density of the carpet to be woven and the dimensions of the repeat. Skill is required to convert the original free-flowing motifs into stepped outlines without sacrificing aesthetics. Once the design repeat has been created in the form of point paper, multiple copies are prepared and pasted together so that the unity of the design can be checked and, for example, any undesirable diagonal emphasis eliminated. A sample of simulated carpet (pegboard sample), prepared by pressing tufts of yarn into a grid, may be produced for the benefit of the customer or the marketing department. The whole process takes several weeks.

The point paper is then used as a basis for cutting jacquard cards for gripper Axminster, Wilton or face-to-face weaving, or in the case of spool-gripper Axminster weaving, for spool setting.

The significance of the point paper and its relation to the jacquard machine can be understood by considering the pile selection mechanism for 8-frame gripper Axminster weaving illustrated in Figure 16.7. Selection of a yarn colour is initiated by the upward pressure of the card cylinder, B, which causes contact between the vertical feelers numbered 2–8. A hole in the jacquard card has allowed feeler No.3 to remain in its lower position. The pusher plate, A, therefore, moves only needle No.3 to the right to push the corresponding lifting peg, D, from its slot in the selector, G, into the path of the ascending lifting bar, F, which has a fixed stroke. The bar thus lifts the yarn carriers, E, to a height at which yarn No.3 is taken by the gripper, C. If yarn No.1 is required, the yarn carrier remains in its default position, so that no feeler (or no hole in the card) is required.

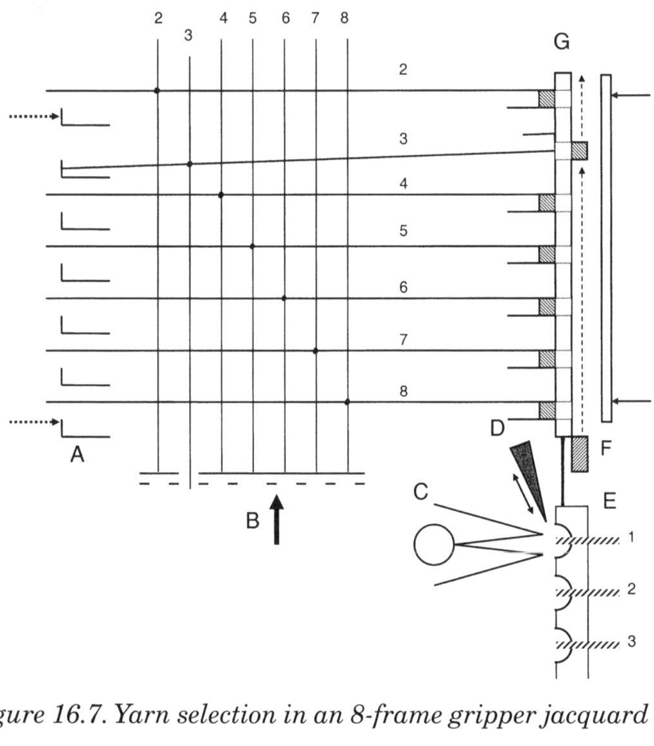

Figure 16.7. Yarn selection in an 8-frame gripper jacquard loom.

Figure 16.8 illustrates the section of a cut jacquard card (top) corresponding to a row of a design taken from the point paper (centre), together with a reference to the eight colours in the frame plan (bottom). Note that frames 6–8 have additional colours planted in them, so that there are 10 colours in all in the design. Note also that when frame No.1 is required no hole is punched. Conventionally, the most frequently used yarn is put in the top frame (No. 1 frame) since the carrier for this frame does not move and therefore less lifting power is required.

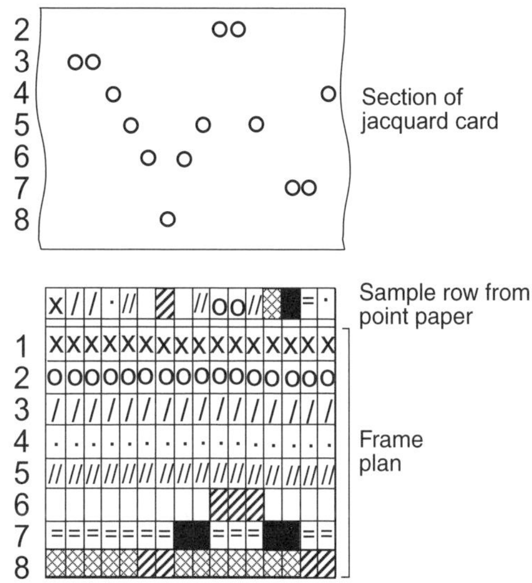

Figure 16.8. The relationship between point paper and the jacquard card.

16.3 COMPUTER AIDED DESIGN

Milliken, USA, first applied CAD facilities in the carpet industry in the 1970s in association with their Millitron computer injection dyeing system (see Section 15.10.4). CAD design systems dedicated to woven carpets were developed in the early 1980s, and quickly became highly sophisticated following the introduction of electronic jacquards for carpet looms and improved computer hardware. Because of the difficulties of visualising the number of designs achievable by changing the creeling sequence in sliding needlebars, there was strong motivation for applying CAD to crossover tufting (see Section 16.5.2) at an early date. Applications of CAD in other areas of tufting quickly followed.

Computer aided design facilities are now considered essential in an effective carpet design studio. Although generalised software for textile designing could be used for carpets, it does not include specialised carpet technology (eg, planting of colours in weaving) or output production data (eg, yarn requirements per colour) and is not easily interfaced with production equipment, so that software dedicated to the particular production system is vital.

Modern systems include scanners for converting sketches or existing carpets into electronic data, electronic sketch pads that can 'paint' in not only colours but background patterns, editing systems that can reduce the number of colours to practical levels and manipulate images, and facilities for viewing multiple repeats, often in perspective. The systems have output facilities that can produce point paper or can be interfaced with jacquard cutting machines or electronic pattern control systems for tufting or weaving machines. Designs can also be printed on paper in styles simulating carpets and generated as photographs or pegboard samples. Information useful to production management is generated by most of the CAD systems in use.

16.3.1 Computing Power

Powerful computers are required for designing carpets, since graphics and full colour both require large amounts of computing power, measured in megabytes (bytes $\times 10^6$) or gigabytes (bytes $\times 10^9$). A byte is a binary number consisting of 8 bits (0 or 1) that can represent the arabic numbers 0–255 (256 numbers in all). Computing power consists essentially of random access memory (RAM) for manipulating data, and hard disk memory for storing data when the machine is switched off. For textile designing, computers having 8–128 Mb of RAM plus 512 Mb to 4 Gb of hard disk memory are required. Computers also incorporate ZIP disks or CD-ROM or CD-R systems that enable designs to be stored remote from the computer or be physically transported to another location. Alternatively, design output may be transmitted electronically.

16.3.2 Input Tools

Mouse – CAD software operates in a windows type environment so that a computer mouse operating on a pad is generally used to load the program, select computer functions and select colours, by clicking the mouse against the appropriate icon or box listed in the window. The mouse could theoretically be used for drawing on the screen, but is excessively sensitive, making it difficult to draw with precision.

Keyboard – A keyboard is necessary for typing information associated with a design (eg, the name and reference number of the design).

Electronic stylus and graphics tablet – An electronic drawing board about 450 mm square plus an electronic stylus are operated by the designer like brush and paper, but the results are seen on the screen and not on the drawing board. Although it does not feel like a brush, some designers prefer using a puck rather than a stylus. A puck, which is an electronic box like a mouse without a ball but having an annular ring with crosswires, rests still when released. It is an ideal tool for editing a design. Using the stylus or puck, a design may be created, starting with a blank screen.

Scanner – An electronic scanner is used to digitise a painted design sketch, item of artwork, general illustration, or even a piece of existing carpet, and transfer the design to the computer screen. If necessary, and this is usually the case with a sketch, the scanner will automatically reduce the number of colours to a practical level. The design displayed on the screen may then progressively be edited into its final form. Alternatively, a motif from the scan may be outlined and selected to be combined with a different background pattern or with other motifs.

16.3.3 Screen

While high quality electron beam systems (VDU monitors) have been favoured for CAD screens, flat LCD screens are becoming increasingly popular because of their display quality and size. However, they are currently relatively expensive in comparison with conventional VDU monitors.

Colours on the screen are created by additive mixing of colours from triads of red, green and blue phosphors. In a VDU monitor an electron gun is provided for each colour and masking systems ensure that each beam impinges only on the correct phosphor. The beams sweep the screen so rapidly that the short glow of each phosphor is renewed with great frequency, providing a steady picture.

Most systems operate by allocating one byte of information to control the intensity of each phosphor, so that $256 \times 256 \times 256 = 16.7$ million possible colours can be displayed, more than can be distinguished by the human eye. However, only one byte is allocated for the number of colours available at each picture cell (pixel). This means that the designer may work with a maximum of 256 colours at a time, but these colours may each be selected from an infinite range.

Screen resolution is an important feature of a CAD system and is defined as the number of pixels (dots) that can be displayed on the whole screen (eg, 1280 $\times$1024 or 1600$\times$1200). The fineness of detail is somewhat different and is defined as the number of pixels per unit length. It is possible for two screens of different sizes to have the same specification for fineness of detail, but the larger screen will have the higher resolution.

16.4 DESIGN OUTPUT

Various possible versions of the completed design may be generated. These fall into two categories: those used to simulate the design for internal purposes or to sell it to customers; and those necessary to facilitate production of the carpet.

16.4.1 Simulated Carpet

On screen

During the final stages of designing, the design on the screen has a grid-like appearance, rather like point paper. The image may be enhanced by means of an 'electronic template' that simulates yarn shadows and yarn hairiness, making it appear very much like a transparent photograph of the carpet surface.

Because a CAD system is not easily transported, it is common to bring potential customers to the design studio to view the image of the carpet on the screen, although the system may be installed at trade fairs to display collections of new designs. Certainly, it is useful within a carpet factory to view designs on the screen as a stimulus to discussions about them.

Photographs

Some CAD systems are fitted with an electron beam system that directly takes Polaroid transparencies or negatives. Such transparencies are useful for lectures about carpet design, and the negatives may be printed via normal photographic services.

However, with the availability of competitively priced computer projection systems it is more common to capture screen images as bit maps and store them to be used as components of a presentation via a computer projection system rather than use slides. Hard copy is more commonly used than photographic prints.

Hard copy

Using high quality ink-jet or thermal printers, images of the design with electronic template can be printed on paper to give very realistic simulations of carpets in miniature. The printed image is built up of dots of (usually) cyan, magenta, yellow and black pigments. For the best quality prints, high definition printers are needed. A printer generating 120 dots per inch (dpi) is suitable only for point paper: for simulated carpets, printers having fineness of detail in the range 280–720 dpi are required. For best results the fineness should be as high as can be afforded, although the finer the detail the greater the number of dots to be printed and therefore the longer the time to print.

Hard copy of good quality is usually about 200 mm wide, and is generated quite slowly (eg, an hour for a single repeat). Faster printing may be achieved using printers having multiple jets.

Note that hard copy is generated by subtractive colour matching, whereas the designs on the screen are based on additive colour matching, so the screen and the hard copy can never appear identical. However, design business in an organisation may become centred around the hard copy, in which case the decision must be made to operate around the colour of the hard copy and regard the colour on the screen simply as a good guide.

Pegboard sample

Pegboard samples of simulated carpet, traditionally made by manually plugging tufts of pile yarn into a grid, can be produced automatically on the basis of digitised designs.

Hardaker's Automatic Carpet Sampler ACS/16 selects tufts from a creel containing up to 16 colours. Samples up to 500×500 mm are produced at a rate of 75 tuft insertions/min on the basis of design data read directly from the CAD facility – there is no need for a conversion program.

The Kibby Pegboard Carpet Sampling Machine is also widely used. The output of a CAD system is downloaded into the hard disk of the machine. The PC of the Kibby machine drives pneumatic and servo-mechanisms that generate the sample. Yarns from the colour palette are mounted on a creel and are led in the specified sequence to their positions in a colour bar. Design data move the bar so that the required colour is positioned for selection by a gripper (Figure 16.9). A tuft length of yarn is drawn out, cut off and positioned by a guide midway across a hole in a perforated plate located behind the gripper mechanism. The tuft is then pushed into the hole. A selection of perforated plates is available with grids corresponding to a variety of carpet constructions. The design, in the form of coloured yarn, is built up one tuft at a time in this way. On completion, the plate with its tufts is removed and a backing fabric coated with hot-melt adhesive is laminated to the base of the tufts in the Modra HMP Backing Processor. The material is withdrawn from the plate and the pile is shorn level to provide a sample that closely resembles carpet.

Tufts are inserted at 110/min, and samples as fine as 400×560 tufts/m^2 (representing densely woven carpet from a 16-pitch loom) can be produced. Around five samples 300×300 mm of typical Axminster constructions are produced in an 8 hour shift.

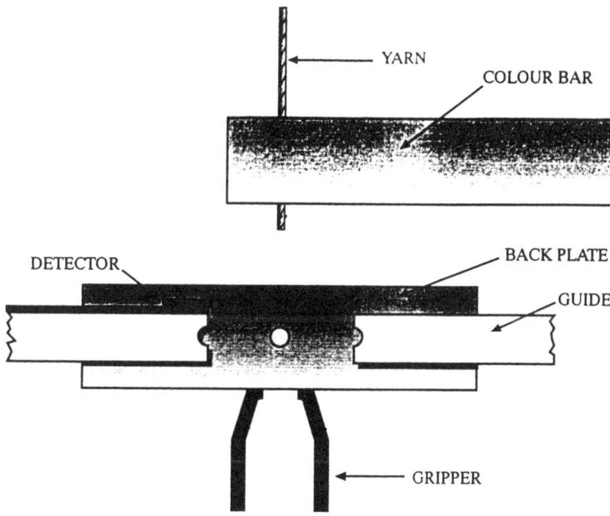

Figure 16.9. Mechanism of the Kibby Pegboard Carpet Sampling Machine. (Courtesy of Wools of New Zealand)

The Mk2 PC Kibby Carpet Sample and Design Machine can provide simulated tufted carpets, as well as the original Axminster styles. A pegboard for tufted carpets has its holes drilled with an element of randomness so as to simulate tufted textures more accurately and the mechanism can produce varied pile heights. Using such equipment, sampling for cut-pile, loop-pile, crossover, scroll and Colortec (see Section 7.4) carpets can be carried out. Separate software, coupled with special pegboards, allows samples simulating Tapistron carpets (see Section 7.4) to be produced.

An alternative to a pegboard sampling machine is the Patternjet of Fine Jets Ltd that can print carpet samples from CAD data (see Section 15.10.4).

Room set

A room may be photographed with a digital camera and the result fed directly into a CAD system. Alternatively, an ordinary photographic print may be scanned and digitised. Next, using special software, the carpet design under consideration is introduced in perspective over the floor area outlined from the digitised photograph displayed on the screen. At this stage, the colourway of the carpet and the colours of the furniture and furnishings can be changed at will.

Room set technology is a powerful tool for interior design and carpet selection.

16.4.2 Production Information

Depending on the technology available in the mill, the CAD system can provide alternative forms of output that can be used, directly or indirectly, to drive carpet production machinery.

Point paper
One form of hard copy that can be generated by a CAD system is point paper, with a numbered grid. Because of the limited dimensions of typical printers, windows of the point paper are usually printed and must be pasted together to provide the complete design.

The point paper is used as a basis for traditional manual cutting of jacquard cards or, in the case of spool-gripper Axminster, is applied at the spool setting table. Point paper is accompanied by other manufacturing information, such as a spool setting table plan.

Jacquard cards
The output of CAD systems in the form of floppy disk may be used to drive electronically controlled card cutting equipment. For example, Hardaker's CCS/2 Card Stamper produces two cards/min in Axminster, Wilton or Vincenzi formats, directly from design data.

Electronic jacquard
Apart from spool-gripper machinery, all types of carpet weaving machinery can be purchased with electronic jacquards (see Section 9.1.4) or can be converted for operation by electronic jacquard. Interface programs that incorporate the specification of the particular loom are available for linking the output of a CAD system to the electronic jacquard. Alternatively, a central unit storing the digitised designs may be networked to a whole shed of looms. Digitised designs may be transmitted electronically to a manufacturing location.

16.5 COMPUTER DESIGNING

16.5.1 Designing Woven Carpets

First, the designer must prepare a technical specification for the carpet and a yarn palette covering the colours to be used.

The technical specification includes the pitch of the loom, the rows of tufts/dm (or inch), the dimensions of the design, whether repeating or not (eg, a rug), and the form of the repeat. This essential information is fed into the design program and the design to be created is identified by name (or number).

The colour palette, consisting typically of 16 colours, is usually selected from the library of yarns available within the company. These may include effect yarns. At some stage, of course, the yarn library itself must be created and this may be done either by matching the colours of standard yarn stocks on the screen or, in the case of a new operation, selecting physical standards for matching (eg, from the Chromatone™ Colour Atlas) or simply mixing colours on the screen to the taste of the designer.

When ready for preparing the design itself, the screen shows a design area with grid or latent grid, the colour palette and the control panel.

The creative step is initiated either by scanning a painted sketch (or other material as described in Section 16.3.2) or by drawing on the screen using the graphics pad and stylus or puck. Rapid sketching is facilitated by aspects of the computer program that generate horizontal, vertical or sloping lines, circles, ellipses or curves through three points. Individual motifs, once created or selected from the scan, may be re-oriented or re-dimensioned at will and may be filed for re-use in another area of the design. In the 'brush' mode, a colour is selected from the palette and the individual areas of the design are coloured in. Work may be overbrushed or areas may be protected from overbrushing. Instead of plain colour, a background of a small repeating design may be brushed in.

Detailed editing, involving a careful consideration of the stepped effect resulting from the grid, is carried out with a window of the design filling the whole design area. A creel planting program may be introduced around this stage. As the design approaches completion, a montage of several repeats of the design is viewed to check the overall effect, and final editing is carried out if necessary. Some programs allow the carpet design to be viewed in the context of a room set.

Figure 16.10 is a print-out of one of the screens used during the designing of woven carpet by the APSO CAD Vantage system. As well as illustrating many of the facilities available to the designer, the figure indicates how the overall design is developed from a number of pre-designed components and draws attention to some of the essential data that are available from the program. On the top right is a palette of 25 colours. Below this is the control panel. The column on the extreme right contains the individual motifs from which the design is composed. Note the area of design to the top left which is composed of small repeats: this has been 'brushed' in quickly using one of the facilities of the software. The design repeat in terms of numbers of tufts and dimensions is shown bottom centre. Adjacent to this the numbers show that the 1 in. bold squares of the grid represent a 7-pitch loom weaving 8 rows/in. To the right of this information is the 'zoom factor', followed by the grid reference of the cursor.

16.5.2 Crossover Tufted Carpets

Because of the difficulty of visualising the numerous designs achievable by changing the creeling sequence of colours, there was strong motivation for applying CAD to crossover tufting (see Section 7.2) at an early date.

Figure 16.11 is a print-out of the key screen in designing a double needlebar crossover tufted carpet. The grid in the left of the design area is for the needlebar movement: the designer has blocked in the required positions of the bars for the various tuft rows in the carpet. The two short bars towards the upper left are for the back and front needlebars, respectively: the designer has selected

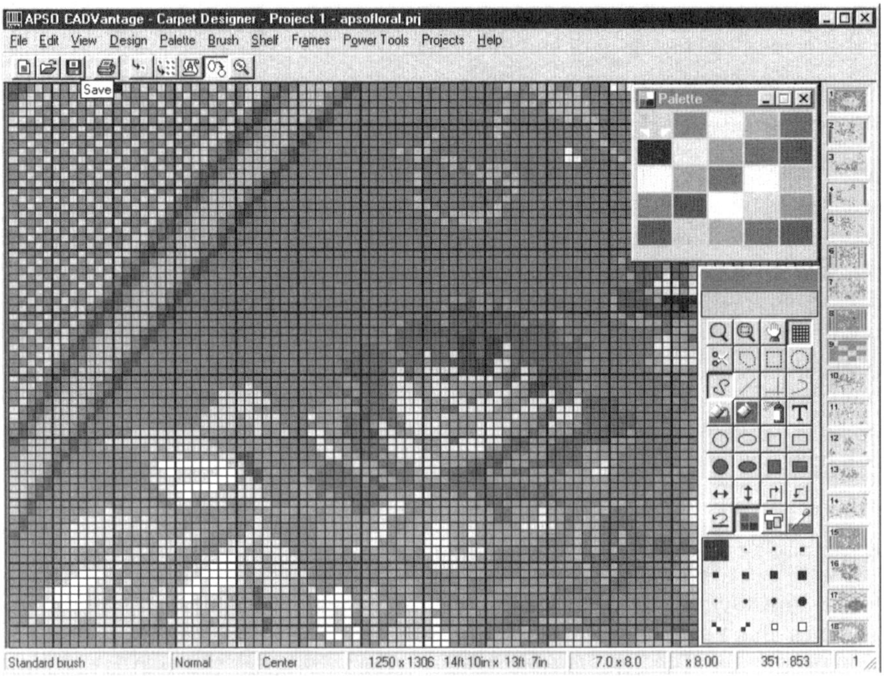

Figure 16.10. Computer screen showing a window of a design for a woven carpet being produced on the APSO CAD Vantage system. (Courtesy of APSO Ltd)

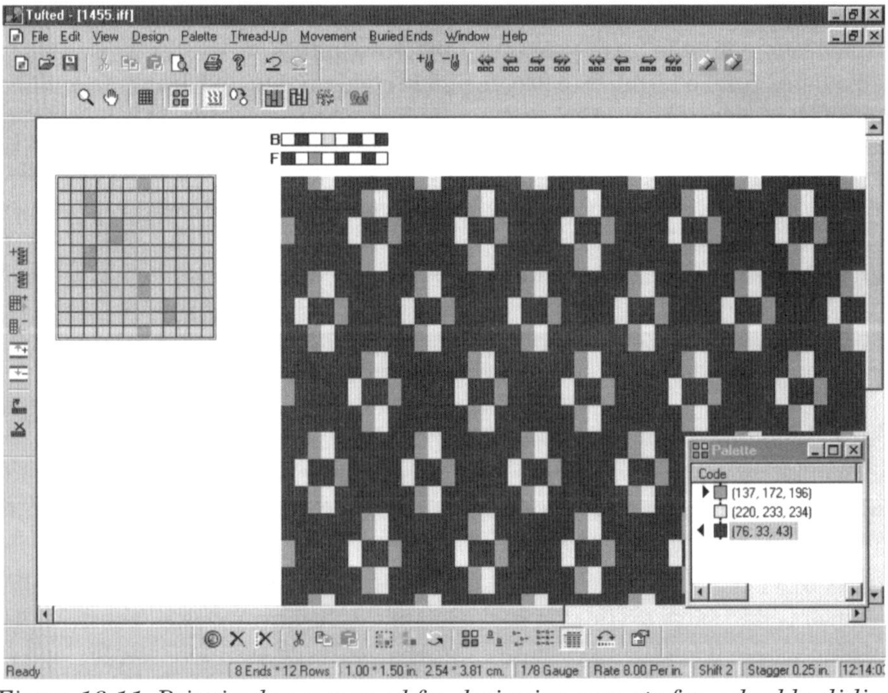

Figure 16.11. Principal screen used for designing carpets for a double sliding needlebar tufting machine, showing needlebar movement, creeling sequence, colour palette and resulting design. (Courtesy of APSO Ltd)

three colours from the palette in the lower right and 'creeled' them on the bars as shown. The large area shows the resulting design. The carpet construction is also recorded at the bottom of the screen.

Figure 16.12 is a print-out of the design in the form of a simulated carpet.

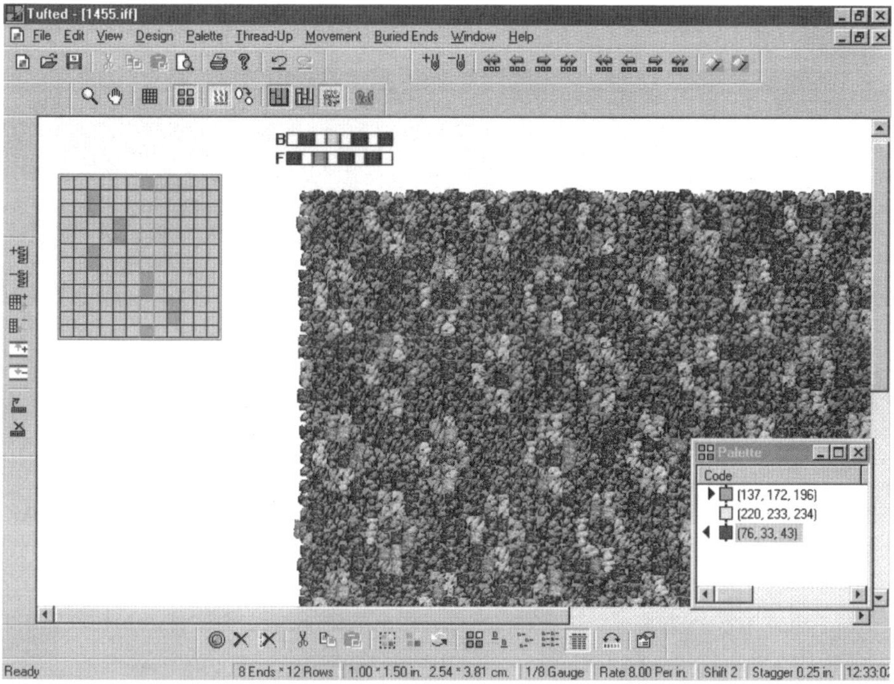

Figure 16.12. Output of Figure 16.11 in the form of simulated carpet. (Courtesy of APSO Ltd)

16.5.3 Buried-Pile Tufted Carpets

When a yarn tensioning patterning system is used to change the loop length in a loop-pile carpet, the high pile of one colour tends to spread out and cover adjacent low pile of a different colour to give 'buried-pile' or 'hidden graphics' patterns (see Chapter 7 for details of the mechanisms). There are two broad types of buried-pile patterning systems. Those based on five, six or seven rollers (often split into sections for convenience of threading) each carrying a large number of yarns, such as Tuftco's Split Rainbow machine or Cobble's Yarn Feed Variator, can only provide simple effects and are generally used to produce small pattern repeats. The other broad type can generate wider repeats based typically on 120 or 126 ends of yarn (or even the full width of the machine), allowing free-flowing curves to be incorporated in the design (scroll patterning).

Figures 16.13 and 16.14 show how a simple buried-pile design is produced, using the same software as in Section 16.5.2. Such a design could be produced on the restricted types of yarn tensioning patterning systems (eg, Split Rainbow or Yarn Feed Variator). Figure 16.13 shows how two colours only from the palette

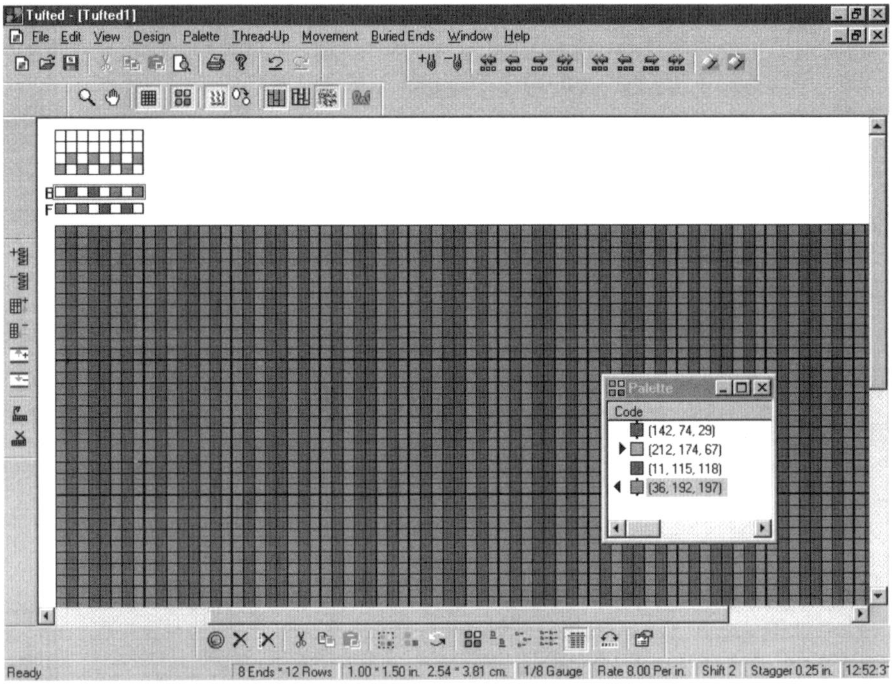

Figure 16.13. Design screen set up for buried-pile carpet in two colours. (Courtesy of APSO Ltd)

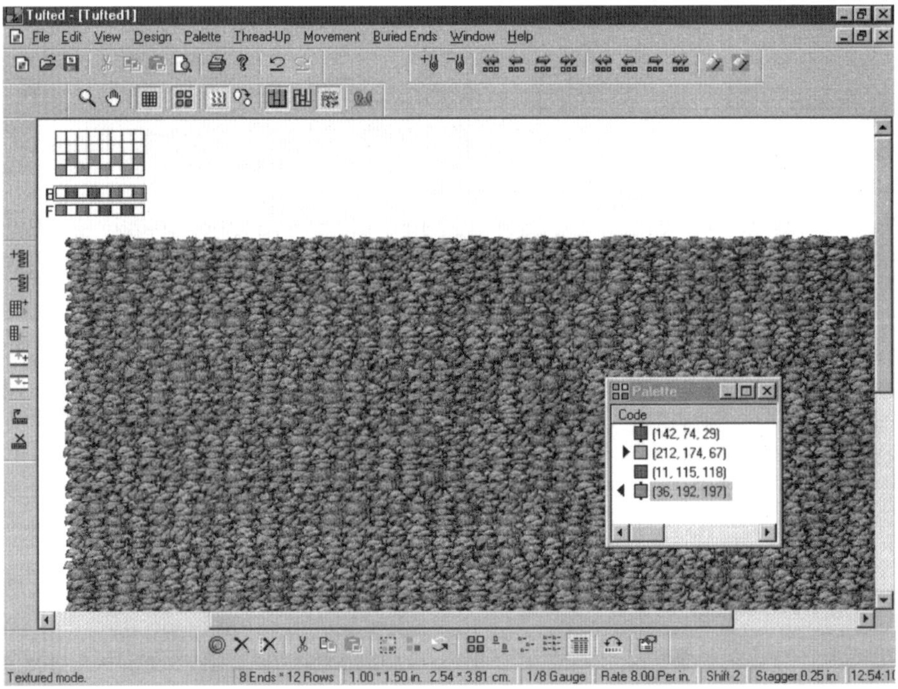

Figure 16.14. Simulated carpet from Figure 16.13. (Courtesy of APSO Ltd)

(lower right window) of four are creeled (large area) and this is how the carpet would appear if all the loops were in high pile. There is no needlebar movement. There is, however, the possibility of alternate ends (in this case on the back or front bars) being low or high pile, and this possibility is next selected. The designer has then instructed the patterning system to alternate the sequence of high and low loops every six stitches to give the very simple design shown in Figure 16.14 (in the form of a simulated carpet produced by the use of an electronic template). Note that the buried ends of yarn are just visible, typical of the effect in a real carpet.

Designing for a cut/loop carpet is carried out in an identical manner: the patterns will be identical, but the appearance of the carpets will be different because of the contrasts of texture as well as colour.

16.5.4 Combination Tufting

The software outlined in Sections 16.5.2 and 16.5.3 can clearly be used to design for machines that can produce buried-pile (or cut/loop) carpets augmented by sliding needlebar effects. A reading of Section 7.3 will reveal that combination machines are often more than simple combinations of tension and crossover tufting: there may be different types of patterning systems operating on the two needlebars, and the needlebars may be in different gauges. Therefore, designing for some combination tufting systems, but by no means all, may be accomplished by introducing the appropriate limitations on relatively simple software.

For some machines, special software and interfaces are necessary. The NedGraphics Windows 95 Tuft program, for example, includes interfaces for many of the speciality tufting machines offered by CMC, Tuftco, Cobble, Chattanooga Tufters and Yamaguchi.

Comprehensive software has been written by APSO for designing carpets to be produced on a double sliding needlebar machine equipped for buried-pile and cut/loop tufting controlled on both bars by an FRS patterning system (see Section 7.1.5). Restrictions can be applied to this software to make it suitable for some other tufting machines.

Figures 16.15 and 16.16 represent two of the screeens used in the design of a four-colour carpet tufted from two colours of yarn using the buried-pile and cut/loop facilities but not the sliding needlebars. The additional colours are, in fact, tonal variants of the two basic colours: the darker tone is displayed by cut pile and the lighter tone by loop pile.

Figure 16.15 includes a small window of the design (that can, if required, have a width repeat equal to the width of the carpet) and the Apply Effects window that illustrates how the four colours in the carpet (Palette Items) are obtained from the Weave Data.

> Colour 1 (light green in the colour version of the illustration) is obtained with both yarns in loop pile (circle code) and with the green in high pile and the other yarn (sand colour in reality) low (small dot code).

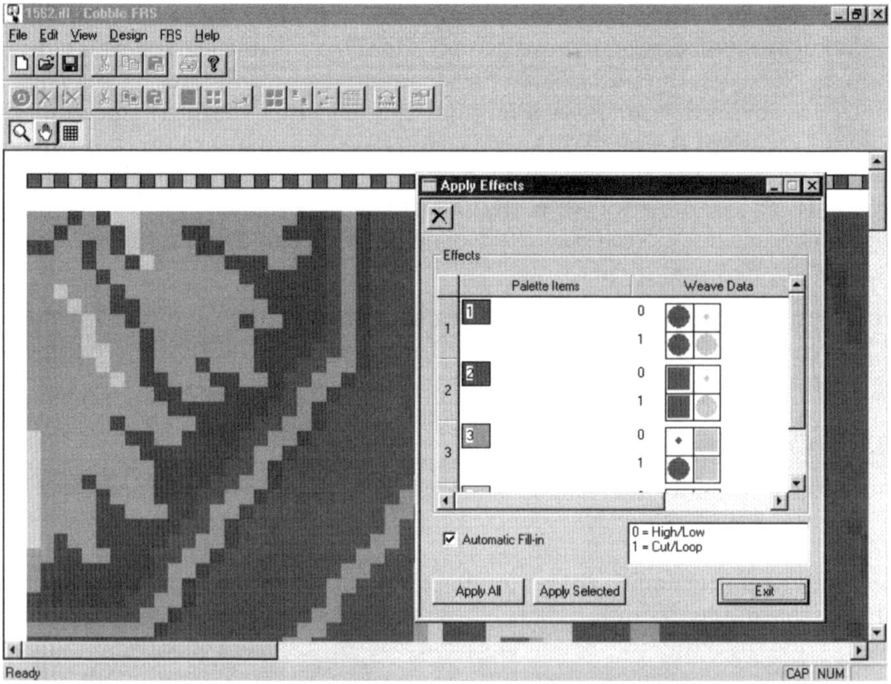

Figure 16.15. A screen in the design sequence for Full Repeat Scroll with window showing how three colours are achieved from only two yarns. (Courtesy of APSO Ltd)

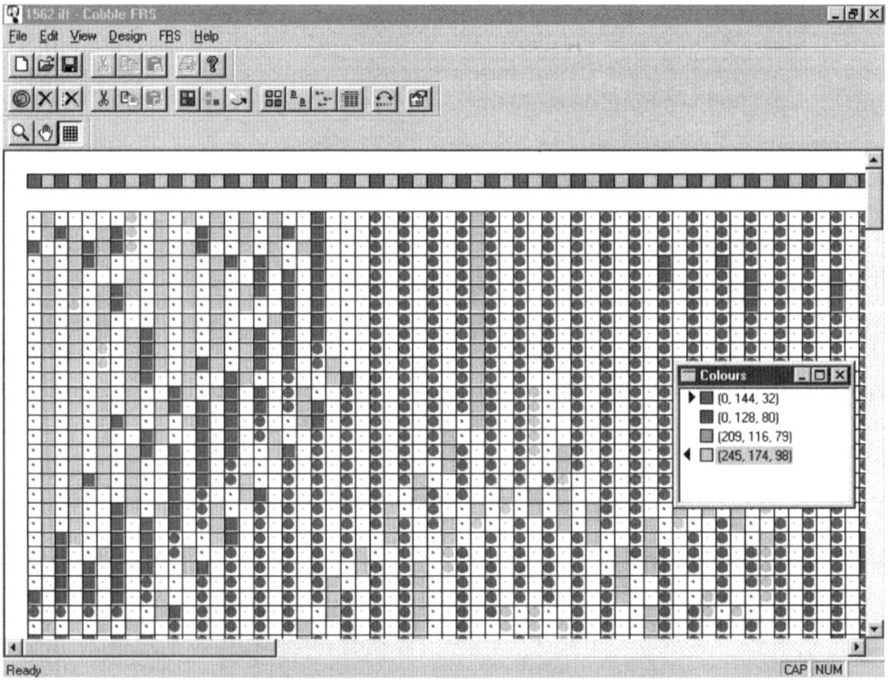

Figure 16.16. Symbol display for the design window shown in Figure 16.15. (Courtesy of APSO Ltd)

Colour 2 (dark green) is obtained with the green yarn in cut pile (square code) and the sand yarn in loop pile; the green being high and the sand low.

Colour 3 (dark sand) is obtained with the sand yarn in high cut pile and the green in low loop pile.

Colour 4 (light sand) is obtained with the sand yarn in high loop pile and the green yarn in low loop pile.

Figure 16.16 is a symbol display of the area of the design considered in Figure 16.15. The high yarns may be green or sand (greys within the limits of a black and white illustration) and may be dark cut (squares) or light loops (circles). Low loops (dots) represent buried pile.

Note that the two yarns are creeled A,B and the tufting is linear (no use of sliding needlebars). In other designs, the sliding needlebars may be used to create major design features, but they are often used for more subtle purposes, such as to fill in large areas of colour. In one area of the symbol display of Figure 16.16, three adjacent lines of tufts have low loops and probably would not be adequately covered in a real carpet. However, if the high colour in such circumstances (indeed, in the more extreme case of a large area) were shifted laterally two gauge widths and back on alternate rows, better cover of the low loops would be achieved.

16.5.5 Screen Printing

Designing for screen printing (see Section 15.10) is similar to designing for woven carpets, in that the starting point may be a painted sketch to be scanned into the computer, or the design may be created on the screen, and there are two to twelve colours. The principal difference is that the grid may be finer, down to the mesh of the screen to be used (50–200). Curves may, therefore, be smoother. Of course, when the intention is to simulate a woven carpet, a grid appropriate to a woven carpet is selected and a stepped design results.

The unique feature of screen printing software is the facility to separate colours and output them in a form that can be used to engrave screens. Account must be taken of the printing technology in order to avoid a gap of white fibre between the colours, or excessive overprinting. Designers use their experience of the interaction of carpet texture, screen mesh (raster), paste formulation, depth of colour, etc, to specify an overlap or cutback between colours, and the software takes account of this when calculating the areas to be engraved on the screen.

Figure 16.17 shows a four-colour print design with the separation of colour 3. The overall design is shown in the top left window of the screen and the areas for colour 3 are in the larger window. The designer has specified a cutback of 0.7 mm for colour 1, a cutback of 0.4 mm for colour 2, and an overlap of 0.7 mm for colour 4. Because separation of colour 3 is in progress, the selected overlap/cutback for this colour has been removed from the box.

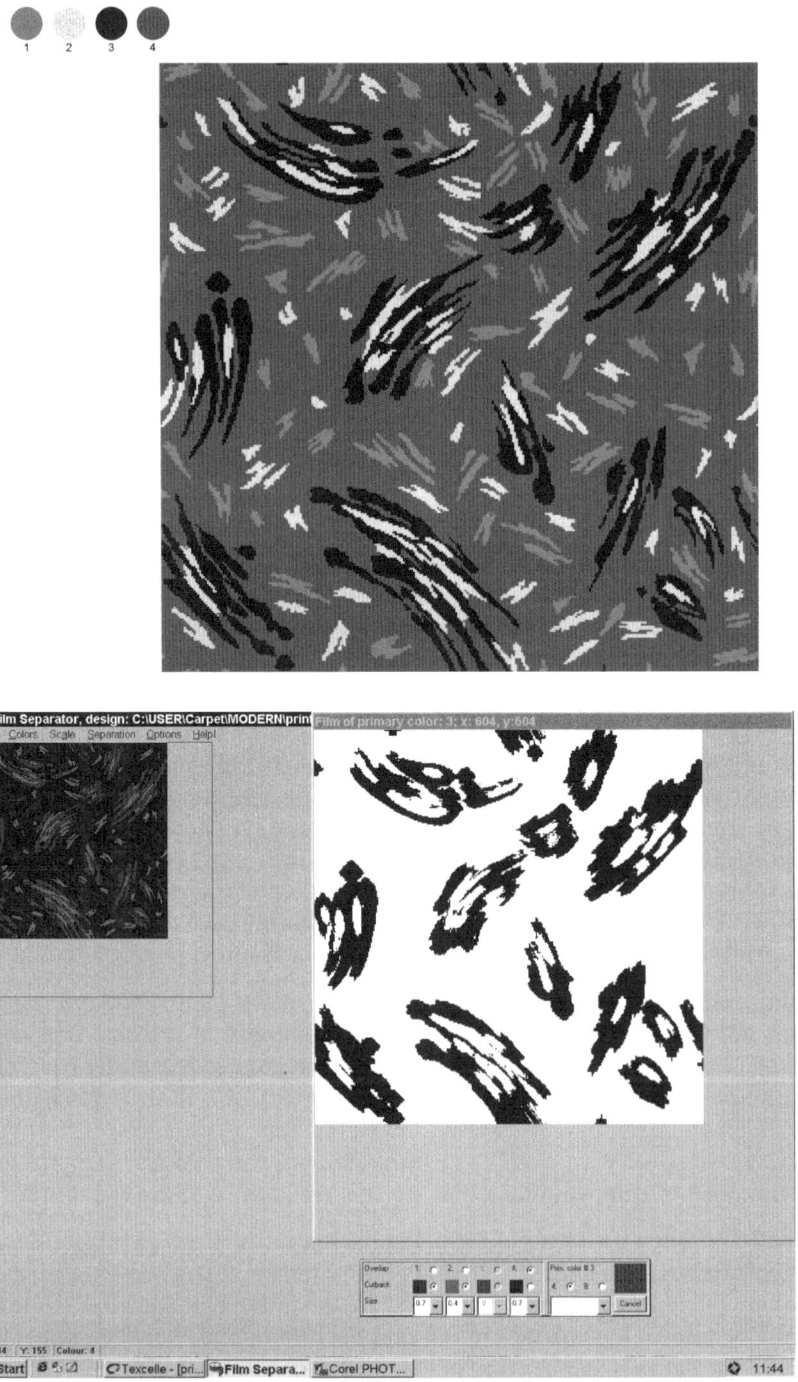

Figure 16.17 Separation of colour 3 from the four-colour print design above, with an indication of the facility to specify overlap or undercut between colours. (Courtesy of NedGraphics Print Ltd)

Each colour is separated in turn and the separations are printed to full scale on acetate film with four register marks outside the design area. These films provide the starting point for the photographic process for engraving the screens.

REFERENCES

1. Joy Turner Luke. *The Munsell Color System: A Language for Color*, Fairchild Publications, New York, 1996.
2. J Ellison. Yarn colour systems for the carpet industry. In: *Proceedings of World Textile Congress on Floorcoverings*, University of Huddersfield, 12–13 July 1995, 235–251.
3. *The Chromatone™ Carpet Colour Atlas*. RGS Chromatone Ltd, Aberdour, Scotland.

17 Finishing of Woven Carpets

Although many finishing processes are common to both tufted and woven carpets, there are others that are not, and the ones that are common to both sectors of the industry often have somewhat different objectives and may be carried out in different sequences.

Finishing is a quite accurate description of the late-stage processing of woven carpets, whereas the backcoating step converts the products of the tufting machine from limp cloths having a tenuous hold on the tufts to a firm stable carpet, so that the finishing of tufted carpets may be regarded as an integral part of carpet manufacture. It is therefore convenient to consider the finishing of these two types of carpet in different chapters.

Chemical finishing is considered in Chapter 15, except for washing which is unique to rugs.

The usual steps in the finishing of woven carpets are: inspection and mending, shearing and backcoating. Optionally, carpets may be given an anti-shading treatment.

Shearing and backcoating are sometimes carried out in-line. The exact finishing routine depends on the type of carpet and the policy of the mill. Some woven carpets do not require backcoating, for example.

17.1 PREPARATION FOR INSPECTION

The irregular surface and presence of debris on a loom-state carpet preclude reliable detection of faults, so carpets must be prepared for the inspection process.

On entering the finishing department, the carpets are passed through measuring rollers and checked against manufacturing details (obvious faults, such as wrong colour, need to be notified quickly to the weaving department). Debris is removed from the back and face of the carpet by brushing, and the pile may be given a light surface shear. More detailed information on shearing is given in Section 17.5. A process that is unique to the lower carpet woven on single-rapier (or single-shuttle) face-to-face looms is back-cleaning – the removal of loose pile yarns (the yarns that would lie dead in the backing in other structures) in a machine that scrapes the back.

17.2 MENDING (PICKING)

First, the carpets are inspected face down as they are transported intermittently, under the control of the operative, over highly illuminated inspection tables. Short lengths of chain, stuffer or weft yarns are sewn in as required at points where breaks are detected.

A second inspection to detect faults in the pile is carried out with the carpet pile-up using surface or back lighting as is considered appropriate for the carpet quality. Surface touch is also used as a means of detecting missing tufts or gross yarn defects. There may be tufts of the wrong colour, overtwisted yarn, soiled places or unsightly knots. Incorrect or faulty tufts are removed and fresh tufts of the correct colour are sewn in and the ends trimmed off. There may be an intermediate brushing and shearing operation to facilitate the next step.

17.3 ANTI-SHADING

Cut-pile carpets, in particular, are susceptible to developing a patchy appearance through the pile leaning in different directions according to the pattern of traffic or crushing of the pile during transport, handling or fitting. Shading or 'watermarking' is an optical effect resulting from the changes in reflection obtained by viewing the sides or ends of the tufts, according to the pile lay [1]. All pile fibres are susceptible; lustrous piles show the effects more than matt yarns; and shading is more apparent in plain carpets than patterned ones. Once shading has occurred, the original appearance of the carpet cannot be restored.

If the pile of the carpet is firmly laid in the manufacturing direction before it leaves the mill, its direction is less likely to be reversed during handling or wear. The WRONZ Trutrac machine augments the natural pile orientation and provides a controlled setting of the pile in this condition. Force is applied under accurately controlled conditions to ensure uniformity of pile lay across and along the carpet. The degree of resistance to shading can be adjusted to suit the carpet style and the particular end use. The Trutrac machine can be conveniently located in the carpet finishing line prior to back-coating, but can be by-passed if required. A typical plant configuration is shown in Figure 17.1. Alternatively, the Trutrac machine can be used off-line.

17.4 BACKCOATING

Backcoating removes all creases and folds from the carpet, improves tuft bind, contributes towards good dimensional stability and prevents broadloom carpets from fraying when they are cut for installation. In loop-pile carpets, tuft bind must be particularly strong to prevent laddering. Most squares are backcoated, the target being an invisible, flexible film that is durable and unaffected by climatic conditions. Carpets having a firm backing structure, such as particularly dense multiframe face-to-face, Wilton or Kardax-weave Axminster squares, may not require an application of finish but are brushed and steamed to relax and burst the pile and flatten the backing structure, and are then dried.

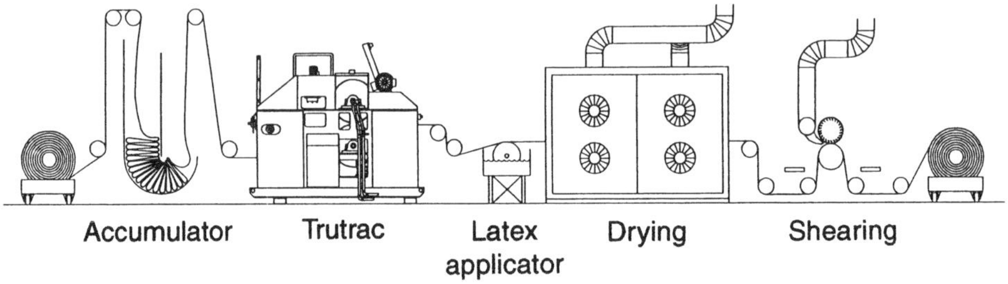

Figure 17.1. Location of the WRONZ Developments Trutrac machine in the finishing sequence for woven carpets.

Steam may be jetted on the face of the carpet or applied from the back. A favoured procedure that gives uniform results is to steam both sides in a simple vertical steamer (Figure 17.2). The Trutrac machine involves the use of steam, rendering a separate steamer unnecessary.

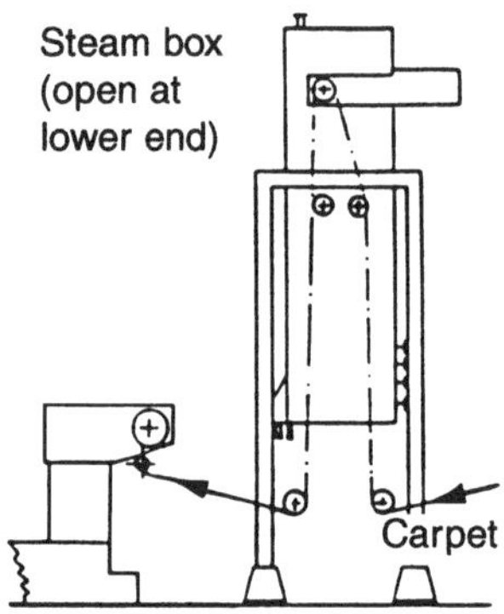

Figure 17.2. Vertical steamer for relaxing and bulking the pile. (Courtesy of Sellers Engineers Ltd)

Plain Wilton and wall-to-wall Axminster carpets are backcoated, usually with self-cross-linking (carboxylated) styrene-butadiene (X-SBR) latex (see Section 18.4.3). Application of the latex formulation is commonly by lick roller (Figure 17.3). Primary control of the level of application is by the viscosity of the formulation. Once a formulation has been established, each batch should be checked in a viscometer. Control may also be exerted in the latex applicator. Increased weight of latex compound, coupled with improved penetration into the backing, is obtained by increasing the speed of the lick roller. Increasing the pressure of the doctor blade (scraper blade) also improves penetration but decreases the level of application. Moving the pressure rollers to increase the arc of contact improves penetration, with little effect on the quantity applied.

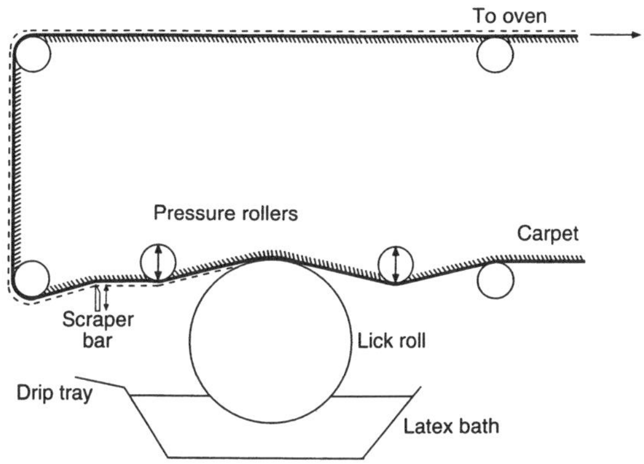

Figure 17.3. Lick roller system for applying latex to the back of a carpet.

In the tufted carpet industry, where backcoating/laminating is a key feature of the manufacturing process, there is a move away from lick rollers to direct application of latex in the form of unstable foam applied at a measured rate (see Section 18.4.2). Some weavers have also adopted this technique.

For residential carpets, X-SBR latex is applied at a level of 140–240 g/m^2 solids. No filler is necessary, but if it is considered desirable to increase the total weight of the carpet, calcium carbonate powder, up to twice the weight of rubber, may be incorporated in the formulation.

The backcoating compound used for contract carpets often depends on the necessity or otherwise to meet a flammability specification. For example, 300 g/m^2 X-SBR with 750 g/m^2 aluminium trihydrate may be applied. For stringent flammability specifications, EVA copolymers may be used as an alternative to X-SBR.

Prior to the use of synthetic latex backcoating compounds, starches were applied to woven carpets. This practice may persist in some long-established product lines.

After applying the latex, the aqueous component of the formulation must be dried off and the latex cross-linked by heating at 150°C. In a factory with a moderate throughput, drying and curing may be accomplished together by passing the carpet over a hot drum (Figure 17.4). For high throughput, a heavy-duty stenter oven designed for carpet applications, perhaps in-line with a heated drum, is required.

17.5 BRUSHING AND SHEARING

Steaming (or the steam generated by drying the latex) will have developed pile bulk in preparation for the final brushing and shearing [2, 3] of the carpet. For wool, additional in-line steaming is sometimes considered desirable to condition the fibres for brushing and shearing.

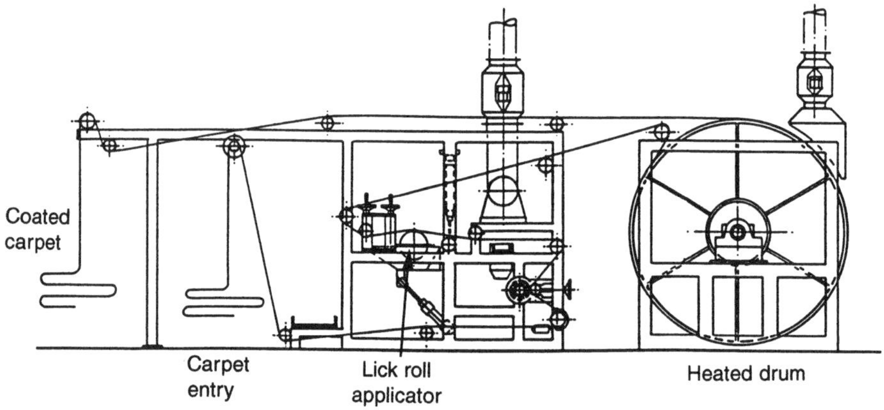

Figure 17.4. Plant for backcoating and drying woven carpets. (Courtesy of Sellers Engineers Ltd)

A rotary face brush with synthetic fibre bristles is an essential component of a shearing line. The carpet is passed over a bed to ensure consistency of brushing. The brush removes surface fibre and raises fibres into the shearing position. Optionally, a pile lifter (a wire brush) may be used to compensate for any crushed pile.

Other in-line features of a modern shearing machine are a metal detector, to avoid damaging the shearing head, and a seam passage control system that pivots the bed away briefly when a seam passes through so that a minimal length of carpet remains unshorn.

The basic components of the shearing head are shown in Figure 17.5. For woven carpets, the bed is usually rotary, as shown in the diagram, but may be a solid bar with a curved edge. As a result of the opening of the pile over the bed, tufts are presented to the stationary cutting blade individually. The sharp edge of this blade is the essential cutting element and the helical blades (there are 24–40 mounted on a cylinder), which have a file-cut edge, gather the surface fibres and transport them to the stationary blade for shearing (A). The helical blades

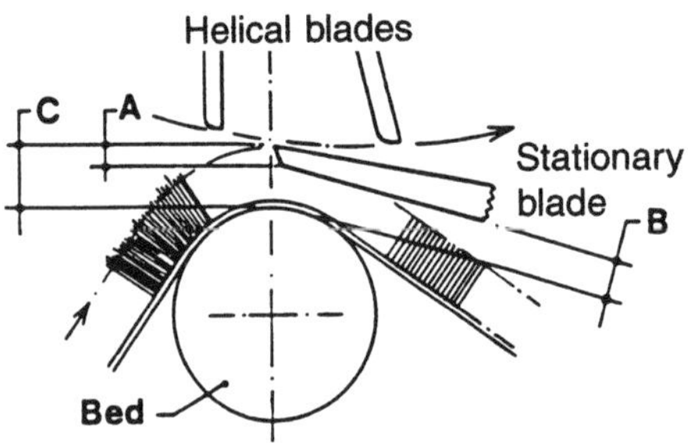

Figure 17.5 The key elements of a carpet shearing machine. (Courtesy of Sellers Engineers Ltd)

are often (incorrectly) called spirals. Fibre dust is removed by suction and collected in bags. The distance C governs the pile height of the finished carpet. The stationary blade is tapered to allow the pile to pass through gap B.

Plain cut-pile carpets are given more attention at shearing than patterned carpets, since the quality of their surface contributes strongly to their attraction. Best results are achieved by shearing lightly several times. To facilitate repetitive shearing, multihead machines are manufactured. Figure 17.6 shows the layout of a double-head shearing machine.

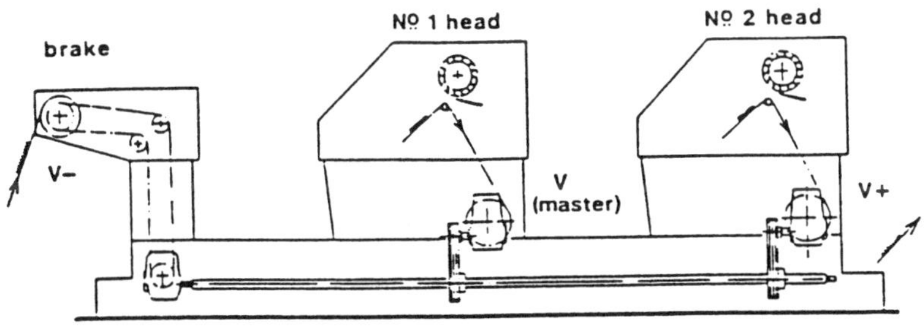

Figure 17.6. Layout of a double-head carpet shearing machine, emphasising carpet tension control based on the speed of the master head. (Courtesy of Sellers Engineers Ltd)

17.6 FINAL INSPECTION/WAREHOUSING

The purpose of the final inspection is to select first-grade carpet and to classify any faulty material as imperfect. The lengths of each are measured.

Squares and other shapes of rug are cut to length and the edges are overlocked, using specialised sewing machines. Usually, the ends are fringed to simulate hand-knotted carpets, but may be hemmed. Broadloom carpet is rolled, pile inside, and transferred to horizontal racks using a forklift truck fitted with a boom. Alternatively, carpet may be stored in a compact automatic warehouse. Carpet may be sold to large retailers or wholesalers as whole rolls. For furniture stores and smaller retailers, a cut-length service is provided: an unroll/re-roll unit unrolls the carpet, measures the required length automatically, makes a square cut, rolls the length and seals it in plastic film or paper for dispatch. The bulk is re-rolled and its residual length recorded.

17.7 CHEMICAL WASHING

Hand-knotted carpets are not finished like broadloom carpets, and are often subjected to a special finishing procedure known as chemical washing. Machine made squares are sometimes chemically washed to develop a finish similar to that of a knotted carpet.

Chemical washing applies only to wool pile carpets. It chemically damages the fibres and should only be carried out on very densely constructed carpets: if the access of chemicals is restricted by the dense pile, the most serious chemical damage is confined to the pile surface.

The objectives of chemical washing are to:

- cleanse the carpet;
- lay the pile, increasing the lustre of the carpet;
- soften the colours;
- generally impart an antique appearance.

In a typical Asian procedure, the back of the carpet is optionally singed, and the carpet is then soaked in caustic soda solution for 30–120 minutes. It is removed, rinsed with water on a concrete floor and worked with stiff brushes or wooden blades to remove loose fibre and to lay the pile in one direction. Sodium hypochlorite solution is then applied and the pile is again worked with brushes or blades. The carpet is rinsed with copious amounts of water, again working the pile. The washed carpets are dried in the sun.

Carpet importing companies in the principal markets for knotted carpets have industrialised the process.

The lustre and faded appearance of a washed carpet are said to simulate that of an antique carpet that has been subjected to many years of foot traffic. Microscopic examination of a washed carpet shows that the lustre is not only provided by the unidirectional pile lay but by the virtually total removal of the scales (cuticle) of the wool pile at the pile surface.

17.8 FINISHING OF FLOKATI RUGS

Traditional Greek flokati rugs (see Section 10.5) are finished "under the waterfalls" [4]. Cold water from the falls is led by flumes into spouts that churn the water in circular processing vats 6–9 m deep. Rugs are swirled in these vats for 2–8 hours (the longer times for the higher qualities) and removed occasionally for turning and inspecting the result. They are dried in the open air and, finally, lint and fluff are combed out and the back is brushed.

The simple shag pile as woven is transformed into a bulky fleece-like structure in which groups of pile tufts are felted into pointed strands. The wool backing undergoes up to 40% linear shrinkage, locking the tufts in place. Around 17% of the fibre is lost during finishing and is trapped downstream for recycling in woollen spinning.

REFERENCES

1. J W S Hearle and G A Carnaby. Carpet shading explained. In: *Proceedings of Tifcon '92, Carpets – The Changing Environment,* The Textile Institute, Manchester, 1992.
2. K Bottom. Shear beauty – towards the perfect surface. *Textile Horizons,* 1986, 6(2), 38–42.

3. H Heinrichs and J Löscher. Shearing machine requirement profile in carpet finishing. *International Textile Bulletin Dyeing / Printing / Finishing*, 1996, 2, 22–23.
4. D Sebastian. *The Complete Handbook of Flokati and Carpet Making*. Nick Kokinos, Athens, 1978.

18 Finishing of Tufted Carpets

The key finishing process for tufted carpets is backcoating, which has several important functions:

- to bind the tufts to the backing fabric;
- to improve fibre bind;
- to provide dimensional stability;
- to add weight to the carpet to improve laying properties; and possibly
- to impart particular performance features.

Backcoating, in effect, transforms the tufted cloth into a practical carpet.

Other essential finishing processes are inspection and mending and preparation for warehousing. Cut-pile carpets are subjected to a shearing process, which must be particularly carefully carried out on plain velours qualities, where the surface finish is one of the main features of the carpet.

18.1 INSPECTION AND MENDING

Tufted fabric is usually inspected immediately after tufting in-line with the machine, the fabric being led vertically in the path of high-level illumination especially for this purpose. There is an arrangement of dancing rollers that can accumulate carpet for a short time, allowing carpet on the inspection frame to be stopped while some faults are repaired.

It is important to thoroughly inspect the first few metres of a new batch of carpet for yarn faults (unlevel yarn, uneven dyeing, uneven setting, stains). A spinner will usually accept the return of unsatisfactory yarn on cone along with a short length of the faulty carpet, but will dispute the wisdom of converting the whole batch into carpet. If there is a stripe that is associated with one particular cone (or a small number of cones), this can be transposed on the creel to be tufted at the edge so that it is later removed harmlessly with the edge trimmings. Some stripes may be associated with the setting and maintenance of the tufting machine, rather than with the yarn, and these, too, should be corrected before continuing the production run. Possible faults include stripes or bands of short tufts caused by excessive yarn tension; stripes or bands of apparently different colour; J-tufting due to blunt knives; dragged filaments or uncut loops due to incorrect settings of knives and loopers; and rills arising from incorrect selection of backing fabric (see Section 5.2). At this stage, the carpet may be tested to check that its constructional details are as specified for the particular quality.

Missing stitches are filled in by the inspector using a mending gun threaded with the appropriate colour of yarn. The mending gun is a single-needle, pneumatically driven tufting machine, similar to those used in the hand tufting industry (see Section 8.2.4). Pile yarn is propelled through a hollow needle by compressed air and the needle is reciprocated by air pressure to stitch into the backing fabric. A pair of spiked wheels, driven by a crank, assists the operator to propel the gun along the tufting line, and the weight of the gun is supported by a counterbalance system. Only loop pile is tufted, the loop length being controlled by the circumference of the yarn feed rollers fitted. When cut-pile carpet is being repaired, loops of generous length are tufted by the mending gun and the surplus is trimmed off with scissors.

The need for mending can be reduced by purchasing yarn of good quality appropriate to the gauge of the tufting machine, maintaining the creel, yarn guides and tufting elements in good condition, selecting the most suitable primary backing fabric and using monitoring and control equipment.

In addition to the inspection at the tufting stage, piece-coloured products need to be inspected for dyeing or printing faults. Little can be done to mend these, except in the case of winch/beck dyeing, where unlevel or off-shade lots may be re-dyed or overdyed to a darker colour. Otherwise, the carpet is simply classified into first grade or sub-standard grades.

A final inspection after backcoating is required to check for faults introduced at shearing or backcoating: the final inspection also represents an opportunity to identify any tufting or yarn faults that might not have been apparent at the first inspection.

18.2 SHEARING

Loop-pile carpets tufted from spun yarns may be subjected to light shearing to reduce surface hairiness; and sculptured piles may be given a heavy shearing to remove the tips of the high pile, creating an effect somewhat similar to level cut/loop. Otherwise, shearing of tufted carpets is concentrated on cut-pile materials. The long-term trend towards fine-gauge tufting has drawn attention to the importance of shearing [1]. The shearing machine represents an opportunity to create a highly attractive surface on a fine-gauge velour: conversely, bad shearing can ruin the quality of the surface, sometimes with dramatic effect.

Although shearing in-line with backcoating is not unknown, the control required of both processes is such that high quality carpets require shearing to be carried out separately. Shearing of fine-gauge plain velours is preferably carried out before backcoating (and is then known as soft shearing) because carpet sheared at this stage will be cut parallel to the shearing bed, giving a uniform pile height. After backcoating, the overall thickness of the carpet presented for shearing will be less uniform than it was before backcoating, and the pile cannot be opened as well to facilitate the shearing action. A further reason for soft shearing is that great benefit is achieved by breaking open the carpet at the shearing point so that the stitches are presented to the stationary blade more individually. Laminated goods are insufficiently flexible to allow such opening of the pile to be achieved. Shearing after backfinishing is usually reserved for the less critical constructions. If shearing is carried out in-line, it should be protected by accumulators so that the shearing section alone may be stopped for short periods.

The principle of shearing is described in Section 17.5. When fine-gauge tufted velours are to be processed, the detailed considerations are a little different. Steaming before shearing is essential for unbacked goods, for example.

Two types of bed are offered by machinery manufacturers, fixed beds or rotary beds, for use with different constructions of carpet. The fixed bed provides optimum opening of the pile and is mainly applicable to unbacked goods. Laminated carpets do not pass easily over a low-radius fixed bed and may be susceptible to a stick/slip situation if some latex compound has penetrated through the secondary fabric: shearing faults in the form of bars across the carpet can result. Rotary beds are, therefore, favoured for secondary backed goods. Flexibility to shear a wide range of carpets is provided by a shearing machine having a combination bed that can be converted quickly to fixed or rotary, as required (Figure 18.1).

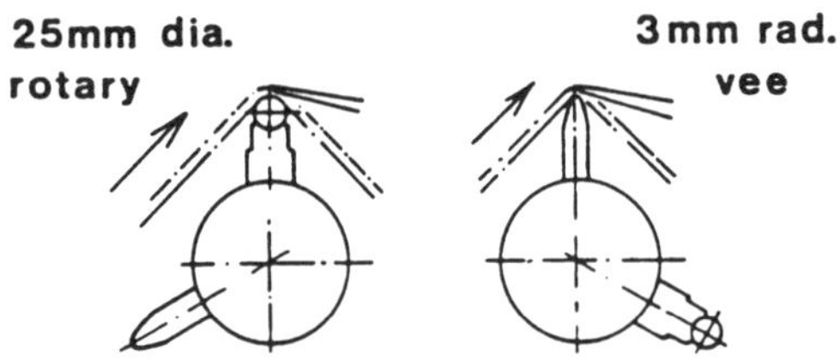

Figure 18.1. Example of a combination bed. (Courtesy of Sellers Engineers Ltd)

A further aid to smooth presentation of the goods for shearing is a pneumatic slipping clutch on the transport roller drive for achieving fine control of tension. The multiple needle penetrations in densely tufted fine-gauge products make the cloths rather sensitive to tension.

It is desirable for there to be a brief spell of time for the pile fibres to spring into the shearing position between the passage of one helical blade and the next: modern shearing machines are, therefore, fitted with fewer blades (24–26) than the traditional machines. The pitch of the blades on the cylinder is correspondingly greater.

The cylinder blades have a file-cut edge (Figure 18.2). During shearing, fibres tend to become bunched into the corners of the file cut so that fine velours require a fine file cut. The rotating blades are polished to reduce the possibility of oily flock adhering to them.

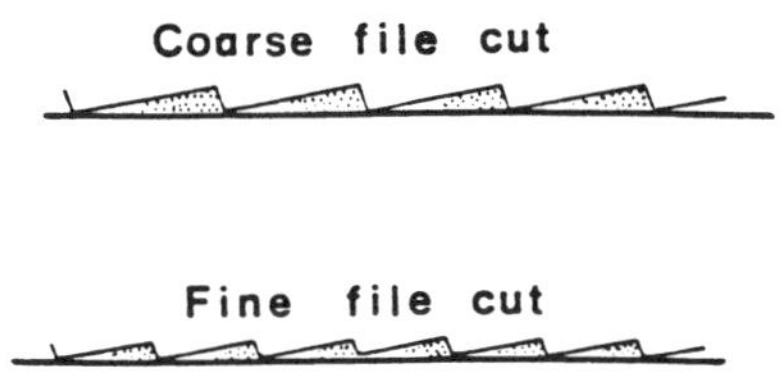

Figure 18.2. Coarse and fine file cut on helical blade. (Courtesy of Sellers Engineers Ltd)

Fibre damage at the point of shearing can affect the handle of the carpet and in the extreme case a frosted appearance from fibrillated fibre ends can result. Such problems can be minimised by keeping the stationary blade cool. Sophisticated systems for surface lubrication of the rotary blades are provided for this purpose: the rate of feed of oil should be the minimum required for the fibre type being processed.

Best results in shearing are obtained by progressive action, so that carpets requiring a particularly good finish are processed on a multi-head line. A three-head line is illustrated in Figure 18.3. Most such lines would incorporate a pre-steamer, as illustrated in Figure 17.2.

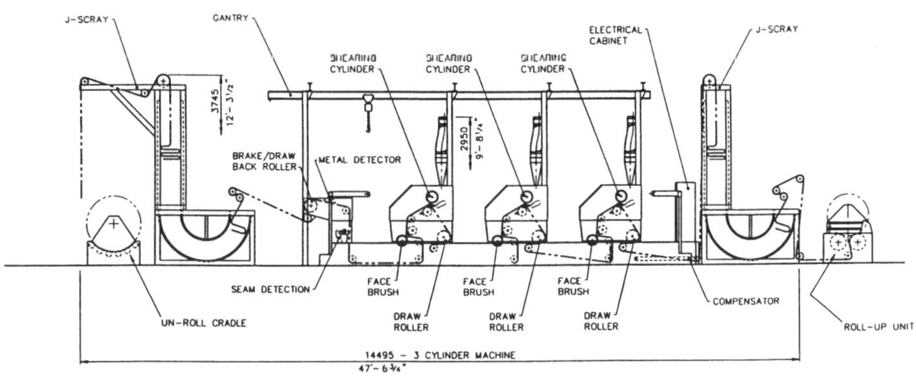

Figure 18.3. Triple-head shearing line. (Courtesy of Sellers Engineers Ltd)

18.3. ELECTRO-POLISHING

A unique machine for finishing lustrous velours is the Pol-rotor unit of Heinz Hergert. The essential component is a heated metal cylinder having grooves roughly parallel to the axis. The carpet is passed pile down over the rapidly rotating drum and is subjected to a beating and lifting action. Carpets finished on this machine have enhanced lustre owing to the fibre parallelisation that is imparted.

18.4 BACKCOATING AND LAMINATING

There are a number of alternative means of binding the pile material and stabilising tufted carpets, most of them based on rubber latex. Environmental considerations (see Section 5.4) have drawn attention to the backcoating process, creating research topics, and the range of possibilities is becoming extended.

As it was once widely conducted, the backcoating process was likely to provide carpets that emitted volatile organic compounds (VOCs) alleged to be associated with 'sick building syndrome'. Attention by suppliers of polymers to eliminate residues of volatile monomers and other low molecular weight organic compounds, coupled with careful control of the backcoating process by carpet manufacturers, has essentially eliminated this problem. Correctly finished carpets inevitably have some smell but, a few days after installation, they should have an undetectable or only a very slight odour. Of products used in the indoor environment, new carpet is one of the lowest emitters of VOCs [2].

The most serious environmental problem relating to carpet finishing is that of disposal. (Disposal of edge trimmings from carpet finishing is a supplementary issue.) Increasing thought is being given to improving the environment by discontinuing the disposal of carpets by landfill (see Section 5.4). Sophisticated systems for recycling are being introduced, but the simplest solution, incineration, is hampered by the presence of chalk filler in the latex which can give rise to dust emissions and undesirable ash. There is, therefore, motivation for the production of carpets that are free from inorganic matter. Accordingly, new backfinishing/laminating processes are being developed.

Most carpets are backfinished using latex, but the range of options is extensive and includes the following.

Unitary latex backing: A single layer of filled latex for carpets intended for direct-stick applications, and for hand-tufted products.

Laminated secondary backing: A secondary fabric layer is laminated with filled latex. If the secondary layer is a thick needlefelt, this may be regarded as an integral underlay.

Foam backing: Application of a precoat to lock the tufts is followed by application of filled latex foam, giving an integral underlay.

Organic fillers: The above options can be modified by substituting an organic product for the inorganic filler.

Speciality backing compounds: To satisfy particular requirements (antistatic, antibacterial, flame retardancy) special latex additives or special polymers may be used.

Polyurethane foam backing: Said to be energy efficient, free from VOCs, and to have good physical properties.

Hot-melt unitary backing: Polyethylene backings, applied by powder scattering, followed by infrared heating, are established for the automobile industry.

Hot-melt laminating offers the possibility of carpets composed entirely of polyolefins.

Laminating using melt-bonding fibres in both primary and secondary backing fabrics also offers the prospect of 100% polyolefin carpets.

18.4.1 Preparation for Backcoating

The entry section of a backcoating range may have the following components.

Unroll unit: In the case of piece-coloured qualities, this may be replaced by a cart of plaited carpet.

Accumulator: Usually a J-box or a double J-box, possibly supplemented by dancing rollers, to allow batches of carpet to be seamed together without stopping the range.

Singeing machine (optional): The purpose of singeing is to reduce the hairiness of the back of the tufted cloth.

Sewing machine: A robust sewing machine mounted on a frame, capable of joining batches of carpet with a firm seam, square to the selvedge, is essential to the smooth operation of a backfinishing line.

Brush / backbeater: Loose fibre, etc, is usually removed by beating the back of the horizontal carpet through a canvas screen using a slatted drum. A typical backbeating unit is shown in Figure 18.4.

Weft control system: Particularly if the carpet has a geometric design, produced either at tufting or by printing, it is important to align the tufted cloth square before backfinishing. A skilled operative may do this by adjusting the relative speeds of the stenter rails and using a system of bow and skew rollers: automatic systems can respond rapidly to changes in carpet batches. For example, the heavy duty Orthomat RFMC-GR of Mahlo scans the stitch rows with visible or infrared light, computes the result, and through a double system of bow and skew rollers can compensate for up to 300 mm of bow and 550 mm of skew.

18.4.2 Latex Applicators

The lick roller type of latex applicator for woven carpets illustrated in Figure 17.3 is also widely used for applying latex compound to tufted carpets. It has the advantage that penetration into the backstitch as well as the application level can be adjusted by the pressure on the carpet and the speed of the roller, although primary control is provided by the viscosity of the compound.

Direct applicators of the type illustrated in Figure 18.5 are increasingly favoured. Latex compound, in the form of an unstable foam, is metered across the carpet

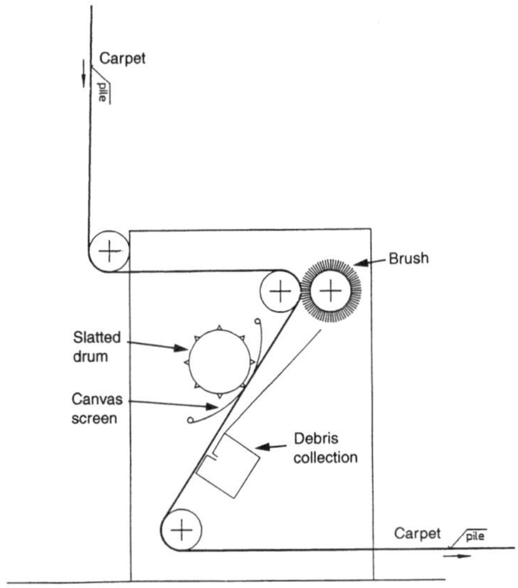

Figure 18.4. Carpet cleaning unit comprising a surface brush and a beating drum acting on the back. To protect the backing, a strip of canvas is mounted between the slatted drum and the carpet.
(Courtesy of Fleissner GmbH and Co)

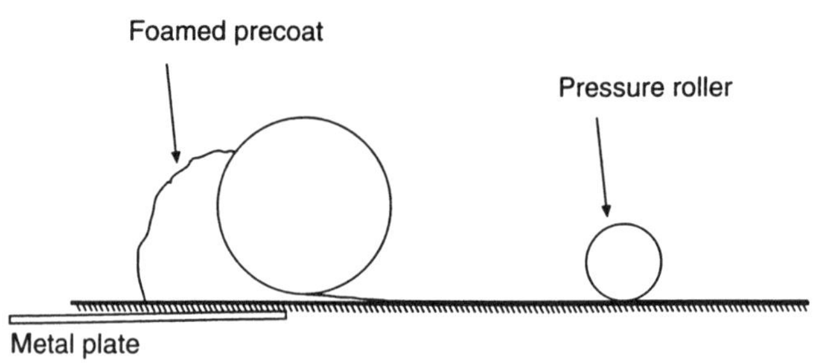

Figure 18.5. Direct application of foamed latex compound.

at the required application rate to form a bank in front of a doctor roller. Collapsing of the foam into the carpet is promoted by the doctor roller, and an optional pressure roller may assist further penetration. The system provides accurate control of the level of application, there is no effluent and air replaces some of the water as a diluent so that less energy is required for evaporation.

18.4.3 Backing Compounds

Most backing compounds are based on X-SBR and products include variants having slightly different properties which are selected for specific purposes or for particular carpet constructions. After the polymerisation process, the latex emulsion should be subjected to steam distillation to remove unreacted monomers and other VOCs that otherwise could give rise to objectionable odours from the finished carpet.

Other polymers, such as styrene/acrylate and EVA copolymers in the form of emulsions similar to latex, may be selected as alternatives to X-SBR products when greater variations in properties are required. EVA is often selected when the carpet is to meet a flammability specification, since it is non-aromatic and has more favourable flammability properties. Waterglass (sodium silicate) blended with a little latex to reduce brittleness has been studied as a backcoating for knitted carpets [3], its merits being flame resistance and environmental friendliness.

Polymer emulsions, including X-SBR, usually have shelf lives of 6 months or more, but the products applied in the carpet industry have additional components and the total compounds are less stable. Common practice is, therefore, for the compounding to be carried out in the carpet mill, in response to immediate demand. For smaller carpet mills, companies exist that provide a compounding service, offering quick response to orders. In addition to latex or other polymers, the compound may contain the following:

- Filler, to add weight to the carpet and extend the latex at minimum cost. Chalk or other finely divided form of calcium carbonate is the most common filler. Hydrated alumina is regarded as a flame-resistant additive. Powdered polyolefin may be used as a combustible organic filler, and ground carpet waste has been recommended. Note that too high a filler loading can adversely affect the properties of the latex.
- Bactericide, to prevent the growth of mildew or other biological species in the carpet.
- Surfactant, to aid penetration into the yarns.
- Thickener, to increase the viscosity if necessary.
- Antistat, to provide conductivity in an antistatic carpet. Carbon black or organic additives are alternatives.
- Water softener, if required.
- Water to the required total solids.

Quantities are specified in terms of parts per hundred of dry rubber (pphr).

Viscosity is an important controlling factor in application, especially lick roller application. As mechanical working can affect viscosity, the compounding procedure should be standardised and the viscosity of each batch checked.

18.4.4 Unitary Latex Backing

A unitary backing must have adequate physical properties to provide dimensional stability, as well as binding the fibres and tufts. The compound is based on a flexible type of X-SBR polymer and the level of filler is low by comparison with other latex compounds.

A typical unitary backing formulation contains 100 pphr X-SBR, 250 pphr filler and water to 75% solids. Application level: 1 kg/m^2 minimum.

The carpet is dried relatively slowly (eg, 110°C) to avoid bubbling the coating and the latex is cured at 150°C. A high production line has a relatively short oven for drying and a longer one for curing. For lower production rates, a single oven adjusted to provide a temperature gradient is used. Before removing the carpet from the stenter pins it must be cooled to avoid shrinkage.

Gun tufted carpets have a special chemical type of unitary backing to facilitate small-scale processing. The latex is pre-vulcanised in emulsion form by the manufacturer, and may be applied by brush or spray and air dried.

18.4.5 Single-pan Laminating with X-SBR

Jute secondary backing may be laminated to simple tufted constructions using the equipment illustrated in Figure 18.6. The filler level is around 400 pphr and the total solids 75%, to apply 800–1000 g/m^2 solids.

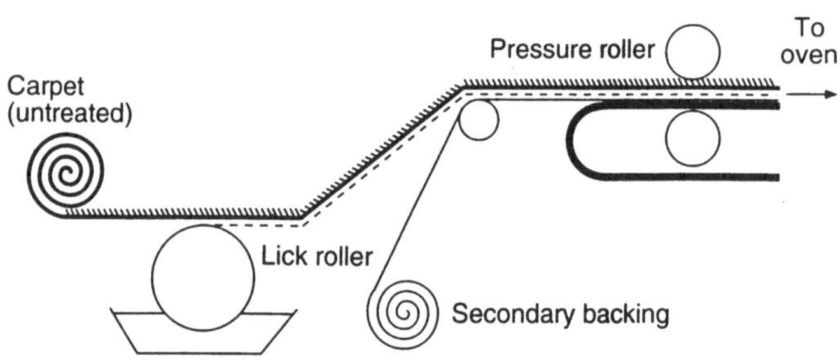

Figure 18.6. Principle of single-pan laminating.

The compound is applied to the back of the carpet and the jute backing is married with it from below. A pair of nip rollers located immediately before the stenter presses the laminate firmly together.

The single-pan system requires a backing compound that can penetrate the tufts, fill the backstitch and adhere the secondary fabric. An adequate compromise between these requirements may be reached in the case of jute, but the system is not effective with polypropylene secondary fabrics.

18.4.6 Double-pan Laminating

An old single-pan line may be upgraded to make it suitable for applying polypropylene secondary fabrics by adding another lick roller applicator immediately following the first one. The first applicator is used to apply a heavily filled, low viscosity precoat to the carpet, and the second to apply a lightly filled adhesive layer, also to the carpet. Other procedures provide better control, however.

18.4.7 Kiss Coating

Good control of laminating can be achieved by applying a heavy (500–1000 g/m^2) low viscosity, highly filled (400–800 pphr) precoat to the carpet to bind the tufts and fill the backstitch, and a lighter (about 250 g/m^2) adhesive coat with less filler (100–250 pphr) to the backing fabric, using the configuration shown in Figure 18.7.

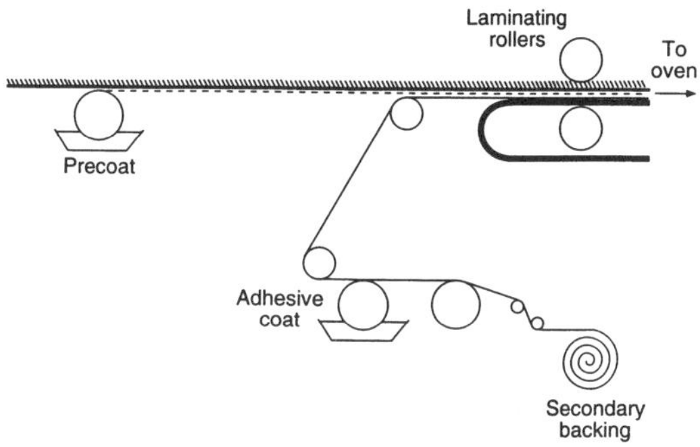

Figure 18.7. Layout of a kiss-coat laminating system.

The composition of the precoat and its level of application very much depend on the construction of the carpet to be finished. If the backstitch is deeply structured, as in crossover tufting, a reduced level of filler should be used and a wetting agent included in the formulation to aid penetration of the yarn. A high level of application is required to fill the structure.

18.4.8 Pile Down Laminating

An alternative system for applying a precoat and an adhesive coat separately is favoured in Europe because it can be incorporated in a range suitable for either foam backing or laminating. In the USA, where foam backing is less common, the configuration for kiss coating (Figure 18.7) was first developed and is more widely used.

Figure 18.8 shows the layout of a pile down laminating system. The precoat is applied by a lick roller while the carpet is pile up; the carpet is turned pile down; the precoat is dried under a bank of infrared heaters; the adhesive coat is applied by direct coating over the dried precoat; the secondary backing is introduced from above and is pressed to the tufted material by laminating rollers before it passes into the oven. As an alternative to using a lick roller for applying the precoat, a direct applicator may be used, with the carpet following a longer path pile down.

18.4.9 Separate Application of Precoat

Knitted carpets, which may be very unstable prior to finishing, may usefully have the precoat applied on a range consisting of applicator and drum dryer, similar to a range for backcoating woven carpets (Figure 17.4). Carpet that has been stabilised in this way is then laminated on a separate range.

18.4.10 Foam Backing

Latex foam may be applied to tufted carpet on a range having the configuration shown in Figure 18.9. A precoat, similar in composition to that used for carpet

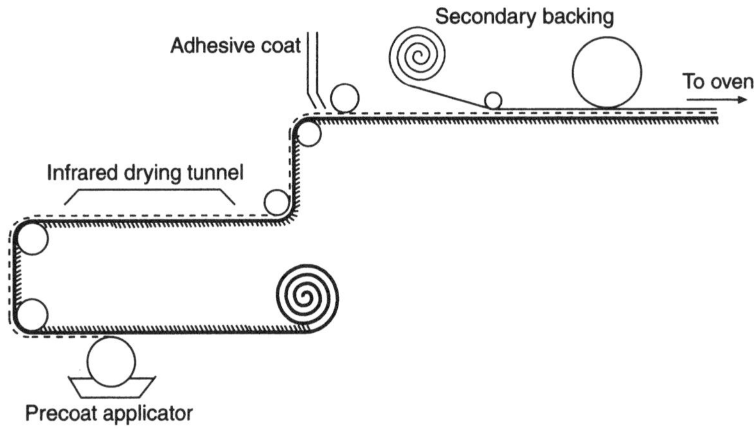

Figure 18.8. Layout of a pile down laminating system.

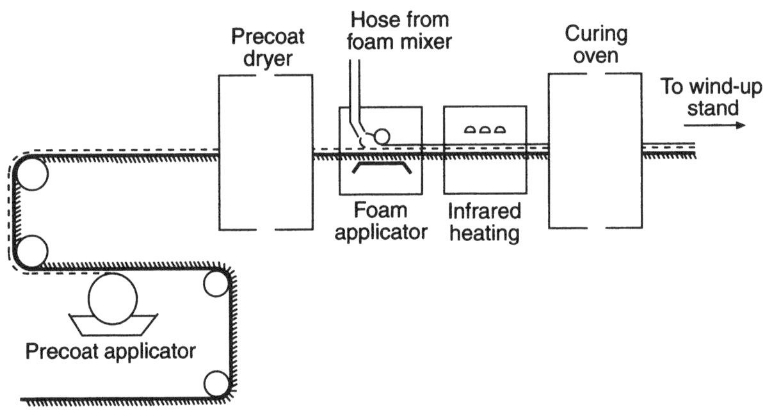

Figure 18.9. Layout of a foam backing range.

to be laminated, is first applied, dried and cured. Latex foam quite different in composition from that used in a direct applicator for laminating is then hosed on to the carpet and doctored level. The surface is cured by infrared heaters to create a skin, the foam is optionally embossed (eg, with the trademark of the carpet or with a simple design) and cured slowly by heating in a continuous oven at 150°C. Vulcanisation must be complete in order to avoid emissions of volatile sulphur compounds from the finished carpet. The backed carpet is cooled before removing from the stenter.

Because X-SBR latices cannot be used in the thick layers of foam required on foam backed carpets, SBR latices are the preferred option. No-gel foam systems are generally used to stabilise the foam during the application procedure, rather than the phase-inversion gelling system used for foam cushions and mattresses (ie, the foam is stabilised by means of thickeners and emulsifiers).

A typical no-gel foam consists of SBR latex, 150 pphr filler and vulcanising paste. The basic compound is foamed with a specified ratio of air to give a product having the required density. The vulcanising paste has a somewhat complex composition, comprising three distinct chemical systems: colloid products to stabilise the foam during the different stages of processing; a curing system containing powdered sulphur (the basis of chemical cross-links) accelerators and activators; and anti-oxidants to scavenge free-radicals from oxygen sources that can degrade the foam during carpet wear. A mechanical test developed by Centexbel may be used to evaluate the standard of vulcanisation [4].

Polyurethane foam can provide a high quality backing applied by a non-aqueous process that requires little energy, but is more expensive than latex foam and is not commonly used in Europe. Polyurethane foam is, however, the most common type of foam backing in the USA, and is briefly described in Section 19.1.3.

18.4.11 Drying/Curing Ovens

Although heated drums and infrared units may be required for special purposes, the type of oven in widespread use in backcoating lines is the dual airflow type illustrated in Figure 15.15. A special need is the facility to direct heat preferentially to the back of the carpet.

18.4.12 Hot-melt Systems

Automobile carpets have, for many years, been backcoated with fusible polyethylene (with or without a secondary fabric) to achieve products that can be moulded into floor wells. The production sequence involves dispersing polyethylene powder on the back of the pre-coated carpet and sintering it by passing under infrared heaters. After cooling, panels of the required shape are cut and the backcoating is re-softened by infrared heaters. While hot, the panels are pressed into moulds having the same shapes as the floor wells.

Dispersion coating has created interest as a means of providing wall-to-wall carpets that are free from filler [5] so that they can be recycled as fuel, without creating excessive ash or dust emissions (the major component of a low-grade conventional tufted carpet can be inorganic filler). Secondary advantages are that edge waste and other carpet waste can be ground up and used as organic filler; no energy is required for evaporating water; and no VOCs are involved.

Powder scattering involves precision metering of granules of fusible polymer from a hopper, flicking them from a metering roller to fall uniformly on the back of the carpet, and fusing the material with infrared heating. The principal polymers considered are LDPE (low density polyethylene), LLDPE (linear low density polyethylene) and EVA.

For laminating a secondary backing fabric, a precoat for anchoring the tufts and an adhesive coat are recommended by Cavitec. The precoat is a mixture of LDPE and an additive formulated for good flow characteristics, and is applied at a level of about 140 g/m^2. The adhesive coat, which also stabilises the carpet, may be LDPE, LLDPE or a mixture of the two, plus optional organic filler (ground carpet waste), applied in-line after the precoat to give a sinter layer of about 330 g/m^2. The secondary backing fabric is pressed into the sinter layer while the thermoplastic is still hot.

An alternative hot-melt technology has the objective of producing carpet entirely from polyolefin [6]. The basis of the process is an Adflex X100G soft polypropylene copolymer from Montell blended with resin and wax (conventional atactic polypropylene has too sharp a melting point to be useful as an adhesive). Bematec developed the Gear-in-Die Coater to apply the polymer, which has a higher viscosity than conventional hot-melt materials. The coater has two die lips: a lay-on and a lay-off lip. Molten polymer is delivered at a constant rate to create a film on the die lips. The moving carpet (pile down) takes the film from the lay-off lip, via a low-pressure meniscus, to create a coating thinner than that on the die lip. Adjustment of the gap between the lay-off lip and the carpet, as well as the relative speeds, is necessary to achieve the desired application level. The secondary backing fabric is laminated immediately.

This process produces 100% polyolefin carpets that have a calorific value equal to that of crude oil. The process may also be used for carpets having nylon or polyester pile.

Thermoplastic elastomers have also become available in a form suitable for powder coating. Kraton polymers (Shell Chemicals) are claimed to be easy to process in coating and thermoforming, and create a strong bond between the carpet fibres and the backing. Their rubbery properties make them suitable as unitary backing. Kraton polymers are said to give a "compatibalising" effect that enables plastic blends from carpet manufacture to be recycled.

18.4.13 Use of Melt-bonding Fibres

An alternative means of producing filler-free carpet is to use primary and secondary fabrics each having a layer of melt-bonding fibre needled into it [7].

Orientation of the primary backing during tufting is such as to locate the backstitch in the fleece of melt-bonding fibre. If required, the tufted cloth can then be dyed or printed, provided the temperature does not exceed 110°C. The tufted cloth is then married with the special ActionBac secondary fabric so that the two layers of melt-bonding fibre are in contact, and they are held on stenter pins to maintain the required dimensions. Lamination is accomplished by heating the backing to 140°C for at least 90 seconds and by the action of cooled nip rollers. Note that time is required for the bonding material to penetrate the pile yarn. The laminate is passed through a cooling zone to fix the dimensions before rolling up.

The special primary fabric carries 100 g/m^2 melt-bonding fibre. Up to 200 g/m^2 may be used on the secondary fabric, the higher amounts being for carpets having deep backstitch, notably crossover styles.

Advantages other than the targeted environmental issues are lower capital investment; improved dimensional stability to water; greater flexibility than latexed carpets; carpets can be folded during transportation and fitting without creasing; and lower weight and volume of the rolls.

Disadvantages, such as distortion of polypropylene and wool-rich piles owing to the relatively high processing temperatures and pressures, and too high a lustre of the backing for retail styles, led to the development by Sellers and Amoco Fabrics of a combined melt-bonding fibre/hot-melt system (Figure 18.10), the principal components of which are:

- let-off system for the carpet;
- stenter frame to control dimensions;
- infrared heaters to activate the melt-bonding fibre;
- cooled nip rollers to assist penetration of binder into the tufts;
- powder applicator;
- further infrared heaters to melt the powder coating;
- secondary backing feed;
- laminating rollers;
- nip rollers to consolidate the laminate;
- cooling zone;
- standard slitters and carpet take-up.

All these latex-free systems give lighter and more flexible carpets than have traditionally been produced, considered to be attractions to carpet fitters.

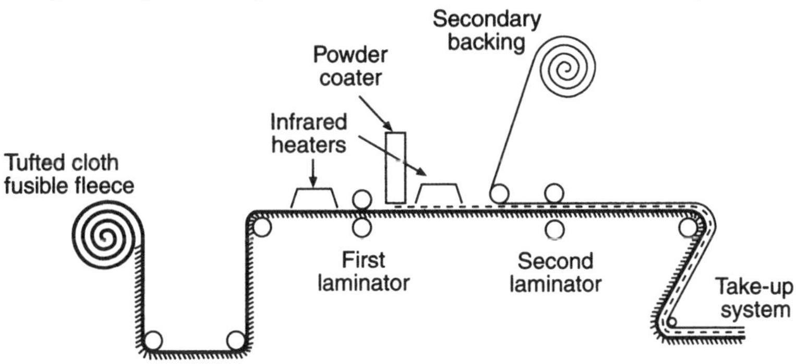

Figure 18.10. Schematic diagram of the Sellers / AdBac™ laminating line.

18.4.14 Post-backfinishing Processes

After cooling, the carpet emerging from any coating/laminating line must be trimmed, inspected, cut into lengths, rolled and packed. To facilitate these operations, at least two accumulators are required. Cutting and rolling may be classified into the production of complete rolls, and the preparation of lengths cut to the requirements of the carpet fitter. The latter may be short lengths required by residential sales outlets or a multiplicity of lengths needed to make up contract orders for large rooms. The presence of faults in the carpet complicates the situation and computers are required to optimise the utilisation of the finished carpet and the cutting process itself [8]. The principal components of post-backfinishing follow.

Anti-shading: Cut-pile carpets that are susceptible to shading may be processed by the WRONZ Trutrac machine to enhance their resistance to pile reversal, as discussed for woven carpets in Section 17.3. The Trutrac machine is normally separate from the backfinishing line, and the anti-shading process is carried out before trimming, inspection and wrapping.

Edge trimming: Tufted carpet is always produced a little wider than required, and wider still when sliding needlebars are used, so the finished carpet needs to be trimmed to the required width. This is accomplished by rotary knives located close to the ends of the stenter chains. Allowance for edge trimming may be minimised by using video control technology that allows the knives to follow the edge of the pile more closely. Selvedge trimmings tend to be voluminous, but can be made more compact by equipment that chops them into short lengths.

Accumulators: Two accumulators are required: one to allow time for inspection and the other to allow cutting, rolling and wrapping to be carried out. If the second accumulator has fixed storage (ie, is off-line), a cut optimisation system may be used. The first accumulator should hold a carpet length equivalent to 5–7 times the line speed (eg, 6×60 = 360 m), and is based on dancing rollers. If cutting and packaging are carried out in-line, a second dancing roller system holding 50–70 m of carpet is required.

Inspection: As the carpet passes over the inspection table, 'hard' (serious) faults are marked (with metal stickers in a modern system). When a computerised system is used 'soft' faults and other quality data may be keyed into it. If desired, the carpet can be turned through 45° after inspecting the face so that the backing can be inspected as well.

Measuring: A prominent supplier of cutting and rolling equipment, ETF, favours a simple measuring wheel, but with a self-calibrating system, based on a follower in contact with the carpet, that activates two photodetectors exactly 1.5 m apart and an electronic feedback system. Maximum deviations of 150 mm/100 linear m are claimed.

Cutting and rolling: Rolls up to 1 m in diameter may be prepared in-line and consist of pre-selected standard lengths cut automatically when programmed to do so by the measuring system, or shorter lengths governed by factors such as mill policy, avoidance of faulty carpet, end of line, etc. Carpets may be rolled with or without cardboard tubes, and pile-inside or pile-outside, depending on the policy of the mill or the configuration of the backing line. Surface drives are most common, but centre-wind systems are preferred for some polypropylene products. Cutting is by double rotating knives having a scissors action.

Wrapping: Rolls may be wrapped in film (polyester or polypropylene) or possibly paper. Film of a width depending on the diameter of the roll is unrolled in the axial direction of the carpet and laid on a belt table in the path of the carpet to be wrapped. If the film is to be attached to the flap of the carpet this may be done with low-tack polyester tape (foam backing), heavier-tack polyester tape (jute or polypropylene secondary backing) or plastic tacks. Sealing of the wrapped roll may be by means of a heat sealing bar. The roll ends may be closed by securing the loose film with bungs in the cardboard centre (if used), by heat seal or by an ultrasonic device.

Optimisation software: Cut-length machines are operated off-line. They utilise belt tables and wrapping systems similar to those for on-line wrapping, and are made having different rates of production for different scales of operation. They can be optimised to achieve maximum production of good quality lengths from large rolls that may contain faults. Similarly, somewhat different optimisation software can be used for the production of standard rolls (with fewer short rolls) or contract lengths from a fixed accumulator unit of at least 150 m capacity located next to the movable accumulator between the inspection table and roll-up area. The functions of ETF optimisation software include:

- reading carpet piece data (from bar code or computer);
- entering quality information (lengths of faults, etc);
- automatic optimisation of lengths (via queuing of items, including faulty areas);
- data storage for further handling by the rolling/packaging installation;
- cutting, rolling and packing according to stored data;
- output of roll data to floppy disk or on-line to central computer;
- initiating printing of roll tickets and shift production data.

Cutting and wrapping of narrow carpet: Some markets require narrower carpet than the standard 4 or 5 m width so that, for example, 2×2 m slitting may be required. This may be accomplished using a modification of an ETF roll-up unit, involving a retractable knife unit. The two narrow rolls are wrapped side-by-side in their separate sheets of film and telescoped apart for end closure.

18.4.15 Backfinishing Lines

No firm rules can be laid down about the overall composition of a backfinishing line, and the enormous size of most lines means that they could only be illustrated by a fold-out section. Folding leaflets provided by machine manufacturers, such as Babcock, illustrate some diverse alternative lines that may be required for particular purposes and which may be assembled from a selection of individual machines.

The following broad considerations provide guidance to the general nature of a finishing line.

- Capacity should be in line with the forecast production of the tufting department, plus a margin for dealing with peak loads and making up down time. Throughput rates may range from 6 to 60 linear m/min. High production ranges may require additional components (eg, while air cooling of hot carpet may be adequate for a slow range, a separate chiller may be needed for high speeds).

- Styles of carpet to be produced, including possible future styles.
- Market level aimed for – lower qualities may require less sophisticated equipment.
- Environmental considerations in the mill and target markets.
- Versatility (eg, laminate, foam or both).
- Space – configurations including a change in direction of the carpet may be considered (ie, part of the range may be double-decked).

18.5 FINISHING OF RUGS

The principal processes are over-edging, fringing, labelling and packing.

Tufted rugs may be over-edged in piece form, using a dancing roller system, to allow the carpet to be transported free of tension during the sewing process. A machine offered by ETF operates on both sides of the carpet simultaneously and is available in working widths from 0.7 to 4.0 m. After over-edging, the carpet is cross-cut into rugs according to the design (or length in the case of plain rugs).

Sewing of fringes on the ends of the rugs to simulate hand-knotted rugs may be carried out on a conveyor system operated intermittently or on an air table. The ETF table has special nozzles grouped in sections (for uniformity of effect) that blow air at an angle to support the carpet and make it easier for the operative to move. Over-edging may be carried out at an air table as an alternative to in-line over-edging.

Automatic rolling/wrapping machines for rugs are based around a belt table that may also be used for final inspection. If specified, the machine can include a thermal labelling unit for applying labels to the back of the rugs. Several rugs may be rolled and packaged together.

18.6 SAMPLING

Presentable samples are required for pattern books to be used by retailers and for general marketing and display purposes.

The ETF carpet sample cutting machine M5522 inputs 2 m wide carpet from roll and automatically cuts the material longitudinally and crosswise to create samples in sizes down to 50×130 mm. The required number of samples can be pre-set on a counter. The cut samples are automatically discharged or taken by conveyor to an edge bevelling machine.

Automatic machines are available from ETF that have four mini-shearing cylinders, enabling all four sides to be bevelled without turning the sample. Dust extraction ducts are provided for each shearing unit. Depending on sample size, 1500–3000 samples may be bevelled an hour. For lower rates of output, single-cylinder bevelling units for manual feeding are available. An interesting option is a fusing unit for securing the bevelled edges of polypropylene pile.

Associated with sampling machinery, various items of equipment are available to facilitate secure mounting of samples in safety, including the preparation of sample books.

REFERENCES

1. K Bottom. Shear beauty – towards the perfect surface. *Textile Horizons*, 1986, 6(2), 38–42.
2. Carpet and Indoor Air Quality. White Paper, DuPont HomePage, www.dupont.com/dcf/dfs/environmental/iaq/html.Internet, 1996.
3. J Weinhold. Carpet Knit – warp knitting technology sets new standards in carpet finishing. Melliand International, 1997, (3) 140–141.
4. D Simoens, H Pottel and E Albrecht. How to avoid producing uncured foam backings. In: *Proceedings of International Man-made Fibre Congress*, Dornbirn, September 1993, Austrian Man-made Fibre Institute, 1993.
5. C Welter. New techniques in the production of environmentally friendly carpets. *Coatings*, 1993 (11).
6. H Wunsch. Novel techniques for production of polyolefin floorcoverings. *International Carpet Yearbook*, 1997, 33–37.
7. J Diggle. Adbac™ – a latex and filler free backing system. In: *Proceedings of Tifcon '98*, The Textile Institute, Manchester, 1998.
8. R J J Koppens. Innovations around extensions of backing ranges. In: *Proceedings of World Textile Congress on Floorcoverings*, University of Huddersfield, July 1995, 220–234.

19 Carpet Tiles

An apocryphal belief in the industry is that carpet tiles were invented by a Dutch pig farmer. Each winter he would improve the comfort of his stone-flagged home by covering the floor with a layer of pig bristle, stuck down with bitumen. One spring, when removing the insulating layer with a spade, he found that he could lift quite large sheets, so he cut the sheets of bristle and bitumen into tiles for re-laying the following winter.

Today's mature tile industry is focused on contract applications rather than the residential market, and the ability to be lifted and re-laid features among some of the following advantages.

- Easy access to services in raised sub-floors.
- Easy access to flat cable laid on top of sub-floors.
- Damaged or stained tiles can be replaced, if the installation is not too old.
- Easier transport, compared with broadloom, to locations in high-rise buildings.
- Styling possibilities through combinations of different coloured tiles or plain and patterned tiles (examples are given in the colour section).

Tiles may be based on needled, tufted, bonded, flocked or woven carpets and are made for all market levels. There are two general types: heavy grades (at least 4 kg/m^2) that can be laid loose (self-laying grades), and lighter grades that need to be fixed to the floor with adhesive, either permanently or, more usually, using a release adhesive. In most cases the backing must be specially engineered to enable a carpet to meet the general requirements for tiles, which are as follows.

- High mass per unit area, especially for self-laying tiles.
- Particularly good dimensional stability. This requirement extends to resistance to curling and arching, as well as shrinkage.
- Controlled flexibility to assist laying; but excessive flexibility can lead to tiles rucking and moving.
- Adequate thickness to cover minor imperfections in the floor.

- Accurate dimensions and clean cutting, including freedom from fraying (tufted and woven styles).

There are other specific requirements related to particular applications.

19.1 BACKING POLYMERS

Backing materials are used to provide mass, thickness and the required flexibility. Dimensional stability is usually achieved by including at least one glass fibre scrim, rather than by the backing polymer alone.

Bitumen and PVC are the most common backing polymers. Bitumen is the cheaper of the two, but care has to be taken to avoid soiling walls and the tile surface during installation, and the installer is advised to have white spirit available for stain removal. PVC is cleaner and can be made more flexible and elastic, without introducing the risk of plastic flow. Both materials have been criticised for behaviour in a conflagration: bitumen provides fuel and PVC generates toxic fumes. However, the backings of installed tiles are not exposed, so that any fire hazard is determined by the pile material until a fire has reached the stage of being totally uncontrollable. An advantage of bitumen is that it can slowly creep to accommodate minor floor irregularities.

19.1.1 Bitumen

Bitumens, sometimes known as asphalts, are the residues from petroleum fractionation and consist of complex high molecular weight hydrocarbons dispersed in oily hydrocarbons, with some free carbon. The grades used for tiles are oxidised materials that are virtually solid at room temperature. Melting points are not sharp. To achieve the required application properties, a blend of 90/100°C and 110/120°C material is used. Bitumen is delivered hot from tankers.

Compounding is usually carried out by the tile manufacturer in heated tanks. The compound may contain up to 100 pphr chalk filler, antistats and polymers such as polypropylene, LDPE, EVA or other property modifiers. (The finished backing must not be brittle, must flex, but should not creep.) Typically, 3–4 kg/m^2 of compound is applied to the carpet substrate. The melt at 160°C may be applied by reverse coating, direct coating of the carpet, or a combination of the two. The carpet substrate may be needled, woven or tufted, pre-coated in the normal way with SBR latex.

A reverse coating line offered by Sellers Engineers (Figure 19.1) is designed to laminate a secondary backing fabric, two coats of bitumen, a glass fibre scrim and the carpet, to give the construction illustrated in Figure 19.2, with in-line cutting of tiles. The secondary backing fabric is essential to prevent bitumen from staining the pile of adjacent tiles during storage and transport, and may be printed with arrows to mark the direction of manufacture and the brand name. Nonwoven polypropylene/polyester fabrics, around 120 g/m^2 are typically used as secondary backing material. The glass fibre scrim, used to provide dimensional stability, is advantageously an open-weave construction of about 30 g/m^2.

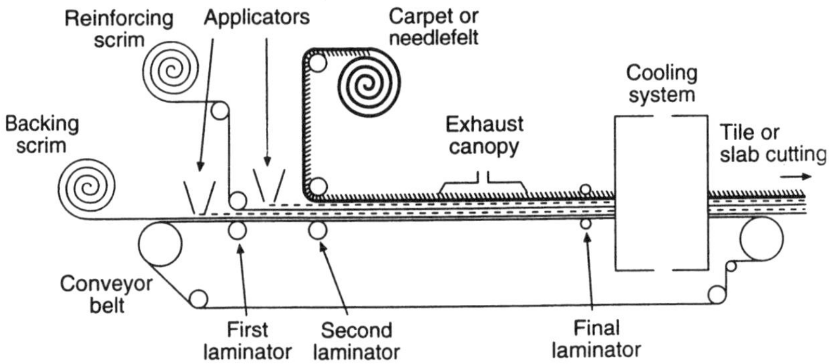

Figure 19.1. Sellers' reverse coating line for applying bitumen to a carpet substrate.

The line starts with a double fabric-unroll unit, one for the backing fabric and one for the reinforcing glass fibre scrim, each with a position for a spare roll. While the scrim is skied, the secondary backing is tensioned and laid on a stainless steel conveyor belt. Bitumen is flowed on from an applicator and levelled with a doctor blade. The tensioned scrim is then introduced at laminating rollers. A second layer of bitumen is applied and the tensioned carpet is led in from a gantry at a second set of laminating rollers. An exhaust canopy removes fumes and accomplishes preliminary cooling. After a final pressing between nip rollers, the material is jetted with chilled air in at least one powerful cooling unit. The laminate is then skied under minimum tension (driven overhead rollers) and brushed before passing to an automatic tile cutting machine, or is cut into slabs and stored flat to cool thoroughly and anneal over a period of around 3 days prior to cutting.

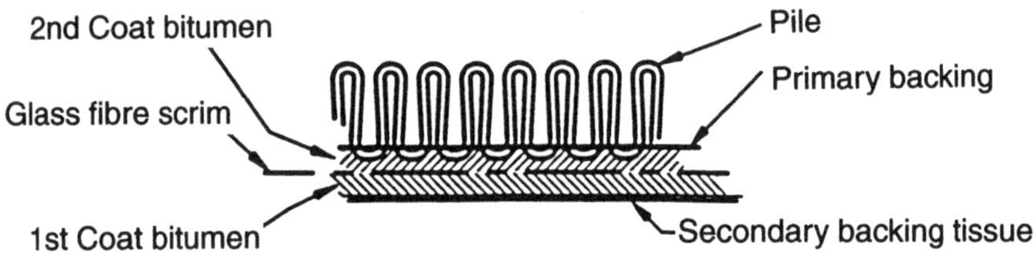

Figure 19.2. Cross-section of a tufted tile having a bitumen base. (Courtesy of Sellers Engineers Ltd)

It is possible to apply bitumen in a single step by direct backcoating of the carpet, provided an open-weave glass fibre scrim that allows bitumen to penetrate through it is used. However, it is not so easy to achieve a level result, and double application techniques are favoured for reliability. With a double direct coating technique it is possible to apply a layer of APP as an alternative to a secondary backing fabric.

19.1.2 Polyvinylchloride

PVC is applied in the form of a plastisol, a paste of PVC and plasticiser which fuse together under the influence of heat to give a rubbery sheet. Fillers and other additives are incorporated in the paste. As with bitumen, application weights are 3–4 kg/m^2, except for electrostatically flocked carpets (see Section 14.2), which are often lightweight products.

The carpet substrates may be bonded or flocked (which already have a PVC layer), tufted or woven. Some needled products are coated with PVC. As temperatures in the fusing oven may be as high as 180°C, there are restrictions on the materials used in the carpet substrate. Tufted carpets should have a polyester primary backing fabric, rather than the usual polypropylene. The precoat should be EVA since the usual SBR latex has poor adhesion to PVC. EVA emulsions that can be applied by the same techniques as latex are available for this purpose. A PVC precoat is a satisfactory alternative.

As with bitumen, application may be by direct coating, reverse coating or a combination of the two. However, a secondary backing fabric is optional since the fused polymer is clean to handle and arrows, logos, patterns, etc, can be embossed on its surface with hot rollers.

Typically, a pre-coated pile carpet is directly spread with PVC paste; a glass fibre scrim is applied and laminated with nip rollers; and a second layer of PVC applied. The laminate then passes to a gelling unit, which may be a hot air oven or infrared heater.

A fusion-bonded tile will often have three layers of PVC and two glass fibre layers, one the original fabric on which the pile was formed and one incorporated to improve dimensional stability.

When the reverse coating method is used, a secondary backing layer is used to support the first layer of PVC so that the final tile has a fabric lower surface. For example, Bonar Floors' heavier grade of flocked tile has a nonwoven polyester/cellulose secondary backing, a layer of PVC, a glass fibre scrim, another PVC layer, the glass fibre primary fabric, the adhesive and the pile. The total weight is 4.2 kg/m^2 and the total thickness is 5.3 mm. A lighter grade has only the applied PVC layer, the primary glass fibre backing, the adhesive and the pile: the weight is 2.4 kg/m^2.

19.1.3 Other Polymers

Atactic polypropylene

Atactic polypropylene (APP) is an amorphous polymer with properties that are somewhat analogous to those of bitumen, but it is a purer substance and is white and clean. It may be applied in a similar way to bitumen by reverse coating or direct coating. The temperature of the melt is about 150°C.

The basic APP is formulated with filler and stabilisers and with property modifiers such as isotactic polypropylene and elastomers. No plasticiser is included, since plasticisers can interact adversely with some adhesives.

Vinyl acetate/ethylene

Vinyl acetate/ethylene copolymer (EVA) is used when it is necessary to meet particular flammability specifications.

EVA compounded with filler, anti-oxidant and other polymer may be applied as a hot melt. It may be presumed that the process is not straightforward, however, since the principal application technique is Du Pont's Keldax proprietary laminating system. The EVA is extruded into a sheet and is concurrently coated with Surlyn ionomer resin to improve its adhesive properties. A glass fibre scrim is incorporated and the backing assembly is laminated to the carpet using an adhesive.

Polyurethane

Polyurethane is claimed to be environmentally friendly and little energy is used during application. Because of its relatively high cost, it has a very small market share in many countries.

Polyurethane elastomer is formed by reacting a di-isocyanate with a polyhydric alcohol at relatively low temperature. The final stage of synthesis is accomplished during the laminating step. The components are mixed continuously and the compound is spread on a belt having a release finish. A glass fibre scrim is introduced and then the carpet. The laminate is passed through an oven at 60°C to complete the process of polymerisation and adhesion.

Low density polyethylene

Low density polyethylene (LDPE) is used to a minor extent and for lightweight tiles, since current techniques allow only low weights to be applied. Recyclability is the principal motivating factor. A coating of sintered polyethylene may be applied, starting with granules of LDPE. Pre-coated carpet is led pile down under a powder scattering machine that meters on the polyethylene, and then under infrared heaters. The granules fuse to themselves and to the precoat to give a stable backing material.

19.2 TILE CUTTING AND FINISHING

The traditional method of cutting tiles was by die stamping. Modern blade cutting machines operate continuously and in many cases can be operated in-line with the laminating range. Less waste is produced than with die stamping.

The Sellers Tile Cutting Machine has guided blades working both lengthwise and in cross-cut. Operating speeds may be up to 6 m of laminate/min. Square tiles in the size range 400–600 mm can be cut and, if necessary, rectangular tiles 400–600 mm wide and 400–2000 mm long: most commonly tiles are 500×500 mm, or 18×18 inches in the USA. Tiles 600×600 mm are offered by some companies, to coincide with the dimensions of raised flooring modules. Milliken's Affinity range of 600×600 mm tiles is designed specifically for installation over metal access panels. Each tile has a single magnetic strip, enabling it to be laid firmly, and lifted cleanly, without residual adhesive.

When cutting bitumen tiles in-line with the laminating range it is vital that the bitumen is thoroughly cooled, otherwise the straight edge of the cut will deteriorate through flow of the material. There is also the risk of warm bitumen sticking to the knives, or even cold bitumen sticking to blunt knives that have been heated by friction. Knives must be maintained sharp whatever the backing polymer, and must be regularly inspected for build-up of deposits.

Carpets for tile manufacture often do not need shearing (face-to-face bonded carpets) or have been sheared before laminating (cut-pile tufted carpets). In some cases, it is desirable to shear the material in tile form, to improve the pile surface and ensure that all tiles are the same thickness. Narrow width shearing machines that shear tiles on a conveyor belt are available for this purpose.

After cutting and finishing, tiles are stacked in boxes, pile-to-pile and back-to-back, unless required to be printed.

19.3 PATTERNING

19.3.1 Tile Laying

Tiles having a cut pile are especially likely to show apparent colour difference when viewed from different angles, owing to the pile lying in the direction of manufacture. When laying the tiles, it is therefore important to consult the arrow on the back to ensure that they are laid in the intended direction.

To ensure that tiles are installed at right angles, it is advisable not to rely on walls being accurately at 90°. Instead, a setting out point that would utilise a maximum sized perimeter tile should be selected and two chalk lines at right angles drawn through this point, checking the angle by measuring a triangle 3×4×5 m. Four tiles are installed around the setting point (with release adhesive or loose, as appropriate) and are followed with rows of tiles along the two chalk lines. The area is completed one quadrant at a time, adding tiles in 'stair step' formation (Figure 19.3). Other details for achieving quality installations are provided by the tile manufacturer.

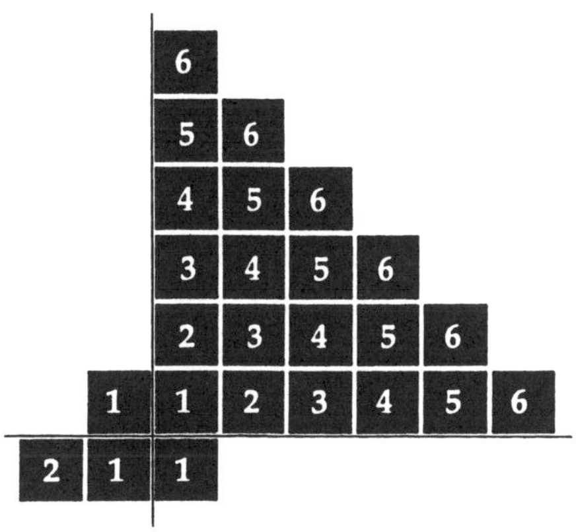

Figure 19.3. Sequence for installation of tiles. (Courtesy of The Carpet Tile Company)

If all the arrows are aligned parallel, the installation will simulate broadloom carpet. If the arrows are alternately in opposite directions, a tonal chequerboard effect can be achieved. For a more definite chequerboard pattern, two different colours of tile are used. Patterns other than chequerboard can be made, according to the overall plan for interior design: for example, different work stations can be defined by lines of tiles or by different coloured areas. More elaborate results may be obtained by using combinations of plain and patterned tiles.

19.3.2 Die Cutting

Single designs that greatly increase the patterning scope of tiles may be made by outlining a motif on two tiles of different colours and stamping the outline with a die cutter. The items cut out are exchanged between the two tiles and the inlaid items are assembled with adhesive tape. Sometimes, one of the two may be rejected for aesthetic reasons (eg, a company emblem may only be acceptable in a particular colour).

The flexibility of this design technique can be greatly improved by using laser cutting or water-jet cutting, instead of die stamping. More intricate designs with more use of colour can be obtained by these techniques (see Plates 28 and 29).

19.3.3 Yarn Effects

Additional colour, sometimes with small random patterns, can be provided in tufted and bonded tiles by using pile yarns with colour effects, such as marls, intermingled filaments, space-dyed effects and tweeds or other spinning effects.

The Westbond Flex tile manufacturing process, described in Section 14.1.4, allows clumps of several colours to be spaced randomly over a plain background, but using a technique that can be reproduced.

19.3.4 Tufted and Woven Designs

Crossover and buried-pile designs that are not too definite can be used in tiles to produce an overall effect. Woven carpets can be converted into such tiles. Designs with a linear pattern are normally installed in chequerboard fashion, while large swirls, for example, may sometimes be laid at random. The tile supplier specifies the installation technique.

19.3.5 Precise Designs on Bonded Tiles

Two processes for producing fully patterned bonded tiles are described in Section 14.1.3. The patterning system of the Bondax process is based on a spool-gripper mechanism, so there are no restrictions in colour. A feature of the designs offered is that they match on all four sides. The Axtile process is based on an eight colour electronic jacquard, with an option for a 12 colour system. Its principal target was to reproduce repeating Axminster style designs, but coordinated tiles can be produced to provide more extensive designs, including simulations of bordered squares.

19.3.6 Printing

Low-pile nylon tiles (flocked, needled, tufted) in pale background colours can be overprinted in short runs by a dry system involving heat transfer from impregnated paper (rather as T-shirts are printed). Detailed emblems and even photographic images can be printed on tiles in this way.

Screen printing has a significant market share in the patterning of carpet tiles and is the principal system for flocked tiles. The Westprint system, developed for printing Westbond tiles, utilises foamed print paste, and additives that would normally need to be removed in an afterwash can be avoided in such formulations. Processing consists of simply printing and steaming.

Computer controlled jet printing is the most flexible system for introducing pattern to tiles. Register between tiles is extremely accurate and designs can, if required, have repeat lengths covering several tiles, or be non-repeating. Special versions of the Chromojet and Millitron machines for tile printing are available (Section 15.10.4) and are widely used in the tile industry (Plates 30, 31 and 32).

Interior designers have exploited the possibilities of jet-printed tiles and combinations with plain tiles to great effect.

20 *Testing*

Carpets may be tested for a variety of reasons:

- As part of quality control or quality management procedures associated with carpet manufacture.
- To specify the product in objective terms, either by its manufacturer or its purchaser.
- To measure the consumer properties of carpets – those properties that confer an immediate improvement on the interior of a building (eg, acoustic and thermal properties).
- To predict the deterioration of carpets in wear.
- To identify the source of faults that may have arisen during carpet manufacture.
- To provide evidence relating to consumer complaints or commercial disputes.

Apart from testing the carpets themselves, a competent carpet manufacturing organisation will also have in place procedures for checking the quality of its raw materials. Such testing is not considered in this chapter.

Tests should be:

- accurate (ie, they should measure what they are intended to measure and performance tests should be relevant to real use);
- precise (ie, give reproducible results between laboratories as well as within laboratories);
- sensitive (ie, should discriminate clearly between different products);
- economic (ie, not too slow or expensive).

Different carpet products often need to be subjected to different ranges of tests. Thus, a carpet intended for use as a sports surface will have quite different requirements from one to be used in a ship, and a carpet to be installed on a domestic staircase will have different requirements again.

A systematic tactivisual examination (plus smelling) of a sample can give guidance on the test schedule required to identify potential failure modes.

Performance testing is particularly difficult. 'Hot-house' testing is necessary in order to achieve results in a useful time frame and at a reasonable cost. Attempts to simulate the action of foot traffic, which is intermittent and allows considerable time for the pile to recover from one impact of the shoe before another takes place, have been unsuccessful. And there are many types of traffic: straight walking, turning action, stair traffic and wheeled traffic (eg, trolleys of various kinds, wheelchairs). Moreover, different aspects of the appearance of the carpet, its thickness and its mass can be affected by the traffic. Expectations of carpet performance in common domestic circumstances may vary considerably from one country to another. In Germany, for example, a dense utilitarian loop-pile carpet may be selected as the ideal residential wall-to-wall material on which to lay a valuable oriental rug, whereas in the USA, a carpet with a thick cut-pile showing footmarks may rather be taken as evidence of luxurious living.

The range of tests and their varied interpretation, coupled with the difficulties of carrying them out with precision and accuracy, mean that carpet testing is a dynamic and contentious area of research, worthy of a book in its own right. Moreover, a detailed, unambiguous description of each test method would be too unwieldy to attempt in this book. This chapter can do little more than summarise the current situation of carpet testing, pointing out some of the pitfalls.

It is a cliché that concrete would perform better than carpet in many carpet performance tests, but it is one worth repeating to draw attention to the need for a balanced approach to the subject. The properties of a carpet may be viewed as a chain (Figure 20.1) each link of which should be sound if the carpet is to fulfil consumer expectations [1]. Attempts to combine the results of several tests in a mathematical formula to give a single number that represents the 'quality' of the carpet are scientifically unsound. A high level of resistance to soiling can never compensate for low resistance to change of texture, for example. However, systems that summate points for each test result have some merit, provided that a minimum number of points for each aspect of performance is achieved. It should also be appreciated that the need for testing very much depends on the application of the product: an aircraft carpet must clearly satisfy well defined technical criteria, whereas a gun tufted rug is assessed chiefly according to its aesthetic merits. However, mainstream carpet products, whether intended for residential or contract applications, usually need to be tested and the results presented in a clear way, so as to provide the customer with useful information.

20.1 SIGNIFICANCE OF TEST RESULTS

For an understanding of the significance of test results it is important to know how precise and accurate they are. Figure 20.2 distinguishes between precision and accuracy through the analogy of target shooting.

The accuracy of a test may be assessed by considering how closely it measures the 'true' value. A simple example is measurement of total thickness of a carpet:

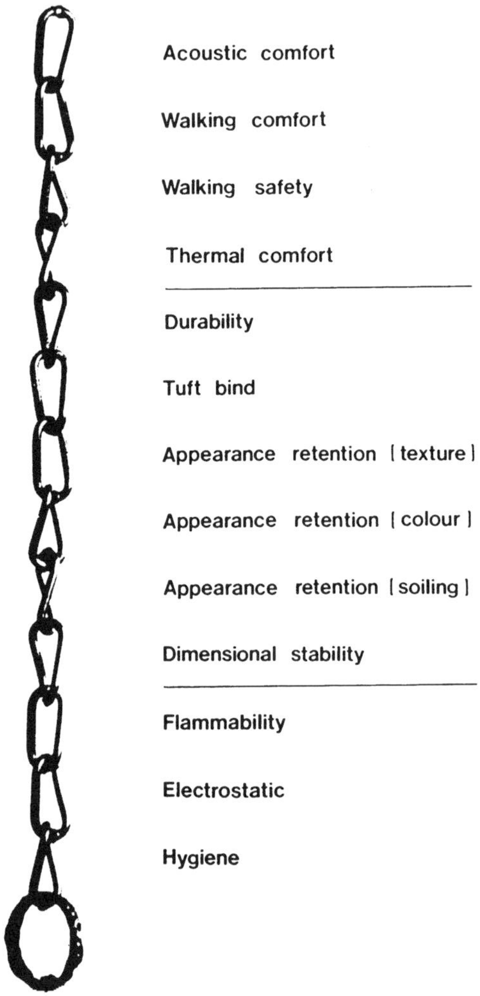

Figure 20.1. The chain of carpet properties (abbreviated) classified into three groups [1].

is the pressure under which the thickness is measured just sufficient to eliminate the effect of protruding fibres without compressing the pile itself? Other cases may require a careful study of the test method, especially when surrogate values are being measured. For example, when assessing the insect resistance of a wool carpet it is usual to measure the level of insect-resist agent in the pile rather than to carry out difficult and slow biological assay tests. There must be confidence in the correlation work that was carried out between different levels of insect-resist agent and the results of biological assays. Moreover, to be satisfied that the insect-resist agent is likely to be effective over the life of the carpet, its level should have been measured after relevant durability tests (cleaning and light fastness).

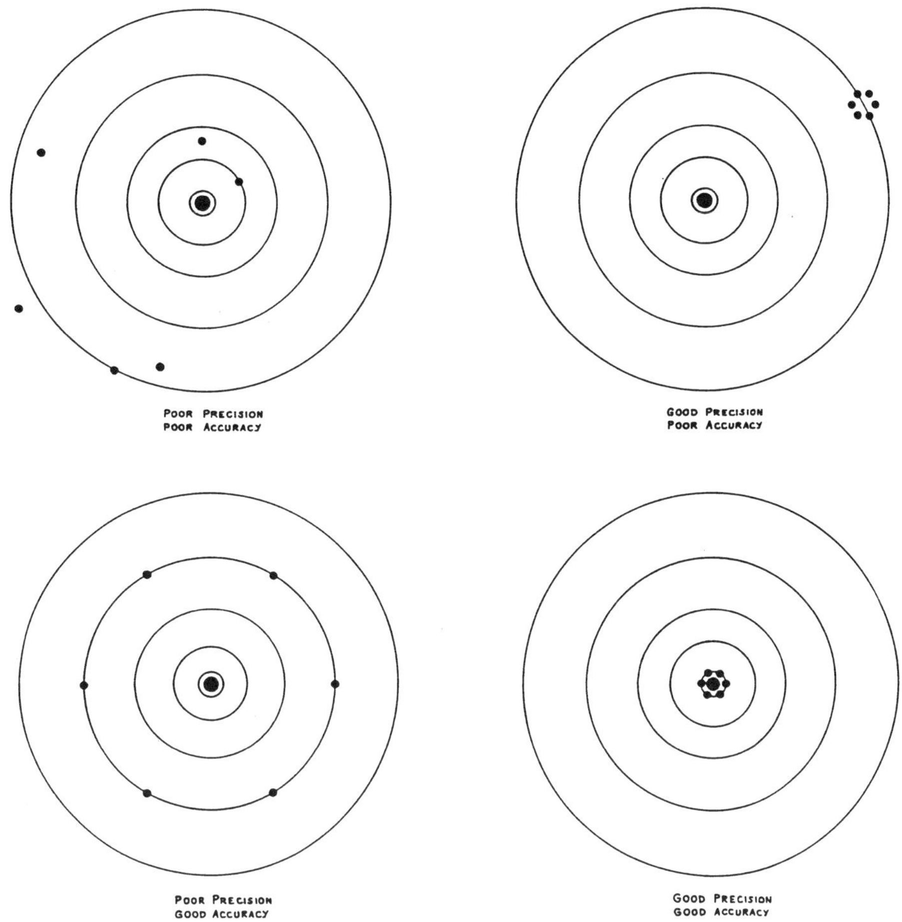

Figure 20.2. Precision and accuracy in target shooting.

The precision of a test can be measured. Errors may arise from the following.

- Sampling – Clearly, a sample of carpet should be representative of the bulk (eg, it should not be cut from an area directly against a seam). It is desirable, but not always possible, to take several samples to account for small variations in properties that can exist over the area of the carpet.
- Instruments – The instruments used for carrying out the test must be well maintained and calibrated. The conditioned atmosphere, if required, must be similarly controlled. Equipment should operate to recognised standards.
- Operators. The operators must be well trained, must carry out the test in the manner specified, and should be encouraged to report unusual aspects.

It is not easy to achieve the same results in different laboratories, nationally and internationally, and 'round-robin' testing or surveys by accreditation authorities may be necessary in particular circumstances.

Despite the best precautions, repeat tests on the same material will not give exactly the same results, owing to random fluctuations of variables. However, they should lie within agreed tolerances.

Test variations may be expressed in mathematical terms such as:

standard deviation $s = \sqrt{\frac{\Sigma d^2}{n-1}}$

where d is the deviation of a test result, x, from the mean, $\overline{x}$, of n individual test results; and

coefficient of variation $CV(\%) = \frac{s}{\overline{x}} \times 100$

Using such parameters, the statistical significance of differences between test results from two carpets can be ascertained. Rigorous application of statistical methods to the results of some carpet performance tests may reveal that they may not be as clear as would appear from a superficial consideration.

National and international standards organisations are working towards achieving better accuracy and precision in testing for the benefit of the industry. The principal abbreviations pertaining to standard test methods are as follows.

ISO	International Organisation for Standardisation
EN	European Test Method (CEN, European Committee for Standardisation)
ECE	European Colour Fastness Establishment
ASTM	American Society for Testing and Materials
AATCC	American Association of Textile Chemists and Colorists
BS	British Standards Institution
SDC	Society of Dyers and Colourists
DIN	German Institute for Normalisation
JIS	Japanese Industrial Standards
AS/NZS	Joint Standards for Standards Australia and Standards New Zealand

In addition to the official bodies, some fibre suppliers (including wool interests) have test methods associated with labelling schemes for products made from their fibres and these are usually operated internationally. For example, Wools of New Zealand (WNZ) and The Woolmark Company (formerly the International Wool Secretariat (IWS)) administer similar testing schemes that are appropriate for 100% wool and wool-rich carpets. National carpet industry bodies are sometimes active in developing test methods and standards of performance.

Their procedures are often proposed as national standards and the national standards may later be considered as international standards. For specific applications, bodies outside the carpet industry may become involved in the development of standards (eg, aircraft manufacturers in standards for carpets for aircraft, or sports organisations in standards for artificial turf). Where flammability standards are imposed, the carpet manufacturer is advised to ascertain exactly what test criteria the products are required to fulfil, since in some countries the city fire chief may set the standards for that city alone. It may be argued that carpets installed in public or high-rise buildings should be regarded as building materials. However, it is unacceptable that all building materials – wood, concrete, steel, plastics, floorcoverings – be subjected to the same tests, and carpet manufacturers should be prepared to counter such arguments.

Agreements are in place to rationalise large areas of this complex situation regarding international standards. Thus, CEN was formed to develop pan-European standards. One consequence is that the British Standards Institution has an official obligation to publish all–European Standards (ENs), which has led to a series of BS EN numbers. Similar obligations apply to the members of ISO as a result of the Vienna Agreement. Combined prefixes to standards' numbers identify the national and/or European standards that have been superseded eg, BS EN ISO 105-A01 *Textiles. Test for Colour Fastness. General Principles of Testing*. Another example of an international agreement is that made in 1992 in which Standards New Zealand and Standards Australia recorded their joint commitment to prepare and market joint standards, the primary objective being to remove barriers to trans-Tasman trade, with an ultimate goal of linking standardisation work first within the Asia-Pacific region, then globally.

Considerable progress towards such unification has been made through the ISO, although the progress may not have been as rapid as many in the industry would wish. It is necessarily more difficult to reconcile ideas from a wider geographic catchment than within one country and no doubt some local standards will be retained for many years. However, there is already equivalence between many national and international standard test methods.

There is widespread belief that international standardisation of carpet test methods is a worthwhile attainable goal. However, the diversity of cultures in the carpet markets and in the industry itself is such that expectations of, and therefore levels of performance in, those test methods may not all be identical in the different regions.

20.2 CARPET CONSTRUCTION AND COMPOSITION

With the exception of pile fibre content, tests in this category are usually performed on the same sample, but with replicate tests for each item.

20.2.1 Pile Composition

A systematic scheme of chemical testing developed by The Textile Institute [2] is available for identifying a wide range of textile fibres. Within the carpet industry, common pile fibres can be identified more simply by a combination of a heating/burning test in a flame, microscopic examination, tests for nitrogen (heating with soda lime generates ammonia – alkaline to litmus) and chlorine (green flame from hot copper wire test), and a flotation test in cold water with a little wetting agent.

Nylon: Burns slowly with an odour of celery; usually extinguishing to form a hard grey bead. Various fibre cross-sections. Positive test for nitrogen.

Acrylic: Burns readily with a luminous smoky flame, melting to a hard black bead. Round or dog bone cross-section. Positive test for nitrogen.

Modacrylic: Burns less readily than acrylic. Positive tests for nitrogen and chlorine.

Polypropylene: Melts readily; molten material burns slowly like wax; forms a white bead. Various cross-sections. No nitrogen or chlorine. Floats.

Polyester: Burns slowly with a sweetish odour; usually self-extinguishing, melting to a black bead. Round cross-section. No nitrogen or chlorine. Sinks.

Wool: Burns with melting and foaming to give a friable black ash; characteristic smell of burning hair. Variable fibre diameter, surface scale structure.

Cotton: Burns after removal from flame with a smell of burning paper; afterglow; grey feathery ash. Fine and convoluted fibres; collapsed tube cross-section.

Sisal, coir and jute: Burn rather like cotton. Characteristic cross-sections (see Sections 2.8, 2.9, 2.10).

Wool/synthetic blend proportion: Shear off a sample of pile; condition at 65% RH and 20°C; weigh. Dissolve the wool in hot alkaline hypochlorite; filter off the residue, wash and dry; re-condition and weigh. Express weight loss as wool content at standard regain (%).

20.2.2 Total Carpet Mass and Surface Pile Mass (ISO 8543)

Specimens (usually from replicate) 200 mm square are cut and conditioned at 65% RH and 20°C. The total carpet mass in g/m^2 is determined by weighing.

The pile is removed with a band knife; smaller (150 mm square) specimens are cut to avoid any shearing damage and are conditioned before weighing. Difficulties may be experienced in deciding when all the pile has been removed. Care is necessary to avoid removing latex and other backing material. In particular, problems may arise with bonded and tufted carpets having a non-woven primary backing. The band knife must be set up precisely. The surface pile mass (conditioned weight, not standard regain) is calculated by subtraction.

The total pile mass of woven carpets is sometimes determined by dissecting a weighed sample and weighing the pile component. This cannot be done with finished tufted carpets because of the presence of latex. An alternative procedure involves dissolving the fibre.

20.2.3 Pile Thickness (ISO 1765)

A thickness gauge (usually with digital read-out) at a pressure of 2 kPa (just sufficient to buckle protruding fibres without compressing the pile) is used to measure the total thickness of the carpet. The thicknesses of the specimens shorn of pile used in the measurement of surface pile mass are also measured, and the pile thickness is determined by difference.

Using a thickness gauge and a range of weights enables the pile thickness to be measured at pressures up to 200 kPa. This compresses the pile to an extent similar to that resulting from a person walking on the carpet. A compression/recovery cycle can be followed using the range of weights, and an optional microprocessor provides automatic calculation and analysis of the data (BS 4098).

The same test is used for underlays over the narrower range 2–100 kPa (BS 5808).

The WRONZ Carpet Pile-Thickness Gauge provides a non-destructive direct method for measuring pile thickness (ISO 10834). It has six needles, loaded by the weight of the instrument, that penetrate the pile until they make contact with the dense backing, and a flat probe that locates the pile surface and is connected to a dial or digital gauge.

The instant results from the WRONZ gauge make it useful for setting carpet manufacturing and shearing machinery and for quality control purposes. The ability to measure the pile thickness of installed carpets has been a boon to those studying carpet wear.

The WIRA Carpet Tile Thickness Gauge requires the tile to be mounted on a rigid steel sample plate with raised edge guides. Readings are taken at each corner.

Figure 20.3 distinguishes between pile height and pile thickness. Pile height is measured by inserting flat metal gauges having 1 mm steps into the pile, an operation that also erects the pile, and finding a gauge that most closely matches the pile height (WNZ TM20). Skill is required to determine the gauge that best matches the pile height.

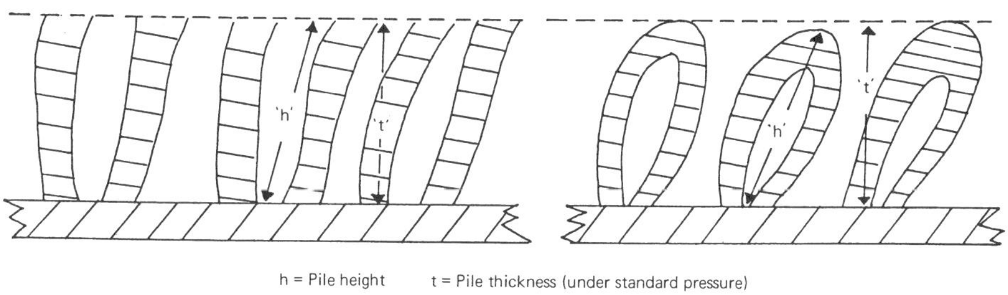

Figure 20.3. Pile height and pile thickness.

20.2.4 Stitches and Rows (gauge or pitch) (ISO 1763)

The numbers of tufts per dm in the machine direction and the weft direction are counted. The result may be reported simply as stitches and rows per dm or the number of rows may be converted into gauge (tufted carpets) or pitch (woven carpets). In the case of multiple tufts inserted in the same penetration point, a comment should be made on the test report as to how the tuft was defined. Definitions exist in ISO 2424.

20.2.5 Tile Squareness (BS 5921)

The WIRA Carpet Tile Thickness Gauge can be fitted with a series of dial or digital gauges to measure the dimensional accuracy of tile cutting at the same time as the consistency of tile thickness is being assessed. Alternatively, the WIRA Tile Squareness Gauge is available as a separate unit. The TFI Areameter is an opto-electronic device based on image analysis, designed for the same purpose.

20.3 CONSUMER PROPERTIES OF CARPETS

During a period when excessive attention was being given to measuring the durability of carpets, Herzog, then director of the Austrian Textile Institute, focused attention on the need to market carpets in terms of their immediate usefulness to the consumer [3]. His list of relevant factors has since been extended to include the following: psychological usefulness (aesthetics, prestige), walking comfort, safety, acoustic comfort, thermal comfort and control of indoor air quality.

Aesthetic properties are not amenable to measurement, but objective information about all the other 'useful' properties of carpets is available [4]. Few of these consumer properties are measured on a regular basis, the notable exceptions being safety from electrostatic shock and safety from fire. However, information about the other aspects is available from research on selected ranges of carpets, and the approximate status of particular carpets can be interpolated from the existing data on the basis of fibre content, surface pile mass and pile thickness. Thus, for example, it may be necessary to measure the acoustic properties of carpets considered for installation in a concert hall or an open plan office area. An experienced architect should be able to adequately assess the influence of carpet on the acoustic comfort of a school classroom on the basis of the constructional details of the carpet, considered in conjunction with existing acoustic data on carpets. Perhaps the most neglected area of testing for consumer properties is walking comfort, where different properties are required for different applications. In general, information about consumer properties of carpets is used in carpet marketing, but not as widely as might be expected.

20.3.1 Walking Comfort

The comfortable sensation experienced when a person walks on carpet depends largely on the carpet cushioning the shock to the body experienced when the foot strikes the floor. Different criteria should be applied to different use areas. A firm carpet that resists compression should be used to prevent muscle fatigue in areas where people are highly active and apply high loadings to the carpet, whereas softer carpets are more comfortable in areas where lower loadings are applied.

Herzog defined a measure of comfort based on an "effective work path" formed from the quotient of work of compression and mean pressure, using loading stages of 1, 2, 5, 10, 20 and 50 kg/cm^2. The hierarchy of locations in terms of required level of walking comfort index is:

- sport;
- corridors, stores, stairs, restaurants;
- entrance halls;
- living rooms;
- bedrooms.

WRONZ developed a more direct test based on the rate of deceleration of a foot (simulated by a pneumatically loaded cylinder) when it strikes the surface of a carpet [5].

All tests for comfort should include tests after a fatiguing treatment when the carpet has 'bedded down', since carpets exist in a compressed form for the greater part of their lives [6]. Such tests have been used principally for research purposes, although a 'comfort index' is included in some carpet classification schemes.

20.3.2 Safety

Safety from falling

The difference between the frictional properties of carpet in general and hard floors in general is so great that it is clear that people are less likely to slip and fall when walking on carpets. If they do fall, the cushioning effect of the carpet means that they are less likely to be injured. The subject is not well researched but is attracting increasing interest from the risk management departments of large companies in the USA because of concern about large claims for compensation for falling injuries sustained by customers at retail and by employees [7]. The cushioning effect of textile sports surfaces (indeed, all sports surfaces) is a very important aspect of safety in sport, and is widely measured, for example, by using the equipment illustrated in Figure 20.4.

A simple test for measuring the surface friction properties of carpets has been developed [8] based on an Instron tensile tester drawing a specially designed sled across the carpet. The effects of carpet construction on friction are relatively small but topical treatments and variations in fibre type give quite large changes.

Figure 20.4. The Clegg Impact Hammer, used for measuring the surface hardness / cushioning of sports surfaces (natural and synthetic) in relation to the risk of injury from falling and shin-splint syndrome. (Courtesy of New Zealand Sports Turf Institute)

The Tortoise Test (UK) and the Pendulum Test (continental Europe) are designed to test the frictional properties of all types of floorcoverings and they distinguish clearly between carpets and other floorcoverings.

Friction burns to players can be a significant hazard from carpets used as sports surfaces. A German test [9] measures the temperature rise generated by surface friction and can detect significant differences between materials. Watering the surface before the game and at half-time is standard practice in hockey, advantageously controlling ball roll and rebound, as well as reducing the incidence of friction burns. Friction, as related to traction (and sliding tackles in soccer), between player and surface is an important aspect and values for a number of combinations of surface and shoe type have been summarised [10]. There are many specialised tests and specifications relating to different aspects of the performance of textile sports surfaces[11].

Safety from electric shock
Unpleasant electrical shocks may be experienced when persons build up a charge as a result of walking on nylon or wool carpets, in particular (unless specially treated as described in Section 21.2), and subsequently earth themselves. Moreover, there is a danger of such discharges disrupting unprotected computers and other electronic equipment.

Nylon and wool are near the donator end of the triboelectric series and many shoe sole materials are near the acceptor end, so that in dry atmospheres, static charges of opposite sign are generated on the two materials when the shoe is separated from the pile.

Stroll tests

The most commonly used methods of evaluating the electrostatic properties of carpets are based on body voltage graphs. An operator walks on conditioned carpet in a prescribed manner (eg, stepping or scuffing walking) wearing standard shoes and holding an electrode connected to a recording static voltmeter. The charge on the shoe sole is transferred inductively and then by conduction to the walker's body. When the next footfall occurs (on another part of the carpet) the body voltage decreases but not to zero. At the next lifting of the foot, additional body voltage is generated, and so on. Continued walking causes the body voltage to rise in a saw-tooth fashion until, after about half a minute, a plateau is reached as charge dissipation mechanisms balance those of static charge generation (Figure 20.5). Peak voltages around 20 000 V have been observed on untreated carpets in environments having low humidity.

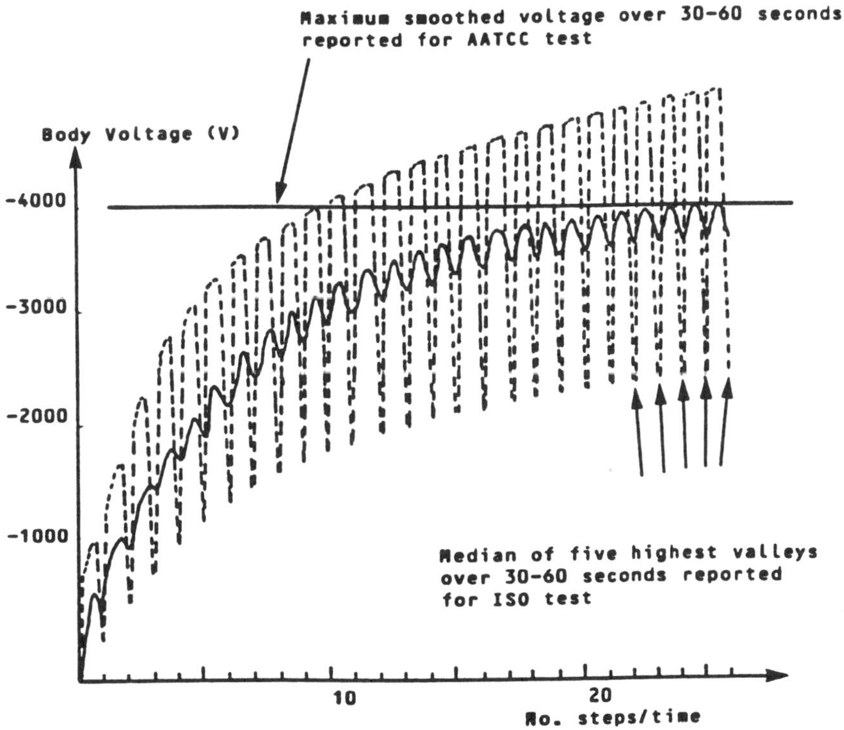

Figure 20.5. Typical curves for generation of body voltage by walking on carpets.

AATCC TM 134 *Electrostatic Propensity of Carpets* is carried out at 20% RH and 21°C. Body voltage is recorded by damping the response of the static voltmeter to a time constant of about 3 seconds and the maximum smoothed voltage (see Figure 20.5) is recorded.

ISO 6356 acknowledges the difficulty of achieving reproducible results between laboratories when testing the same carpets. The test is carried out at 25% RH and 23°C and the median of the five highest valleys in the undamped body voltage curve (ie, the points where two feet are on the carpet) is recorded, as shown in Figure 20.5.

The Estameter (Figure 20.6) is a device for measuring 'body voltages' generated by simulated walking of a standard shoe sole material over the carpet under test. At the time of writing, this test has not been standardised internationally.

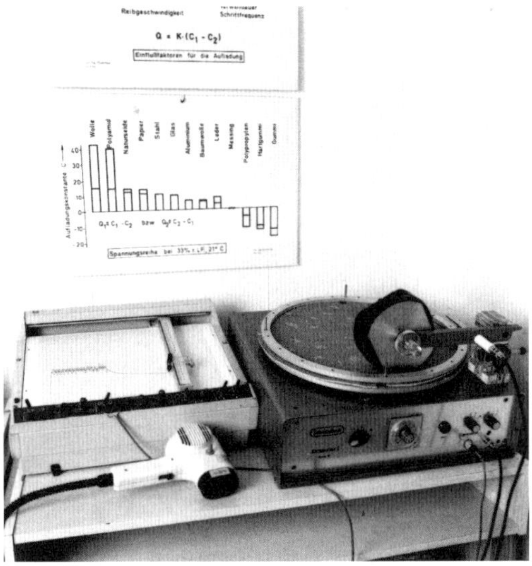

Figure 20.6. The Estameter, used for measuring electrical charges generated by simulated walking. (Courtesy of Teppich Forschung International)

Charge decay
Tests involving a measure of charge decay are used by some producers of synthetic pile carpets to demonstrate the merits of their products, but there are no official standards, except in France. Typically, a person standing on an insulating rubber mat is charged to 10 000 V and moves on to the pre-conditioned carpet under test. The charge decay curve is recorded and some function of it (various measures have been recommended) is used as a measure of the electrostatic properties of the carpet.

Resistance testing
Many computer manufacturers have specified the electrical resistance properties of carpets for computer accommodation, which may be surface resistance or bulk (through) resistance or both. Usually a minimum value (to guard against

electric shock from mains supply) as well as a maximum are specified. Electrode designs vary and, remarkably, some test methods do not standardise the atmospheric conditions, which have a critical influence on the results. ISO 10965, *Textile Floorcoverings-Determination of Electrical Resistance,* published in 1998, should clarify the testing situation, but the standard does not attempt to set standards of performance. IEC 61340-4-1, developed by the International Electrotechnical Commission, applies to all floorcoverings and classifies them as electrostatically conductive, dissipative or antistatic, but the test methods used have proved unsuitable for thick carpets. A classification document may be developed by ISO in the future. The ISO test measures both horizontal resistance (surface resistance) and vertical resistance (surface to back).

Australian Standard AS 2834 is specifically directed towards computer accommodation and includes a test method for surface resistance using NFPA electrodes 1 m apart and a test method for bulk resistance, both carried out at 40% RH and 20°C. Acceptable carpets must give results within the range 5×10^{5} and 2×10^{10} ohms.

Safety from fire

A publication produced for the benefit of partners of Wools of New Zealand lists 84 national and transportation standards pertaining to flammability tests for carpets [12]. Test methods and standards are continually being refined and attempts are being made to simplify this complex situation by the development of international standards. Therefore, carpet manufacturers need to regularly update their information. Despite this dynamic situation, the following broad principles can be discerned.

Fire tests relevant to carpets may be classified into three categories:

- ease of ignition;
- behaviour in a conflagration;
- evolution of heat, smoke and toxic fumes.

Ease of ignition

The most widely used test for ease of ignition is the Methenamine Tablet Test (ASTM D2859). A specified methenamine tablet is placed on the surface of the bone-dry carpet sample and ignited. The radius of char is measured. It is essentially a pass/fail test. ISO 6925 is a similar test but gives the option of testing either a conditioned or dried sample.

All carpets and rugs sold in the USA must pass the ASTM test. The test has also been used widely in other parts of the world and is favoured by CEN/TC134/SC1 (Textile Floorcoverings) but has been criticised by CEN/TC137/FRG (Fire Regulators Group) [13].

Another horizontal test for ease of ignition is the hot metal nut test (BS 4790), originally intended to simulate the effect of a hot coal falling on to a carpet. A stainless steel nut is heated to 900°C and placed in the centre of a sample 300× 300 mm. The greatest radius of char is measured. This test may be used to assess the possible effects of a burning cigarette.

In the Kleinbrenner test of DIN 54332 (developed for drapes) the carpet is mounted vertically and a small flame impinges only on the pile surface.

Research into the spread of fires has shown that a certain combination of carpet and wooden stairs can be susceptible to extremely rapid escalation of fire [13]. The Methenamine Tablet Test does not predict this hazard and it seems that a more energetic source of ignition and a special carpet configuration may be needed for this purpose.

Behaviour in a conflagration
A wide range of tests exist that subject carpets to intense flame or radiant heat, or radiant heat with flame ignition. Most of these tests were developed for building materials in general and some of them have the test material mounted vertically or even inverted. The applicability to carpets of testing in anything other than a horizontal configuration has been disputed.

The most suitable test of this kind for carpets is arguably the NBS Flooring Radiant Panel Test (ASTM E648). A 1 m long sample of carpet is mounted horizontally under a radiant panel inclined at 30° from the horizontal so that the carpet is subjected to a heat flux gradient ranging from 1.0 to 0.1 W/cm^2. A pilot burner is placed at the intense end of the gradient. The distance burned is noted and is reported as the Critical Radiant Flux at that point. Carpets for different classes of building and building areas are required to have different limits for Critical Radiant Flux. Radiant panel tests are often used in transportation applications of carpets.

Some flammability tests for carpets may be considered to test both the ease of ignition and behaviour in a fire. An example is the test for aircraft carpets FAR 25.853b. Specimens 12×3 in. (305×76 mm) are mounted vertically in a cabinet and a small gas flame is applied to the lower end so that both pile and backing are ignited. If the carpet is not self-extinguishing, the duration of burn and glow and length of burn are reported. Any flaming droplets are noted. Testing must also be carried out after subjecting samples to multiple cleaning cycles.

Evolution of heat, smoke and toxic gases
Dense smoke commonly impedes escape from fire and the principal cause of fire deaths is inhalation of the hot toxic products of combustion.

Because of the need to evacuate passengers and crew quickly after a survivable crash, considerable attention has been given to testing for evolution of smoke and toxic gases from burning aircraft carpets. Regulatory authorities, such as the US Federal Aviation Administration and the UK Civil Aviation Authority, do not include such tests in their specifications because of problems in achieving reproducible results and the consequent difficulties of defending test results in public. However, most aircraft manufacturers set standards based on the National Bureau of Standards Smoke Box, ASTM E662 (or a modification of that test), operated in each mode (flaming mode and non-flaming mode). Limits are set for the time to reach maximum permitted smoke density and concentrations of hydrogen fluoride, hydrogen chloride, hydrogen cyanide, sulphur dioxide plus hydrogen sulphide, carbon monoxide, and nitric oxide plus nitrogen dioxide.

NZS/AS 1530.3 is unique in that it simultaneously measures ignitability, flame propagation, heat release and smoke release. The test is based around a vertical radiant panel and the test sample with a pilot flame is initially opposite but some distance away. The sample is moved progressively towards the radiant panel until ignition occurs. Indices for the four parameters are calculated. Limits are set for selected indices for different classes of building.

The Euroclassification of contract floorcoverings (1999 proposals) for flammability by CEN TC127 recommends the vertical test of ISO 11925 Part 2 for ease of ignition (but with generous performance requirements), the NBS Flooring Radiant Panel Test for the measurement of critical radiant flux and evolution of smoke, and heat release tests that virtually no organic material will pass.

20.3.3 Acoustic Comfort

Carpets perform as well in acoustic tests as most dedicated acoustic materials and are often more economical.They may be used as components of the technical design of special rooms such as broadcasting studios, concert halls and large open plan offices. One of the prime functions of carpets in more general areas is to transform a room from one echoing sounds to one more comfortable to the ears. The role of carpets in reducing noise levels in a wide range of contract applications, such as school classrooms, airport lounges and restaurants, is widely appreciated, although not understood in detail. A simple illustration (Figure 20.7) shows the advantage of using carpet for preventing the generation of noise in a classroom. Carpets are often used on walls or panelling in open plan offices to absorb airborne sounds.

Four aspects of acoustics may be influenced by carpets:

- impact sound insulation;
- surface noise control;
- airborne sound reduction;
- transmission of sound into adjacent rooms.

Although carpets on walls have a minor influence on transmission of sound to an adjacent room, there is no substitute for a solid wall between rooms.

Impact sound insulation

Two ISO test methods are relevant: ISO 717-2 and ISO 140-8. Measurement is by the reduction in transmission of sound (in decibel ratio) in a reverberation chamber when an area on the concrete roof of the chamber with or without the sample is impacted by a tapping machine. The degree of reduction depends very much on the frequency of the sound in the range 125–4000 Hz and the results are usually presented in graphical form.

Thick, heavy carpets with underlay perform best, by absorbing the energy of impact of the simulated foot.

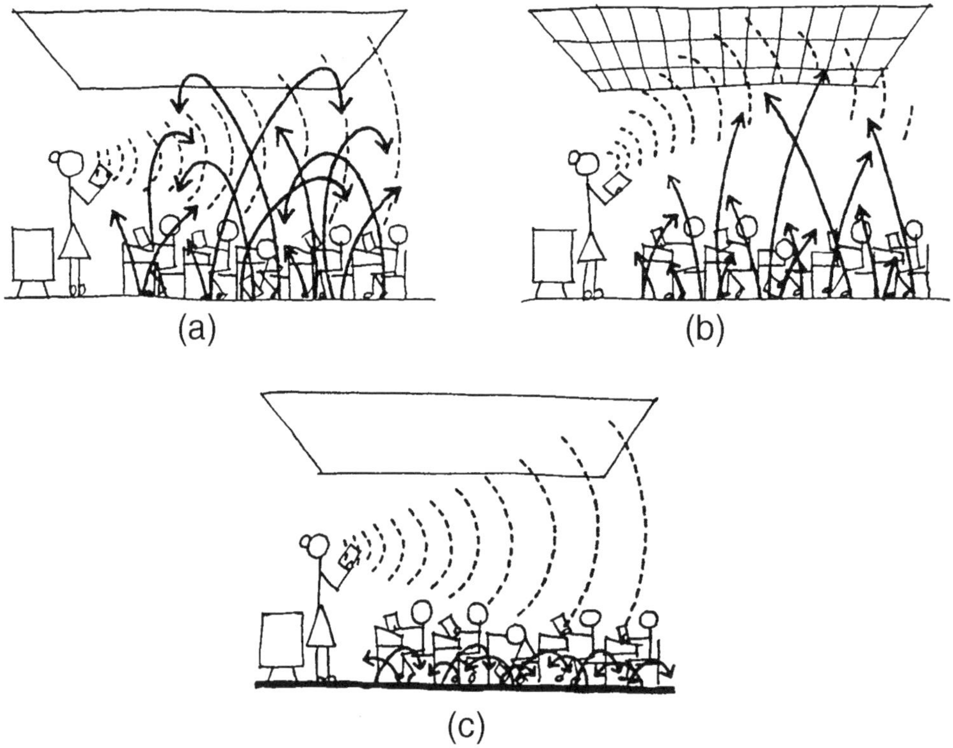

Figure 20.7. Control of classroom noise. (a) Noise from children is perpetuated by hard reflective surfaces. (b) An acoustic ceiling attenuates the teacher's voice as well as noise from the children. (c) Carpet absorbs noise from the children and a hard ceiling usefully reflects the voice of the teacher. (Source: The American Carpet Institute)

Surface noise control

Surface noise control is not measured by standard tests. Strictly, impact sound isolation refers to the insulation of rooms below the floor from sound generated above, whereas carpet users are also interested in the noise generated in the room where the foot traffic is. When comparing carpets, impact sound isolation may be used as a measure of noise generation within the room. The need to consider surface noise control arises when comparing the acoustic properties of different types of floorcoverings. Thus, a cushioned vinyl product may have virtually the same impact sound isolation values as a carpet, but footsteps will be heard in the room itself as a light tapping noise on the vinyl, while such noise is absent when the carpet is used.

Airborne sound reduction

Sound created from any source persists for a long time in rooms with hard floors (eg, indoor swimming facilities, traditional churches) and all rooms are characterised by a reverberation time. The reverberation time is the time required for sound to die away to one millionth of its initial intensity (ie, to drop 60 dB in sound level). Carpets on the floor or the wall may reduce a reverberation time from, say, 2 to 0.5 seconds. With ISO 354, sound absorption coefficients are measured by generating sounds of frequencies ranging from 125 to 4000 Hz within a standard reverberation chamber and measuring the reverberation times with and without carpet present.

Sound absorption coefficients are calculated from the result, taking into account the area of the carpet sample and the dimensions of the chamber: a sound absorption coefficient of 1.0 would indicate complete absorption of sound. The noise reduction coefficient is a single number obtained by averaging the sound absorption coefficients for 250, 500, 1000 and 2000 Hz.

20.3.4 Thermal Comfort

The thermal insulation properties of carpets are valuable in two respects: energy saving and apparent comfort to the occupants of the building. Carpets also contribute positively to the operation of underfloor heating systems, but the thermal characteristics of the carpet must be matched to those of the system.

Energy savings from installing carpets are highly dependent on building design: single storey buildings having ventilated floors benefit most. Savings of 5-13% of heating costs from using wool carpets have been recorded in practical situations [14]. Carpet increases the sense of comfort by reducing heat loss from the body to the floor by radiation and by contact with the feet.

The thermal resistance of carpets can be measured by ISO 8302. Thickness alone is an inadequate guide to thermal resistance, since heat may be transmitted in three ways: convection through movement of air within the carpet; conduction through fibres; and radiation. Thick carpets of moderate density perform best; and wool has a lower thermal conductivity than other pile fibres. Underlays contribute to thermal resistance, but not necessarily additively.

20.3.5 Indoor Air Quality

Although there has been publicity about possible negative effects of carpets on indoor air quality, and research has been necessary to counter this [15], carpets can have a positive effect on the quality of indoor air. For example, measurements of the common city air contaminants formaldehyde, nitrogen dioxide and sulphur dioxide are markedly lower over wool carpets [16]. Additionally, wool carpets have been shown to exert an important buffering effect on changes in air humidity that may originate from human breath and perspiration, cooking and washing, which could otherwise give condensation problems in unventilated buildings [4].

20.4 PERFORMANCE IN USE

Performance testing of carpets is usually taken to mean evaluating the extent of the deterioration of carpets that may be expected to take place during use.

Over the years, the emphasis of test development has moved from wear life to loss of appearance, since consumer dissatisfaction is now focused on deterioration of a carpet's good looks – perhaps a tribute to general improvements in durability that have been achieved by the industry.

When considering carpet performance tests, the type of wear associated with the intended location should also be considered. Types of abrasive wear include:

- walking (straight, turning and on stairs),
- wheeling trolleys,
- the action of castor chairs,
- dragging objects over the carpet,
- the action of cleaning machinery.

In performance testing, the following aspects must be considered.

20.4.1 Durability

Wear life

The first comprehensive controlled real wear trials carried out on carpets [17] showed that wear life is a function of pile mass, P, and surface pile density, P/t (ie, of P^2/t overall), and this function has been confirmed over many years. One predictor of wear life that has been used is carpet construction and composition (ie, P^2/t divided by a fibre factor). Fibre factors used in EN 1307 (see Section 20.6.1) are given in Table 20.1.

Table 20.1. Fibre factors for calculating wear life.

Pile fibre	Fibre factor	Pile fibre	Fibre factor
Nylon BCF	1.0	Wool	1.9
Nylon staple	1.2	Cotton	2.0
Polypropylene BCF	1.2	Modified viscose staple	2.2
Polypropylene staple	1.4	Modacrylic staple	2.4
Polyester staple	1.6	Blends	pro rata
Acrylic staple	1.8		

Schiefer's classical research programme on carpet wear[17] also resulted in a laboratory abrasion tester, which was subsequently developed into the WIRA Carpet Abrasion Machine (IWS/WNZ TM 122). Carpet specimens in the form of small discs are abraded against larger discs of abrasive fabric. Specimen and abradant are each clamped to the ends of vertical shafts which are offset. The shafts rotate in the same direction and at the same speed, providing constant relative velocity at all parts of the specimen, and the loading is 0.56 g/cm^2. Two alternative measures of resistance to abrasion may be used: fibre loss over 5000 rubs (IWS/WNZ TM 283) or number of rubs to expose the backing (IWS/WNZ TM 122).

IWS/WNZ TM 283 for wool and 80/20 wool/nylon carpets is called a wool fibre integrity test and the modern purpose is to determine if the wool in a carpet is damaged or not. If the wool is found to be undamaged, the wear index (P^2/t divided by a fibre factor) is used as a predictor of wear life.

The Lisson Tretrad machine of ISO 12951-1 (Figure 20.8) is an alternative abrasion tester widely used in continental Europe. The suitability of a carpet for use on stairs can be evaluated by a modification that incorporates a stair nosing (ISO 13746). However, the machine exerts a very severe shear action on the pile, tending to remove fibre and tufts from carpet having thick pile, and is not generally favoured outside of continental (Western) Europe.

Figure 20.8. The Lisson Tretrad. (Courtesy of Teppich Forschung International)

Tuft withdrawal force

Actual loss of tufts, or laddering, of loop-pile carpets can be a more immediate problem than abrasive wear, especially when carpets are installed on stairs, so it is vital to check carpets for adequate tuft bind.

In ISO 4919, the force required to withdraw pile yarn from the backing is measured on a tensile tester at a set crosshead speed (or a dedicated WIRA instrument having constant rate of loading may be used); 25 replicate tests are carried out. Note that a tuft is considered to be the material in one dent in the ISO test, even if two yarns are tufted in one needle or woven in one dent, whereas some test methods consider such yarns to constitute two separate tufts. The WIRA test (constant rate of loading) does not always give identical results to ISO 4919 (constant rate of extension).

Fibre bind

Inadequate penetration of latex into the base of the tufts and/or poor yarn construction can result in excessive fibre shedding from cut-pile carpets in the early stages of wear, or cobwebbing in loop-pile carpets.

The 'Velcro test' can be used to assess fibre bind by passing a Velcro-covered roller over the pile surface. The action is difficult to standardise, but the test is used by some carpet manufacturers in-house. A Crockmeter type drive action has been added to this test in South Africa to remove the effect of operator pressure.

EN 1963 uses a shortened Lisson test to assess fibre bind, primarily for loop pile. A modified Martindale test and the Usometer (see Section 20.4.3) are also used for this purpose.

Delamination strength

Inadequate control of backcoating of tufted and knitted carpets may usually be detected by measuring delamination strength (ISO 11857). Strips 50 mm wide from warp and weft (five from each) are partially delaminated manually and the pile and backing components are gripped between the jaws of a tensile tester. The average force required to continue the separation by a peeling action is reported in newtons.

This test is not related to any wear action and is essentially a quality control test for backcoating. Because it does not incorporate fatigue or ageing, it does not detect problems of delamination that can arise when the backing is flexed during wear.

20.4.2 Dimensional Stability

Standard tests for dimensional stability (eg, ISO 2551) focus on the shrinkage that may occur when carpets are subjected to the action of heat and water. Such tests are particularly relevant to carpet tiles.

There is a need for a test that simulates the 'pastry rolling' effect of traffic that can lead to an increase in dimensions and consequent rucking of wall-to-wall installations. In large areas, quite small percentage changes in dimensions can translate into significant length changes.

20.4.3 Change in Appearance

The following definitions originated by ISO TC38/SC12/WG1 are relevant to this section and are given in ISO 2424.

Change in structure / textural change – A visible change in configuration of loops and tufts at the surface of a textile floorcovering.

Loss of tuft definition – The bursting, opening and untwisting of the pile yarn and/or decrimping of the fibres in the surface.

Crushing / flattening – Loss of thickness under the action of a dynamic or static load.

Felting/matting – Loss of pile definition due to entanglement and compression of the pile fibres.

Hairiness/filamentation – The protrusion of fibres above the normal level of the pile surface, not removable by brushing or suction.

Cobwebbing – An extreme form of hairiness/filamentation in which the fibres are entangled to form an interlaced web attachment to the pile surface.

Pilling – An extreme form of hairiness/filamentation in which the fibres are entangled to form small aggregates, which may or may not include fibres from other sources, attached to the pile surface.

Sprouting – The release and appearance during use of extra long tuft legs which were accidentally trapped within the pile during manufacture.

Change in pattern definition – Change in the colour appearance of patterned textile floorcoverings due to mechanical action. A change of pattern definition may be caused by a change in the clarity of the contour lines.

Change in colour – The change or apparent change in colour, assessed using a set of grey scale cards, which may result from one or more of the following: change in orientation of the pile (shading); whitening/chalking; fading; glossing; colour bleeding; staining; soiling.

The overall change in appearance is the difference between a fatigued and an unfatigued specimen, taking into account the following factors. (Note: Changes in structure, thickness, colour and/or pattern may contribute to the overall change in appearance. It is not always possible to distinguish clearly between the factors, since each may have an interaction with the others.)

Thickness loss

ISO 3416 is designed to determine the effect of heavy furniture. The WIRA Carpet Static Loading Tester applies a load of 700 kPa for 24 hours. Thickness recovery is measured after 2 minutes, 1 hour and 24 hours.

ISO 3415 is commonly known as the chair leg test. A pressure of 220 kPa is applied for 2 hours and recovery is measured after 15, 30 and 60 minutes.

The WIRA Dynamic Loading Machine was designed to measure the consolidation of pile that is produced by prolonged foot traffic. Although the test is still used in some organisations, similar flattening is produced in the Hexapod test, which has the primary objective of measuring change in texture. Because the Hexapod produces more information from a single test, it is more commonly used.

Change in texture

Figure 20.9 is the Hexapod Tumbler Tester (ISO 10361).The equipment is also illustrated along with other carpet test equipment in Figure 20.10. Carpet samples are mounted as lining in a drum (over underlay if required) and a 3.8 kg tumbler having six equally spaced polyurethane studs is placed inside. The drum is rotated, with reversals every 15 minutes, for 1500 revolutions (short-term change equivalent to 9–12 months residential wear) and 8000 revolutions (long-term change equivalent to 4–5 years residential wear). Samples are removed for cleaning every 2000 cycles.

The treated samples are assessed for changes in colour, texture and pattern by the procedure of ISO 9405 (ie, comparing against standards and assigned into one of five grades, 1 worst to 5 best, with half grades being permitted). The effects on pattern (if any) in increasing or reducing the visible effects of colour or texture changes in use are taken into account. Fibre shedding is also observed.

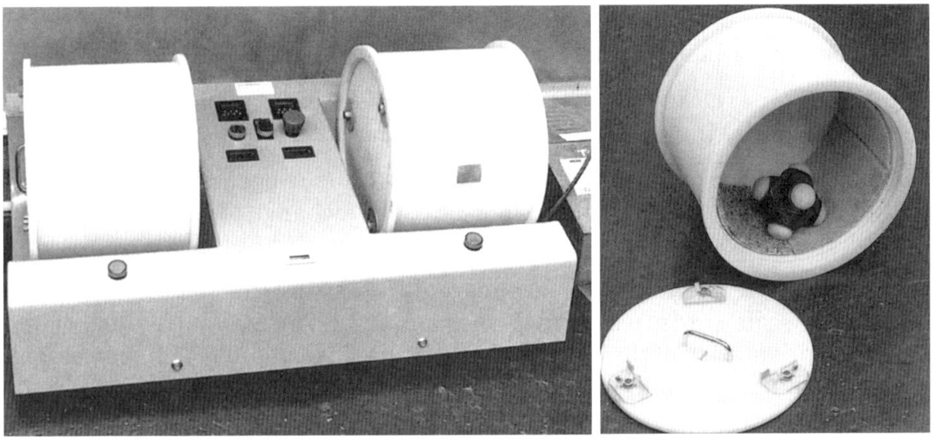

Figure 20.9. WRONZ Developments Hexapod Tumbler Tester.

Figure 20.10. Carpet testing equipment available from WIRA Instrumentation, left to right: WIRA Tuft Withdrawal Tensiometer, WIRA Hexapod Tumbler with Soil Applicator, WIRA Carpet Abrasion Machine, WIRA Carpet Static Loading Tester, WRONZ Digital Carpet Pile Thickness Gauge and WIRA Dynamic Loading Machine.

Actual tested carpet samples in constructions comparable to the carpet under test are used as standards, although photographic standards are increasingly replacing physical standards. Additionally, thickness loss can be measured and expressed as absolute thickness loss or as a percentage of total carpet thickness or pile thickness, as an alternative to the dynamic loading test, but this is not part of the ISO procedure.

The Vetterman Drum (ISO 10361), illustrated in Figure 20.11, performs in a similar manner to the Hexapod Tumbler, except that shed fibres are continuously removed by vacuum extraction. The tumbler is a 7.8 kg steel ball that has 14 rubber discs protruding from it. The Hexapod Tumbler and the Vetterman Drum are regarded by ISO TC38/SC12 as having equal status, but are operated with significantly different test durations (cycles). Figure 20.12 shows a panel of judges assessing change in texture resulting from subjecting carpets to the Vetterman drum test.

Figure 20.11. The Vetterman Drum (Courtesy of Teppich Forschung International)

Image analysis has been developed as a tool for quantifying carpet texture, and has potential for use in objectively measuring texture changes that occur as a result of wear or carpet testing [18].

Pilling and fuzzing: The Usometer (WNZ TM 253) is used to indicate the risk of pilling and fuzzing of loop-pile carpets only. Test specimens are installed on two inside faces of a cubic box and two weighted rubber tubes are placed in the box. The test involves two boxes which are rotated for a specified number of revolutions so that the tumblers scuff the surfaces of the carpets. The tested

Figure 20.12. Assessment of texture change by a panel of judges, using five standards of comparison. (Courtesy of Teppich Forschung International)

specimens are compared with the original carpets and subjectively graded for pilling/fuzzing on a 1–5 scale (5 best).

Texture change due to hot water spillage (WNZ TM 257): This test, which essentially assesses the level of twist set in cut pile, is sometimes known as 'the cup of tea' test. A quantity (75 ml) of pure water at 50°C is poured on the carpet sample, the surplus is blotted up and the carpet is lightly rubbed and allowed to dry. The change in texture is assessed on a 1–5 scale (5 best).

Texture change due to castor chairs: The Baumberg Rollstuhl test of EN 985 subjects a large disc of carpet to the action of three loaded chair castors while it is rotated. The test has not been accepted by ISO. Alternatively, the suitability of carpets for use in castor chair locations may be characterised by minimum levels of P^2/t and tuft bind, and superior performance in the Hexapod/Vetterman Drum and delamination tests.

Pile reversal (shading): A test method developed by WRONZ utilises a large revolving drum having the sample under test mounted on the outside, and a braked smaller drum exerting a shear action on the pile (Figure 20.13). The braking force required to reverse the direction of pile lay is measured. A low braking force indicates a high propensity to shading.

Appearance change (colour)

Changes in colour may occur as a result of pile flattening, pile reversal or fibre fibrillation. Such changes are noted in the tests for change in texture. Separate colour changes that are related to the coloration of the pile material may occur during use.

Figure 20.13. WRONZ Pile Reversal Tester. (Courtesy of WRONZ Developments)

Colour fastness to light (ISO 105-B02): Test samples, along with SDC blue scales, are exposed to simulated daylight from a xenon arc lamp under specified conditions of temperature, humidity and illumination. Assessment is in terms of the number of the blue standard (there is a scale of 1–8) that fades at the same rate as the sample under test. This test is not acceptable for assessing photobleaching.

Colour fastness to rubbing (ISO 105-X12): This is essentially a test for surface dyestuff. Using a Crockmeter, carpets are rubbed in a standard manner with cotton cloth mounted on a 25 mm diameter finger. Two tests are carried out, one with dry and one with wet cotton. The results are assessed against standard grey scales, 1 representing severe staining and 5 no staining.

Colour fastness to shampoo cleaning (AS/NZS 2111.19.2): The carpet sample under test, with its pile in contact with two standard undyed fabrics (cotton and wool), is immersed in standard shampoo solution for 15 minutes at 37°C with the materials pressed together. The shampoo solution is poured off, but the samples are kept under pressure for a further 90 minutes at 37°C. The fabric and samples are separated, and the carpct is hydroextracted and dried below 60°C. Change of shade and staining of the adjacent fabrics are assessed using standard 1–5 grey scales.

Colour fastness to water (ISO 105-E01): This is a general test for wet fastness, similar to the shampoo test, but a little more severe. Wet carpets and standard adjacent fabrics are sandwiched between glass plates for 4 hours at 37°C.

Colour change due to soiling

Carpets may change colour as a result of soiling to an extent that depends on the level of traffic, prevalence of soil in the location, type of soil (colour, composition, inorganic, organic), pile fibre type, presence of residues on the pile fibre, use of soil/stain-resist finish, carpet construction and initial colour of the carpet. The colour of the carpet may be more important than most other factors. In some locations, the carpet specifier (architect or interior designer) would be wise to consider the colour of the carpet in conjunction with the type of soiling likely to be experienced in the locality.

Two international methods are: ISO 11378, Parts 1 and 2.

Part 1: Kappasoil test. The Kappasoil machine consists of a circular table 600 mm in diameter rotating at 18 rpm and reversing every 250 revolutions, to which the test specimens are attached. Between two and twelve specimens, mounted in sectors, are exposed in each test run. Two cones resting radially on the table run over the test specimens; and a soil dispenser placed above the table evenly dispenses 1.4 g of soiling compound at intervals of 250 revolutions over a test period of 5000 revolutions. One of four alternative artificial soils may be used: ISO Artificial Soil; Teppichschmutz; 3M Carpet Soil; and CEN Standard Soil. The specimens are suction cleaned before and after testing, and the colour difference produced by soiling is either measured as ΔE or ΔL (IS0 105), or assessed by large size grey scales. The scope of the test can be extended by cleaning the soiled specimens to quantify the relative efficiencies of cleaning products or processes.

Part 2: Modified Hexapod Drum Test. The test specimens are mounted as lining, along with a light coloured reference sample, in the Hexapod Drum and 250 g of soiled polymer pellets plus 1 kg of 9.5 mm chrome alloy steel balls are introduced. The drum is rotated for 30 minutes and the samples are removed. As in Part 1, the samples are suction cleaned and measured (or assessed) before and after soiling. While AATCC soil is preferred, any of the four soils listed for Part 1 may also be used.

In the case of wool pile carpets, the presence of oily residues, as measured by the dichloromethane extract test, is sometimes used as a surrogate test for propensity to soiling. When used on wool/synthetic carpets the test can give a false negative by extracting material that does not promote soiling. In addition, it cannot predict stripy soiling that can result from irregularly distributed residues.

Susceptibility of carpets to staining

AATCC Test Method 175 (ISO 13750), determines the resistance to staining by acid food colours. A 50 mm ring is placed over the surface of the carpet and 20 ml of a staining solution (1 mg/l FD&C Red 40) is poured in. The bottom of a matching stain cup is pressed through the ring into the pile and moved up and down five times to facilitate wetting of the pile. The ring and cup are removed and the wet staining solution is left in position for 24 hours. After rinsing until the rinse water is clear, and drying, the level of staining is rated against the standard AATCC Red 40 Stain Scale. This is a 1–10 scale where 1 is severe staining and 10 is no staining.

A staining technique developed by the Rengoringsteknisk Institut, Denmark, is more severe in that six staining materials have been standardised: coffee, cocoa, Fanta (carbonated orange drink), tomato ketchup, carbon black (toner for photocopying) and blackboard chalk; other substances can be used as appropriate to local requirements.

The staining apparatus is an acrylic disc with concentric ribs ducted together, coupled to a 20 ml syringe (Figure 20.14). Staining liquid is injected into the carpet by depressing the plunger with the left hand, simultaneously pressing the distribution head against the carpet sample and turning it clockwise with the right hand. A well-defined uniform stain 80 mm in diameter is produced. The stain is allowed to dry at room temperature for 3 days and is then subjected to a standard cleaning procedure. Residual staining is assessed against standard grey scales or by means of a reflectance colorimeter.

Figure 20.14. Rengoringsteknisk Institut staining appartus.

20.5 CHEMICAL TESTING

20.5.1 Environmental Control

GuT

GuT (the Association for Environmentally Friendly Carpets) was founded in 1990 and the majority of European carpet manufacturers are members. The principal objective of the association is to provide the consumer with products that are manufactured using the best environmental practices, that promote the well-being of the consumer, and that at the end of the product's life can be disposed of free from any problems. A further objective is to inform the consumer on environmental aspects of carpets, in part through use of the environmental viability label. Members of GuT are committed to following the best current environmental practices. GuT's activities for future improvements include studies on the disposal and recycling of carpets. Most European laboratories concerned with carpet testing are involved with GuT through the associated testing and production monitoring requirements.

To qualify for the GuT label (Figure 20.15) carpets must be subjected to tests for contaminants, emissions of VOCs, health damaging pesticides and odours.

Figure 20.15. Example of a GuT label. (Courtesy of Teppich Forschung International)

Contaminants not permitted in manufacture, and not detectable in the product include: volatile fluorochlorinated hydrocarbons, asbestos, dye carriers, azo-dyestuffs with carcinogenic amine contaminants, pentachlorophenol, vinyl acetate, formaldehyde, hazardous pesticides, butadiene and vinyl chloride.

Emissions of VOCs are measured by ventilating a 0.5 m^3 box containing 0.2 m^2 of carpet with purified air at a rate of 500 l/h until equilibrium is reached. An air sample of 1000 l is then drawn off and its organic constituents are adsorbed in activated charcoal. The adsorbed material is extracted with carbon disulphide and analysed using a combination of gas chromatography and mass spectrometry. The upper limits quoted by GuT are given in Table 20.2.

The list of banned pesticides includes all chlorinated hydrocarbons, plus Minex, Malathion, Parathion-methyl and Parathion-ethyl. The pyrethroids commonly used for insect-resist treatment of wool are permitted.

Samples tested may exhibit only a 'new' odour of low intensity, typical of the product.

Table 20.2. GuT upper limits for VOC emissions.

Compound	(mg/m^3)
Toluene	0.1
Styrene	0.005
Vinylcyclohexene	0.002
4-phenylcyclohexene	0.03
Aromatic hydrocarbons	0.3
Volatile organic substances	0.5

Environmental Choice New Zealand

Environmental Choice New Zealand has wider environmental objectives than carpets, but includes a Requirements Specification EC-24-95 that covers wool-rich pile carpets.

The emission standards are similar to those of GuT, but testing includes carpet performance requirements, and there are standards for all manufacturing operations, for example:

- Raw wool scouring must comply with strict effluent requirements.
- Synthetic fibre must be sourced from facilities having formal programmes to reduce emissions to air and water.
- Insect-resist treatment may be carried out only at certain stages of manufacture.
- The COD level in dyehouse effluent must not exceed 55 kg/t of fibre dyed.
- Dye recipes must not contain more than 200 mg of chromium/kg of wool dyed.
- Fluorocarbon finishes and sulphonated phenolic stain blockers must not be used.
- Cotton used in the construction of carpets must be unbleached.
- Washdown water from latex equipment must be discharged to an approved effluent plant.

20.5.2 Testing Insect Resistance (wool)

Biological assay testing (ISO 3998) is inapplicable directly as a measure of the required resistance of wool carpets to insect attack. This is because new carpets should contain a sufficient level of insect-resist agent to allow for removal of a proportion during the life of the product by photodegradation and cleaning. The biological effects against the principal insects concerned of all the currently used insect-resist agents applied at a range of levels have been measured. Data are also available for losses of insect-resist agent in cleaning and photodegradation. Commercial requirements for insect resistance are therefore determined by chemical assay. The methods used are high performance liquid chromatography (HPLC) or packed column gas liquid chromatography (packed GLC).

20.5.3 Testing for Visible Faults

Common faults in carpets include stripes, streaks and bands. Such faults are often visible only in large areas under good lighting conditions. An examination of a typical size of sample as used for other tests is unlikely to lead to a valid conclusion.

A band across the direction of manufacture is most likely to be a stop mark or other fault arising at tufting or weaving.

Although stripes, streaks or bands in the direction of manufacture may be perceived as colour differences, there are several possible causes:

- differences in dye uptake,
- differences in texture,
- contamination.

Texture changes that may have been produced during handling or storage should be eliminated by suction cleaning a portion of the sample before examining and describing the fault (thorough cleaning of the whole sample may remove some of the contamination that could be a possible cause of the fault). For many tests, it is necessary to remove the rogue material and compare it with material from the bulk of the pile. This requires skill and possibly two persons – one to observe and direct, and the other to dissect the carpet.

Inherent differences in dyed colour can be checked in the case of spun yarns by hand carding the separated materials and measuring the colour difference. Supplementary information can be obtained by extracting the dye from weighed samples, making up the extracts to standard volumes, the colours of which can be compared using a transmission colorimeter.

Apparent differences in colour owing to different orientation of the tufts may be related to the backing. Check for an incorrect choice of primary tufting fabric (see Section 5.2) or for missing stuffer yarns (woven carpets).

Faults originating at tufting, as described in Chapter 6, may be identified by persons experienced in viewing such faults. Measurements of tuft leg length can assist in identifying faults such as J-tufting and tight ends.

Texture stripes or streaks can result from variations in yarn twist, bulk, set or regain (wool). Counting the number of turns of twist in a large number of tufts to ascertain the average levels of twist in the stripe and background material and weighing these tufts to ascertain the bulk can provide useful information. There may be a twisting fault or a twist-setting fault. Knowledge of the setting process used can guide the investigator to further tests (eg, a variation in set may give rise to a variation in colour).

Soiling stripes are often caused by organic residues that attract soil. Comparing the levels of extractable matter on the rogue and standard materials will usually reveal this fault.

Staining can arise from machinery or from contaminated wet processing liquors. Such stains can often be identified by measuring the metal content of the yarns, notably iron and aluminium.

Detailed procedures for identifying faults are described by Brown [19, 20].

20.6 CARPET CLASSIFICATION SCHEMES

Purchasers select carpets primarily on the basis of colour, texture and design, which they can see and assess. They are also concerned about the performance of the carpet in use – aspects such as wear life, retention of appearance and ease of cleaning, which are invisible. The main purpose of carpet classification schemes is to provide the consumer with reassurance that the carpet will perform satisfactorily in its intended location, usually by means of some form of label. Although some large and regular purchasers of carpets on contract may have their own views on carpet performance, many contract purchasers do not have specialised technical knowledge about carpets. Classification schemes are, therefore, relevant to the contract as well as residential markets. A further purpose of international carpet classification schemes is to facilitate international trade in carpets. Exporters of carpets will clearly find it more convenient to be able to label their carpets according to an international or regional scheme than to have to test them to qualify for a number of national schemes.

Many countries have national schemes; there is a European carpet classification scheme and ISO is developing what should be a truly international scheme. Additionally, several fibre suppliers and carpet manufacturers operate labelling programmes that are essentially carpet classification schemes.

Over the years, communication of the various classes (ie, the design of labels) has changed and no doubt there will be further changes as experience in the marketplace grows and the objectivity of carpet testing improves. Some schemes have classified carpets in terms of a location guide (eg, suitable for bedrooms, living rooms, work areas, etc), while others have focused on traffic levels (eg, two persons, large family, etc). Recently, the emphasis seems to have been on the general quality of the carpet, communicated via a numerical scale or a descriptive scale (normal wear, heavy wear, etc), perhaps with some guidance on location (eg, suitable for stairs).

20.6.1 European Carpet Classification Scheme (EN 1307)

A directive in 1989 from the European Commission relating to safety and health, and with implications for textile floorcoverings, stimulated a voluntary agreement between eight national carpet manufacturer associations to develop a CEN carpet classification scheme. In parallel with the evolution of the scheme, the European Carpet Association arranged for nine testing institutes to undergo training and assessment that led to mutual acceptance of test results. The resulting scheme EN 1307 applies to all wall-to-wall carpets, but has modified requirements for electrostatically flocked carpets. An important feature of the scheme is that it recognises the different expectations from the widely different styles of carpet, and sets different requirements for wear and appearance retention for L and N categories of carpets, informally known as luxury and normal style carpets. Carpets (M category) falling between these two principal categories may be classified for wear according to the requirements for either L or N category carpets, but must meet the same requirements for appearance retention as N category carpets.

The definitions of the three categories are:

- Category L: thick, heavy carpets where the pile mass per unit area above the backing is at least 600 g/m^2 and pile thickness at least 6 mm, or where the pile mass per unit area above the backing is at least 900 g/m^2.
- Category M: all other carpets that fall outside the specifications for categories L and N.
- Category N: thin, low weight carpets where mass per unit area above the backing is 500 g/m^2 or less and pile thickness is 5 mm or less.

All carpets are required to meet basic standards for colour fastness to light, rubbing and water, fibre bind (synthetic carpets) and wool fibre integrity (wool and wool/nylon carpets).
Additional claims may be made for:

- castor chair suitability,
- antistatic properties,
- acoustical properties,
- thermal resistance,
- suitability for use in incidental humid conditions,
- suitability for stairs,
- fraying behaviour.

Between these two groups of properties that are tested for on a pass/fail basis are three properties – wear, appearance retention and luxury – that are used to classify pile textile floorcoverings according to their suitability for use under different circumstances.

The scheme is to be reviewed at five-year intervals (starting in 2002) so that it may be progressively refined.

Classification for wear
Category L carpets are classified into four categories on the basis of wear index (P^2/t divided by a fibre factor, see Table 20.1).

Category N carpets are also classified into four wear categories, but on the basis of an index combining effective pile mass and the results of the Lisson test. To achieve Class 3 rating, the carpets must additionally meet one of three alternative construction requirements (pile mass, tuft density or pile density); and to achieve Class 4 rating the carpets must meet one of two additional requirements, each of which includes a minimum fibre linear density and different minimum pile densities for knitted and woven carpets, and for 'other' carpets.

Classification for appearance retention

All three categories of carpet (L, M and N) are classified according to the results of Vettermann Drum or Hexapod testing, with minimum requirements for appearance retention after short- and long-term testing. Within this scheme, there are variations in the severity of test and the appearance ratings required for the various classes.

Category L carpets with pile of 100% wool or 80/20 wool/nylon have different ratings for Classes 3 and 4 (compared with M and N style carpets) and the required ratings also take into account surface pile density and whether a carpet is patterned or plain. The table of requirements for category L carpets containing other fibres is different again.

For category M and N carpets, the four appearance classes are the same, irrespective of pile composition. However, the test requirements are more severe than those for category L wool carpets in order to align the results for all categories with real wear.

Classification for luxury

There are four alternative ways of classifying carpets for luxury, all based on some aspect or aspects of carpet construction. Moreover, there are two alternative systems of rating, one based on five luxury classes and the other having only three luxury classes.

Grading

EN 1307 has adopted the principle that all carpets should meet minimum standards of wear performance and appearance retention, but points are assigned to the performance in each test, and a grade (measure of overall excellence) is assigned on the basis of the total points attained.

The tests are:

- Wear Index (P^2/t divided by a fibre factor) or Lisson ITR factor.
- Appearance retention (short- and long-term) by Hexapod or Vetterman Drum.
- Surface pile density in g/cm^3.
- Tuft density in $tufts/dm^2$.
- Luxury Rating Class.

There are five grades for domestic carpets and four grades for contract carpets. The overall grade of the carpet is based on the lower of the appearance and wear values, with luxury reported separately.

Labelling

Pending the development of a CEN labelling system, some of the national carpet manufacturer associations adapted or modified their labelling systems (eg, PIT of the Netherlands, ETG of Germany, BCMA of the UK, and UPEC of France). These interim arrangements are largely based on the test methods used for EN 1307, but there may be different requirements for the grades. Figure 20.16 illustrates the ETG labelling system.

Figure 20.16. Example of an ETG label. (Courtesy of Teppich Forschung International)

20.6.2 Other European Classification Schemes

A reading of Section 20.6.1 will reveal that the CEN classification scheme incorporates considerable flexibility in terms of test methods used and ratings applied for the three broad categories of carpet (L, M and N). This minimised the expense of re-equipping testing laboratories and retraining staff in the period of introduction of the scheme and was designed to cause minimum disruption in terms of re-classification of established lines of carpet.

These goals were probably largely achieved for the countries on the European continent. For example, the four wear/appearance retention classes have been adopted for the widely used Europaeische Teppich Gemeinschaft (ETG) scheme and are readily applicable to typical low, dense pile tufted carpets of continental manufacture. The CEN scheme provides the flexibility to classify different styles of carpet (mostly imported) more fairly. The ETG labelling scheme was based essentially on the CEN classification scheme.

In Britain, the traditional classification of carpets was into five domestic and four contract grades, so considerable disruption of the existing situation was required to change to the CEN scheme. Some British manufacturers continued to use these grades towards the end of the twentieth century.

20.6.3 Australian Carpet Classification Scheme

The Australian scheme categorises carpets on the basis of points awarded. Up to 81 points are awarded on the basis of an objective scoring system, with 19 discretionary points awarded by a panel of judges based largely on appearance and retention of appearance (including assessment of the results of Hexapod testing). In addition, carpets must pass minimum standards for colour fastness, oily residues, tuft anchorage, delamination, quality of foam backing (if used), twist level (certain yarns) and insect resistance (wool).

The tests used for allocating the points (maximum) are as follows:

- Surface pile mass/pile height (16 points). Tables of guideline pile masses for different fibres and tufting gauges are available. Higher surface pile masses are expected as gauges become coarser. A formula for calculating the points incorporates the square root of the pile height.
- Volume density (20 points) and tuft density (9 points).
- Dynamic and static loading (5 points each).
- Propensity for soiling (16 points). Calculated on the basis of accepted soiling propensity characteristics of the major carpet fibres plus one point each for fluorocarbon and stain-resist finishes.
- Abrasion resistance (10 points). Calculated on the basis of surface pile mass relative to the guideline pile mass and a fibre factor.

The points required for each category are:

Residential	Light duty	37–46
	Medium duty	47–57
	Heavy duty	58–68
	Extra heavy duty	69 and above
Contract	Light duty	60–69
	Medium duty	70–77
	Heavy duty	78 and above
	Extra heavy duty	Panel's discretion

20.6.4 United States of America

The principal motivation for US involvement in carpet classification is to facilitate international trade, and the industry, through the Carpet and Rug Institute, is contributing positively to the development of an ISO classification scheme.

A North American Standard for the classification of wall-to-wall pile carpets and tiles is under development. The probable basis is a table of minimum requirements for colour fastness, tuft bind, filament damage, delamination strength and flammability; plus classification on the basis of the Vettermann Drum or Hexapod tests.

Independent of the official activities, a number of commercial organisations in the USA are operating labelling/quality programmes that may be regarded as carpet classification schemes.

20.6.5 ISO Carpet Classification

The textile floorcoverings sub-committee ISO TC38/SC12 has developed a document *Guidance on Drafting Classification Schemes for Pile Carpets*. The principal problem in developing an international scheme is to avoid bias against particular styles of carpet (eg, low, dense European styles or thick, luxurious American styles). However, ISO has made substantial progress in the internationalisation of test methods – an essential step towards the development of an international carpet classification scheme. Note that ISO TC38 (Textiles)

SC12 (Floorcoverings) has been replaced by ISO TC219 (Floorcoverings), which deals with textiles, laminates and resilient floorcoverings.

REFERENCES

1. P G H Bakker and G H Crawshaw. The chain of carpet properties – a framework for research on performance features of carpets. In: *Proceedings of 6th International Wool Textile Research Conference*, Pretoria, 1980, Volume IV, 305–325.
2. *Identification of Textile Materials*. The Textile Institute, Manchester.
3. W Herzog. Textile floorcovering: the "Usefulness Index" and its testing. Part I: Walking comfort; Part II: The action of walking. *Textile Institute and Industry*, 1971 9(5), 126–128; 9(6), 153–157.
4. W Meade. *Consumer Properties of Carpets*. WRONZ Technical Bulletin, Wool Research Organisation of New Zealand, Christchurch, January 1998.
5. I D McFarlane and J M Watson. *The Use of an Accelerometer in the Measurement of the Comfort of Soft Floorcoverings*, WRONZ Communication No. C94, Wool Research Organisation of New Zealand, Christchurch, 1984.
6. D G Feather. The comfort factor of carpets. *Wool Science Review No. 45*, May 1973, 39–48.
7. G Wentworth Smith. The research, development and marketing of ergonomic flooring surfaces. In: *Proceedings Carpet Yarn Twist and Heatsetting*, Clemson University Office of Off-Campus Distance Learning and Continuing Education, Clemson, USA, 1999.
8. E J Wood and P K Johnson. *Friction in Carpets*, unpublished WRONZ report, Wool Research Organisation of New Zealand, Christchurch, 1992.
9. *Sportfunctionelle Eigenschaften von Kunstoffrasenbelagen*. Report B1/81, Institute for Sports Science, Germany.
10. G H Crawshaw. *Textile Sports Surfaces and Artificial Grass*, Elsevier Advanced Technology, Oxford, 1989.
11. *BS 7044, Artificial Sports Surfaces, Part 2, Methods of Test.*
12. *Carpet Manufacture 6.11: Flammability Standards in the Marketplace.* Carpet Technical Information, Wools of New Zealand, 1995.
13. J M Brown. Recent developments in carpet specification, grading and standards. In: *Proceedings of Tifcon '98*, The Textile Institute, Manchester, 1998.
14. *Facts About Contract Carpeting*. International Wool Secretariat, London, 1979.
15. *Carpet and Indoor Air Quality*. White Paper. DuPont HomePage, www.dupont.com/dcf/dfs/environmental/iaq.html, 1996.

16. S Causer. *Removal of Indoor Air Contaminants by Wool Carpets.* WRONZ Technical Bulletin, Wool Research Organisation of New Zealand, Christchurch, August 1996.
17. H F Schiefer. *Journal of Research of National Bureau of Standards*, 1942, 29, 333–379.
18. E J Wood. Objective measurement of carpet appearance by image analysis. Part 1: Principles and methodology; Part 2: Applications. *Melliand Textilberichte*, 1996, 77(7/8) 452–459 (E99–102); 77(12), 845–850 (E182–184).
19. E M Brown. *Identification of Carpet Faults*, Cleaning Research Institute, Otley, UK, 1995.
20. E M Brown. *Diagnostic Techniques for the Investigation of Carpet Complaints.* Cleaning Research Institute, Otley, UK, 1995.

21 Speciality Carpets

Carpets produced according to the general principles described in earlier chapters may be destined for applications that require special properties, so that additional processes may be required, or the carpets may need to be engineered specifically for the purpose. A few such examples are outlined in this chapter.

21.1 CARPETS FOR HEALTHCARE ESTABLISHMENTS

Carpets are widely used in private hospitals and are increasingly being installed in public hospitals. Their most important function is to de-institutionalise the environment, reducing the level of stress on patients and contributing positively to the welfare of staff. Reduced noise levels, improved aesthetic appearance, comfort underfoot, safety from falls and an impression of warmth all contribute to the improved environment.

Carpets are not universally applicable in healthcare establishments, which vary widely in their function, while the floor areas within a hospital can be very different, flowing from reception and waiting areas to operating theatres. However, large areas of hospitals, including many types of wards, can benefit from the installation of carpets.

21.1.1 Requirements for Carpets in Hospitals

Many of the performance requirements for carpets in hospitals – durability, appearance retention, colour fastness, acoustics, flammability and electrostatic properties – are not very different from those for carpets intended for other types of public buildings. However, there are additional special requirements. An example specification [1] includes the following aspects.

Designs and colourways should not cause disorientation to patients and staff; and should be compatible with a wide range of colour schemes and furnishings. They should minimise the visual effects of soiling, staining and tracking.

The carpet should be amenable to the movement of trolleys, wheelchairs, walking aids, etc. In the relevant test, wheels of standard sizes, composition and loading are pulled over the carpet with a tensiometer and the impedance measured. When changing an installation from sheet vinyl to carpet, it may be desirable to increase the diameter of wheels on some trolleys. Procedures for evacuating patients during an emergency may include dragging the patient (with mattress) across the floor on a nylon sling. To test for ease of evacuation, the force required to drag a weighted sample of sling material is measured in, opposite to and across the direction of manufacture.

Backing materials should be rot-proof and the coating impervious to spillages and wet cleaning. The purpose of this requirement is to ensure that bacterial growth does not occur under the carpet. The backing should be able to withstand the action of castors.

The carpet should withstand normal cleaning procedures, but more frequent and rigorous cleaning, including spot cleaning, than for usual carpet installations may be expected (eg, daily suction cleaning and monthly wet cleaning).

The carpet should not react with spillages such as urine, vomit, blood, medication, beverages and cleaning solutions. Tests relevant to this aspect include ease of removal of protein (horse serum labelled with acid dye) and urea (detected by HPLC analysis of the aqueous extract) after a standard wet cleaning procedure. For some installations the carpet must be resistant to disinfectants, including hypochlorite.

The rate of drying in a standard atmosphere after a standard wet cleaning procedure is measured. An excessively long drying time indicates a risk of bacterial growth, as well as inconvenience due to a cleaned area being out of service.

It is important that the carpet should not provide a viable medium for pathogens that could contribute to cross-infection. To test for this, a pre-cleaned carpet sample is inoculated with a standard culture of bacteria and sampled at intervals up to 24 hours with contact plates. The contact plates containing nutrient are incubated and the colonies are counted. Additionally, a plug from the contaminated carpet sample is removed after 24 hours and an extract is incubated in nutrient broth. The colonies are counted.

Direct stick methods of installation are preferred to avoid the risk of 'bow waves' developing from wheeled traffic. Impervious seams should be created, using adhesives that do not support microbial growth or emit offensive odours.

21.1.2 Hygienic Characteristics of Carpets

A superficial consideration of carpets by many lay people and some medical people can lead to the belief that carpets harbour germs, may be responsible for allergies and cannot easily be cleaned. Florence Nightingale in 1859 was probably the first person to record such a view in writing. Scientific studies have led to the opposite view. The most comprehensive reports [2, 3] are rather old, but more recent studies [4, 5] have confirmed the earlier findings.

Bacteria are associated with people rather than with carpets. Human skin flakes account for much of the dust in the air and may contain micro-organisms. Skin flakes also provide nutrition for dust mites, the excrement of which is known to be a major indoor source of allergens. Humans may also emit micro-organisms in the form of aerosols from body orifices. Carpet fibres, being of high diameter, are not normally components of airborne dust.

Bacteria settle on floors, carpeted or otherwise, but since they require moisture if they are to develop, they slowly die. Sterile carpets placed on a floor accumulate a population rapidly and an equilibrium then becomes established between bacteria dying and new bacteria settling on the surface. When sampling immediately above or on the floor surface, bacterial counts tend to be similar for carpets and sheet vinyl, although higher counts have been noted after wet mopping vinyl [6], presumably because the moisture temporarily promotes growth. In the air, bacteria counts tend to be lower over carpets than over vinyl: carpets seem to trap bacteria.

The advantages of using fibres treated with bactericides or fibres with inherent antimicrobial properties in hospital carpets have been debated. Dry bactericide and dry bacteria do not interact: but some specifiers feel that a bactericidal pile fibre provides an additional margin for safety when wet spillages occur.

Allergens from house-dust mites commonly trigger asthma and carpets may be feared as reservoirs for such allergens. However, mites do not thrive in conditions of low humidity (ie, in carpets); rather they are associated with the higher humidity conditions in beds. Many studies have shown that mite numbers and allergen levels in hospitals are typically low or zero. Reference 7 is just one example of relevant medical research.

21.1.3 Suitable Carpet Types

The exposure to rolling traffic and the requirement for minimal dissipation of energy in walking by staff point to the need for a low, dense pile. This coincides with the requirement for high durability. For good retention of appearance and minimising of tracking, loop-pile constructions have considerable merit – the unconstrained ends of cut tufts are more susceptible to appearance change. Tweed or intermingled yarns, coupled with small designs, can hide stains and some other defects. Semi-plain or lightly patterned loop-pile tufted carpets in 1/10 or 1/8 gauge meet the general requirements well. Dense cut-pile carpets, woven or tufted, may be applicable in some areas, however.

Loop-pile carpets must have a high quality lock-coat to avoid laddering and cobwebbing in use, and the lock-coat must be augmented by a non-permeable layer. A PVC unitary backcoating is a common solution to the requirement for impermeability.

Flocked and most bonded carpets have an impermeable PVC layer as an inherent requirement of their processes of manufacture (see Sections 14.1 and 14.2). Electrostatically flocked carpets are promoted for hospital carpets on the basis of their impermeability and the fact that the low pile can be cleaned with the same wet scrubbing equipment as sheet vinyl, rationalising floor maintenance programmes.

Selection of pile fibre for tufted products is inevitably based on compromise, and there is no one ideal choice. Nylon and wool offer high resilience and good general wearing properties, but nylon in particular is susceptible to staining, and existing stain-resist treatments have inadequate durability for healthcare establishments. Both nylon and wool carpets need to be engineered for static control, although the solution for wool is simple (see Section 21.2). Polypropylene does not give rise to electrostatic problems and resists aqueous stains, but is less resilient and more flammable than nylon or wool. Perhaps the newer polyester fibres (PTT) may find a niche in healthcare carpets.

21.2 ANTISTATIC CARPETS

The most common electrostatic problem associated with carpets occurs when a person walking over an untreated carpet becomes electrically charged and experiences a shock when this 'body voltage', which may be around 20 000 V, is discharged. This can occur, for example, when touching a lift call-button or a filing cabinet. Some electronic equipment may be disrupted by discharges of body voltages.

Even hygroscopic wool carpets are susceptible to such problems when used in low humidity environments, such as arid or frosty climates, or in air conditioned buildings having no re-humidification facilities or when installed over heated floors. Charge build-up to a maximum takes place over a walking time of 30 seconds so that, in most climates, there are few problems with residential carpets where walking distances are usually short. For most contract installations consideration should be given to making nylon and wool carpets antistatic. Polypropylene and some other fibres are less likely to give problems.

The incorporation of conductive fibres into the pile and use of conductive primary backing materials or backcoatings control the build-up of static charges by conducting the charge across and through the carpet and by dissipating it through corona discharge. In the latter phenomenon, air is ionised by the concentration of electrical charges at the surface and ends of the conductive fibres, enabling the charge to leak away through the air. Some loop-pile carpets may be made satisfactorily antistatic by incorporating conductive filaments in the pile alone, but most carpets need to be made conductive throughout. A surfeit of conductive fibre is unlikely to be used for reasons of cost, but should not be used for reasons of safety: a carpet should not provide an earth for persons accidentally contacting mains power.

A variety of conductive fibres are described in Section 2.6 and in Chapter 5 references are made to conductive backing fabrics and yarns. Conductive materials are based on stainless steel or conductive synthetic fibres incorporating carbon, silver or copper sulphide.

The leading synthetic fibre manufacturers offer brands of nylon that contain small proportions of conductive fibre. To make pile yarns from unbranded nylon or wool, conductive fibres may be incorporated in the pile material in proportions ranging from 0.1–0.5%. At such low levels it is necessary to pre-blend the conductive fibre with a portion of the main blend if uniform distribution is to be achieved.

Conductive backing materials may be created by incorporating filaments or fine (eg, 20 tex) conductive yarns into the warp at weaving and/or applying conductive latex at finishing. Ordinary X-SBR latex can be made conductive by using additives, such as proprietary conductive chemicals, chopped conductive fibre or conductive carbon dispersions. Carbon is the most effective but is often avoided because of compounding issues and its black colour.

21.2.1 Meeting Body Voltage Specifications

Typical body voltage values (AATCC TM 134, step walking) for cut-pile carpets containing different combinations of conductive materials are given in Table 21.1, and body voltage curves for carpets with and without antistatic fibre are shown in Figure 21.1.

Various specifications are set for limits on body voltage generation, typical values being 2000 or 3500 V. Clearly, it is necessary to use both conductive pile and

Table 21.1. Body voltage generation (V) using standard test shoe soles from carpets containing conductive materials [8].

	Neolite	BAM
No conductive fibre or backing	-19 200	-16 600
Conductive primary only	-11 800	-9 900
Conductive carbon backcoating only	-10 000	-7 600
Conductive pile; conductive primary	-3 600	-2 200
Conductive pile; conductive backcoating	-1 900	-100

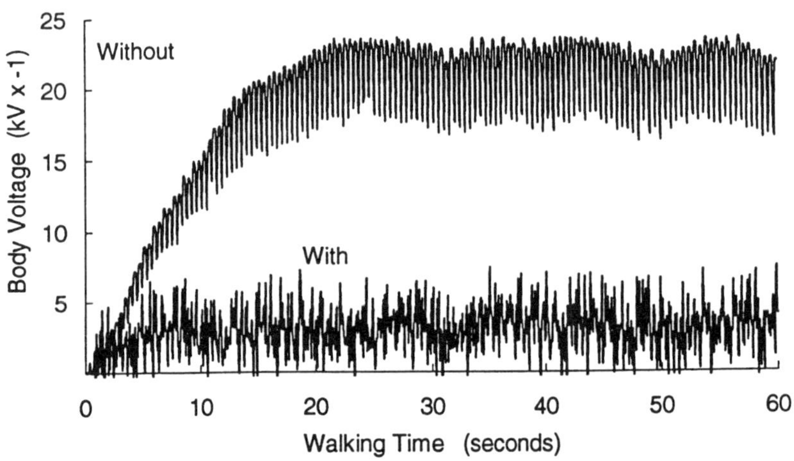

Figure 21.1. Effect on body voltage generation of incorporating 0.3% of antistatic fibre in wool carpet. (Courtesy of WRONZ)

conductive backing (fabric or backcoating) if a carpet is to meet such specifications. There is little margin between the results achieved from using conductive materials and typical specifications, so the engineering of antistatic carpets must be well controlled.

21.2.2 Meeting Resistance Specifications

Carpets to be used for computer accommodation are usually specified in terms of resistance rather than body voltage. For example, IBM specifies a maximum resistance for the flooring system (of which the carpet is one component) of 2×10^{10} ohms and, to minimise electrocution hazards, a minimum resistance of 1.5×10^{5} ohms measured between two points 1 m apart on the carpet. To meet AS 2834, resistances measured at 40% RH and 20°C must lie within the range 5×10^{5} to 2×10^{10} ohms. These limits apply to both the surface resistance (between electrodes 1 m apart) and the bulk resistance (through the carpet). The choice of conductive fibre can also be critical. Good surface conductivity is essential: a fibre having conductive material totally buried in the core is unlikely to be effective. Moreover, the conductive fibre must be adequately distributed throughout the carpet.

Table 21.2. shows the effects of using conductive materials, using more stringent test conditions than required by AS 2834. Again, both conductive pile and conductive backing are required if satisfactory results are to be achieved. It should be noted that there is no linear relationship between body voltage generation and resistance, so that a carpet that may meet a body voltage specification will not necessarily meet a resistance specification.

Table 21.2. Resistances (ohms) of carpets containing conductive materials, at 20% RH and 21°C [8]

	Surface	Bulk
No conductive fibre or backing	3.9×10^{12}	4.8×10^{12}
Conductive pile only	1.1×10^{13}	5.9×10^{11}
Conductive pile; conductive primary	9.3×10^{11}	8.4×10^{12}
Conductive pile; carbon in backcoat	3.6×10^{5}	4.9×10^{5}

21.2.3 Topical Antistatic Treatments

A variety of organic antistats may be applied to the pile material of carpets at levels of 0.3-0.5%, enabling them to meet body voltage specifications. They include anionic surfactants, such as organophosphate esters and poly (ethylene oxide) compounds of various kinds, and cationic surfactants. They operate by transferring small amounts to the shoe soles that contact them so that both the carpet and shoe sole surfaces in effect are given the same triboelectric properties. The conductivity of the carpet remains unchanged, so that carpets treated in this way will not meet specifications based on resistance or charge decay.

The transfer of antistatic agent from carpet to shoe sole implies limited durability, and this is usually the case. Topical antistats are, however, extremely useful when an installed carpet that has not been engineered for static control gives static problems. Such treatments are offered by carpet cleaning companies, and the antistat may be routinely renewed by the cleaning company after each wet cleaning operation.

Unique to wool is a cationic antistat, marketed under the name Disperstat, that is sufficiently durable as to provide satisfactory protection throughout the life of the carpet for all but extremely dense traffic situations. Application procedures for Disperstat are described in Section 15.12.5. Treatment with Disperstat is the most convenient way of making a wool carpet antistatic since no special backing is required. It is suitable for most carpet applications but not for those where conductivity is a requirement.

21.3 HOT CARPETS

A totally different aspect of electricity in carpets is the use of resistance to generate heat. 'Hot carpets' were developed in Japan, where shoes have traditionally been removed before entering the home, to provide comfort to the feet and some background heating. They are usually squares, and may be dimensioned in multiples of tatami sizes.

Commonly, a heated pad situated between the carpet and the underlay is used. To economise on electricity, the pads may be located where people most frequently sit. The carpet above the pad should be less than 9 mm thick and 3250 g/m^2 total mass.

Two general methods of creating hot carpets are used. One is to thread heating wires sinusoidally into the material, in a somewhat similar way to the construction of an electric blanket. The other (less common) is to coat the back of the carpet with carbon dispersed in a backcoating compound to create a coating having the required resistance. Electrodes are attached to opposite corners of the carpet.

The products are designed for 110 V supply and ratings are typically 65–95 W. Groups of mats may be wired in parallel. For convenience, timers and thermostats are included in the circuit.

21.4 LUMINOUS CARPETS

There is a market in the entertainment sector of the hospitality contract market for carpets that fluoresce. The principal objective is to create designs that become prominent and attractive under the ultraviolet light that is a feature of nightclubs, discotheques and some theatres. Fluorescent motifs can also be used to indicate the direction of exits in such darkened environments.

Du Pont's Footlights version of nylon 6.6 has a built-in fluorescent material, so that it fluoresces when dyed to any colour. Footlights yarns may be used as components of a design or for the whole design.

Fluorescent colours may be dyed on wool using acid dyes, such as Rhodamine B (Acid Red 52), Fluorescine (Acid Yellow 73) and a phthalocyanine blue (Direct Blue 189), plus a fluorescent brightening agent. Alternatively, wool that has been treated with an acid-dye-resist agent (eg, Sandospace R) may be dyed with fluorescent basic dyes. The colours achieved have rather poor fastness to light, but this is irrelevant in the intended locations.

Electroluminescent carpets have been patented by WRONZ. Flexible, low voltage, electroluminescent panels are incorporated into the backings of carpets: when they are activated, the light created shines through the primary backing fabric and pile to give illuminated carpets. Colours and patterns can be introduced and switching/sequencing operations can give novel effects. Potential applications are in mats and carpet tiles in entertainment locations, and for safety purposes.

21.5 AIRCRAFT CARPETS

Carpets for the passenger compartments of aircraft have special requirements that may be classified as commercial or technical. Flammability performance is the key safety factor in aircraft carpets. Test methods and limits of performance originate from regulatory bodies and aircraft manufacturing companies. Because of the international nature of the air transport industry, there are moves to standardise flammability testing and performance specifications through the Joint Airworthiness Authority (JAA). Currently, the JAA specifies only a burning test carried out according to the method of Federal Aviation Regulation 25.853b (see Section 20.3.2).

Aircraft manufacturers have, over the years, studied the requirements for carpets in passenger aircraft [9], leading to specifications more stringent than the minimum statutory safety requirements. The principal ones identified are summarised below.

Safety from fire: In addition to the JAA minimum standards for resistance to burning, many aircraft manufacturing companies specify that the carpet must pass the JAA test after repeated cleaning, and may only emit limited smoke and toxic gases when subjected to the action of heat.

Such requirements usually mean that the pile composition is wool. Untreated wool satisfies the basic specification, but wool treated with the Low Smoke Zirpro flame-retarding process (see Section 15.12.2) is often used to provide additional safety. Backing yarns are cellulosic (including FR rayon), FR polyester or glass; backcoatings are usually EVA filled with aluminium trihydrate. Chlorinated fibres or plastics are unacceptable. Nylon carpets have been engineered so that flaming droplets are not formed in the vertical flammability test.

Safety from electric shock: In addition to the possibility of electrical charges being developed by walking, body voltages may result from the interaction of passengers' clothing and upholstery materials. Some in-flight entertainment systems are susceptible to interference from electrostatic effects. To guard against charge generation by walking and to disperse charges from other sources, wool pile aircraft carpets are often engineered with conductive fibres as well as being treated with Disperstat.

Acoustic insulation: An important function of the carpet is to insulate the cabin from sounds. For some models of aircraft, Boeing specifies sound attenuation of 30 dB summated across the octave bands. It may be necessary to use an underlay (pad) to achieve such values because of thickness limitations.

Low weight: The single integrated backing of a woven carpet is lighter than that of the laminated primary/secondary backing system of a tufted carpet, so that more weight is available for surface pile. For example, plain Wilton carpets of total mass 2000 g/m^2 have surface pile mass in the range 600–1050 g/m^2, whereas tufted carpets of the same total mass have surface pile masses in the range of 450–850 g/m^2. The weight requirement, therefore, leads to the use of woven carpets.

Aircraft carpets may be classified as light (less than 1760 g/m^2) or heavy (more than 1760 g/m^2 total weight).

Durability: The requirement for low weight means that the construction of the carpet must be carefully engineered to achieve satisfactory durability. Consideration of the pile mass density factor (P^2/t) (Section 20.4.1) indicates that the permitted surface pile mass is best utilised in low pile thickness. The requirement for appearance retention (see below) tends to lead to the selection of loop-pile carpets. Aircraft carpets, therefore, are mostly loop-pile wireloom carpets, including weaves where the pile is floated over weft only. Wireless looms, such as the Loop Pile Master LPM32 (Section 10.3) or the DuraLite loom (Section 12.6), can equally produce very dense constructions.

Appearance retention: The most damaging aspect of wear is repeated passages of food/beverage carts up and down the aisle of the passenger cabin. The effect on carpet appearance is most severe at the points where the trolleys are reversed. Loop-pile carpets resist this action better than cut-pile.

Dimensional stability: It is vital that the carpet should not expand, and therefore ruck, during use. Limits of 0.5% shrinkage or extension are often set, which are difficult to meet. Cotton or jute backing yarns should be pre-shrunk and the inclusion of a glass stuffer warp is advantageous to prevent extension.

Suitability for installation: Double-sided tape is often used to install carpets in aircraft, and there is the risk that tufts could be pulled from the back when carpets are removed for cleaning. Tuft withdrawal force should, therefore, be particularly high. Scuffing of the carpet by passengers' feet tends to cause fibre shedding; also pointing to the need for good control of backfinishing.

Cleanability: Wet cleaning can restrict the utilisation of the aircraft, so that soil resistance may be a factor determining carpet life. Colour fastness to wet cleaning and to spotting for stain removal should be 4+ on standard grey scales (see Section 20.4.3).

Corrosion: To avoid corrosion of aluminium as a consequence of spillages, or *in situ* wet cleaning, inorganic salts or acids should not be present in aqueous extracts of aircraft carpets.

21.6 OTHER PUBLIC TRANSPORTATION APPLICATIONS

Although not so obviously technical as aircraft carpets, the carpets installed in land and sea transport systems need to be just as safe: it is not easy to escape from a fire in a moving train, nor from a ship at sea.

The flammability specifications set for transportation may originate from national or local bodies, coastguard organisations or the transport companies themselves. Most commonly, the NBS Flooring Radiant Panel Test is used as the basis of criteria for spread of flame; and the NBS Smoke Chamber is used to test for evolution of smoke and toxic gases (see Section 20.3.2).

The International Maritime Organisation in December 1998 introduced a detailed specification for floorcoverings for cruise liners (FTO Code, International Code for Application of Fire Test Procedures). Carpets in areas requiring low flame spread characteristics are to be tested in accordance with IMO Resolution A.653(16)/Resolution MSC 61(67) Annex 1– Parts 2 and 5. The tests cover critical radiant flux at extinguishment, heat for sustained burning, total heat release, and peak heat release rate, as well as smoke and toxic gases.

Durability can be achieved without the strict weight limitations imposed on aircraft carpets. The actual requirements vary tremendously: in some areas of cruise ships, luxury may be more important than durability (although traffic levels can be high), but the intense foot traffic in rapid transit systems in most major cities imposes extreme demands on a carpet. Carpets need to be constructed with a high pile mass density factor (P^2/t), which is the principal factor governing durability (see Section 20.4.1).

In long corridors and large areas, tight limits on dimensional stability need to be set. The actual requirement for a particular order depends on the method of installation (stretched, direct stick).

Rapid transit systems provide arguably the most severe soiling circumstances for any carpet. Less soil is brought into ships, even ferries. Patterned carpets in colours toning with the predominant colour of the local soil and most likely staining materials can make soiling and staining less prominent. The inherent soil resistance of the fibre is the next most important factor: currently available surface finishes are insufficiently durable for such intense traffic. Programmed maintenance is essential if the carpet is to retain its good appearance. At each cleaning step, only a percentage of the original clean appearance can be retrieved, so if the level of appearance due to soiling between cleaning steps is allowed to fall too low, the appearance of the carpet remains at a low level, in line ABCDGHIJ of Figure 21.2 [10]. More frequent cleaning gives the line AM, having a high appearance level.

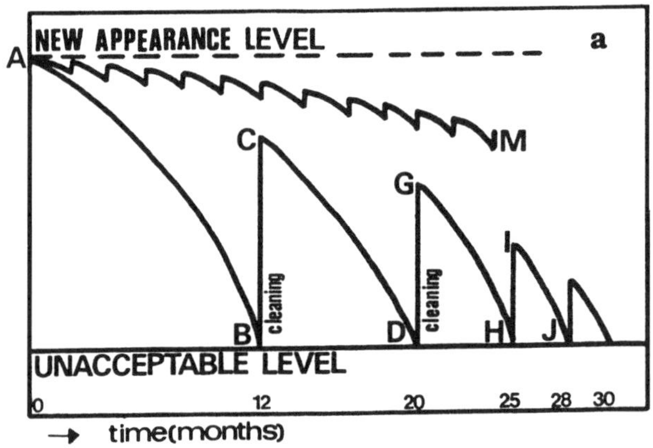

Figure 21.2. Appearance levels of carpets after frequent or infrequent cleaning [10].

21.7 TEXTILE SPORTS SURFACES

Artificial turf is in widespread use for many sports. The successful development of what are, in effect, outdoor carpets, prompted the use of similar products for other outdoor locations such as swimming pool surrounds, verandas, patios, boat deckings, marinas and children's play areas. Grass-like products are also used indoors, for example in car showrooms, exhibition stands and retail display areas. 'Grass' is not always green: it may be offered in brown, tweeds or even blue.

The market for artificial grass in the USA is relatively steady at about 30 million m^2 per annum, although only a minor proportion of this is used for sports. Figures for Europe are less accessible, but the outdoor lifestyle that exists in the USA is not so appropriate in northern Europe so the total market for artificial grass is smaller, although the proportion applied in sports is higher.

21.7.1 General Merits of Artificial Sports Turf

Artificial turf is used in sports for a variety of reasons [11, 12, 13].

Independence of growing conditions
Grass does not grow in northern latitudes in the winter when many sports are played, so that natural turf tends to deteriorate as the season progresses, whereas artificial turf does not; and grass can only be grown under controlled conditions in many arid areas of the world. Monsanto's research programme on artificial turf culminated in a crash development programme when it was found that grass would not grow in the Houston Astrodome when it was completed in 1966. Their programme resulted in the original Astroturf.

Versatility
It is possible for installations of artificial turf to be removable, so that a stadium can be used for events not requiring turf, such as rodeos, mini-automobile races and cattle shows. This was the case with the original Astrodome installation,

where sections of carpet were linked with zip fasteners and could be withdrawn into pits on the perimeter of the action area. Removable installations are common for indoor sports played in multi-purpose halls. In the other direction, carpet-like materials may be installed temporarily to convert ice rinks into areas for other sports or concerts, etc, where their thermal insulation properties are important.

Durability

Real turf is entirely satisfactory for most sports when used for a few games a week, but scarce city playing areas are in high demand and artificial turf can withstand perhaps 50 games a week. Durability is an important motivation for public authorities to install artificial turf for multiple sports facilities. Professional sports organisations in the USA often augment their incomes by renting out their stadiums to amateur sports organisations, and for activities such as rock concerts, opera, boxing matches and ceremonial occasions. When the audience is located on the playing surface, exceptionally high durability is required.

Improved playing characteristics

It is not often the case that artificial turf is preferred to high quality natural turf. One exception is hockey, where the more reliable ball/surface interaction has led to only artificial turf being acceptable at the highest levels of the game. Artificial surfaces are also favoured for tennis and bowls. Methods of measuring ball/surface interaction have been standardised, and have proved vital in the development of satisfactory textile surfaces for many sports (Figure 21.3).

Reduced maintenance costs

Artificial turf requires only brushing or suction cleaning to remove surface debris, occasional repairs, occasional rolling if installed on a dynamic base, and perhaps annual worm/weed/moss killing. Hockey and lawn bowls surfaces are

Figure 21.3. Using a cricket pitch pace and bounce monitor. A cricket ball is fired at about 100 kph at a predetermined angle and the delivery speed, rebound speed and height of bounce are measured. (Courtesy of New Zealand Sports Turf Institute)

an exception in that they require watering. There is no need for cutting or fertilising as is required for grass. Amateur sports organisations often find that they can afford an artificial surface having good playing characteristics, whereas quality turf would require more costly maintenance. However, the cost of depreciating artificial turf over, say, a 10-year period must be considered.

21.7.2 Playing Characteristics

When artificial turf was introduced in the USA it was primarily for American football and for the outfield at baseball. Ball/surface interaction was of minor importance and player/surface interaction was minimised by the clothing worn. Both ball/surface and player/surface interaction are important in many European field sports and for games such as tennis and bowls. The question arose as to whether artificial turf 'played like grass'. As there was a lack of objective information about the playing characteristics of natural turf, this question prompted major research programmes into the subject, principally in the UK, Germany and the Netherlands [11]. The development of artificial turf has, as a result of this research, stimulated some of the improvements in natural sports surfaces that were made in the last 25 years.

The UK Sports Council in the late 1980s published their *Specification for Artificial Sports Surfaces, Part 1,* covering fibrous, polymeric and grass surfaces. It lists broad technical characteristics (required or not) for 28 outdoor and 19 indoor sports, plus 'common space' for multiple indoor sports. BS 7044, *Artificial Sports Surfaces. Part 1. Classification and General Introduction*, is a development of the Sports Council document. BS 7044 Part 2 covers methods of test.

Many sports have not accepted textile surfaces at national and international levels, although they may be widely used at lower levels and for practice. Hockey and bowls have international standards. The International Tennis Federation has published a study towards a standard, and the official bodies for other sports have studies in progress.

It should be noted that the base on which the textile surface is laid is often at least as important as the surface itself in determining playing characteristics and other aspects of performance; and relevant footwear may be required to achieve satisfactory player/surface interaction.

Two general types of base may be used for outdoor installations: an engineered base (an item of civil engineering such as porous asphalt) and a dynamic base (which may be favourable soil, or a sophisticated layer structure of graded particulate material and stones, incorporating geotextiles).

Hockey

Textile sports surfaces have received widespread acceptance in hockey circles: indeed only textile surfaces are accepted at the highest levels of the game. The more consistent ball roll and bounce compared with turf provided motivation for overcoming other problems. The principal problems were the risk of friction burns when players fell, and too fast ball roll. Both problems have been solved by using the surface wet. Standard pitches are equipped with an irrigation system and the pitch is watered before the game and at half-time. Sand-filled surfaces (see Section 21.7.3) are a common alternative to water-filled surfaces.

The performance standards and test procedures required by the International Hockey Federation are: ball rebound and roll; ball-to-surface friction; underfoot friction; impact response; pile/pad deformation; surface colour and gloss; and pitch porosity [14].

Lawn Bowls

As with hockey, the possibility of achieving a consistent level of ball/surface interaction has been a motivation for using textile surfaces for bowls. Player/surface interaction is less relevant. Wetting the surface is, however, used to control the green speed.

The World Bowls Board's performance standard [15] lists six test methods:

WBB-01 Determination of green speed,

WBB-02 Determination of surface draw,

WBB-03 Determination of infiltration rate using a disc permeameter,

WBB-04 Wetting procedure,

WBB-05 Determination of surface cushioning,

BS 7044.2.2 Surface friction.

The first three, plus limits for surface evenness and design levels, are used to set the international performance standard. The last two tests listed are recommended for consideration by potential purchasers of artificial surfaces.

Tennis

The surfaces generally preferred by tennis players are clay and grass. Both require careful maintenance if consistent playing characteristics are to be achieved, and grass is not sufficiently durable for intensive club play. Textile surfaces may be used for outdoor or indoor tennis.

The International Tennis Federation is working towards standardisation of surfaces, both natural and artificial, and has developed a number of test methods, each defining performance criteria [16]. These take into account the tennis balls and shoes that are required to be used. The test methods and criteria are:

Surface pace (ITF CS01/01);

Friction (a) slip resistance (ITF CS02/01); (b) traction (ITF CS03/01);

Energy absorption (ITF CS04/01);

Vertical ball rebound (ITF CS05/01);

Permeability (ITF CS06/01);

Slope (ITF CS07/1);

Evenness (ITF CS08/01).

21.7.3 Products

Textile surfaces have traditionally been used in some sports: the mats of cricket, the canvas of boxing and the baize of billiards. From about 1950, pressed felts began to be used for indoor bowls as permanent installations or roll-up mats. However, this book is concerned with products made by the carpet industry.

The principal techniques of manufacturing artificial turf are tufting, knitting, needling, weaving and electrostatic flocking. Additionally, thermobonded products may be used as components of multilayer structures. Needled products are often used as shock pads under harder sports surfaces and are sometimes laminated to them via a thermobonded layer or by needling to provide an integral shock pad. For athletics surfaces (Figure 21.4), there may be an upper spike-resistant thermobonded layer, a cushion layer of needlefelt and a thin layer of heavy decitex fibres to grip the sub-surface.

Figure 21.4. An installation of artificial turf at Soonchunhyang University, Korea. (Courtesy of Astrograss All Weather Surfaces Ltd)

Pile products may be subdivided into sand-filled products and non-filled products. In the latter, the pile (plus base) is engineered to provide the playing characteristics and the pile is typically dense and heavy. Non-filled products are installed by power stretching or adhesion to the base. Technical motivation (cost was also a major factor) for the development of sand-filled products came from two directions. To reduce the possibility of 'foot lock' (excessively strong interaction of studded footwear with the pile) the density of the pile may be reduced and compensated by filling with sand. Alternatively, a sand-filled pitch may be regarded as a playing surface essentially composed of sand stabilised by the textile material, and provided with grass-like aesthetics by the fibres. Usually, the mass of sand is sufficient to hold the textile material in place and no edge fixing or adhesives are necessary.

Fibres and yarns
Polypropylene and nylon are the only surface fibres used in substantial volumes.

Polypropylene
Special grades of polypropylene are manufactured to withstand weathering. Unprotected polypropylene is highly susceptible to photothermic oxidation that degrades the polymer chains, but careful selection of UV stabilisers can result in fibres that are among the most resistant of all fibres – different stabilisers can offer markedly different levels of protection. Suitably stabilised products give 70–95% strength retention after 10 years' use in geographic areas having up to 100 MJ/cm^2 global radiation per year (most of Europe). Polymer pigmentation has been established as the standard method for colouring polypropylene and the high standards of colour fastness achievable have contributed substantially to its general success in the carpet industry. However, only a few pigments can achieve the standards required for uninterrupted outdoor exposure. The best results are obtained from combinations of phthalocyanine and inorganic pigments, although some manufacturers use organic pigments exclusively for environmental reasons. Polypropylene absorbs virtually no water, a fact that contributes to retention of dimensions, strength and abrasion resistance.

Fibrillated split film is widely used in pile products because it simulates the appearance of grass. Typical yarns are 5000–8000 dtex.

Modified polypropylene yarns have been developed to reduce the risk of friction burns [17]. Polypropylene/polyester copolymers that are softer and have lower surface friction than standard polypropylene are offered for soccer and have even been suggested for rugby. Such 'slide' polypropylenes are more expensive but life-cycle costing has shown that they may be an economical alternative to pitch watering, depending on the local cost of water. Yarns made from finer (90 dtex) monofilament of cylindrical cross-section give reduced friction heating, especially in loop-pile constructions, and are less susceptible to colour change owing to fibrillation. A third possibility that is quite widely practised is to use tapes curled and set at 120°C by the knit-deknit process: contacts with the body are preferentially with the sides of the tapes rather than the cut ends. Knit-deknit or stuffer-box crimping also give a non-directional surface – important for ball/surface interaction in games such as bowls.

Nylon
Nylon was the pile fibre used in the original Astroturf, and represents the second most important pile material in textile sports surfaces. Both nylon 6 and nylon 6.6 are used. Yarns are typically monofilament of rectangular or round cross-section, around 500 dtex. As with polypropylene, the incorporation of effective stabilisers to minimise photothermic oxidation is essential. Coloration is by pigmentation, rather than dyeing, to achieve the required standards of colour fastness.

Tufted products

Apart from the special pile yarn, tufted artificial turf requires a special primary backing fabric and a permeable backcoating. Primary backings are usually woven from polypropylene tape containing high levels of UV stabilisers and are dark in colour (eg, black or brown) to avoid the possibility of undesirable grinning of ecru tape. The weight range is a little heavier than for ordinary carpets (120–150 g/m^2). Backcoating for tuft lock may be with X-SBR or EVA but applied by special techniques (eg, use of air doctors) to retain permeability, allowing rainwater or irrigation water to drain away (Figure 21.5). Satisfactory permeability may alternatively be achieved by puncturing the backing after latexing using special equipment (eg, a hot spiked roller to create heat-fused orifices). Polyurethane foam having a closed cell structure that resists the uptake of water may be applied as an integral shock pad: it must be embossed with drainage channels to satisfy the requirements for permeability.

Figure 21.5. Using a disc permeameter to quantify the drainage properties of an installation of artificial turf. (Courtesy of New Zealand Sports Turf Institute)

Primary backing fabric having a surface layer of melt-bonding fibre is advantageous for providing tuft bind without the risk of creating an impervious layer (see Section 5.2). As an alternative to polymeric shock pads, needlefelts may be laminated to the back, possibly using melt-bonding fibres. Shock pads (required principally to achieve the required ball rebound and impact response over engineered surfaces) are typically 10–15 mm thick. An alternative approach to achieving the required impact response is to cast a thick (35–40 mm) shock pad consisting of rubber and aggregate on to the engineered base.

Cut-pile tufting gauges in the range 5/32–5/16 are used for non-filled products (3/16 gauge is very common) and pile heights are usually in the region of 8–15 mm for straight yarns but may be over 30 mm for curled yarns. Surface pile weights are in the range 1000–2000 g/m^2.

Seaming may be by means of an adhesive system (contact adhesive or two-part polyurethane) or sewing.

Sand-filled products are in the lighter surface pile weight range at 900-1250 g/m^2 and are tufted on 5/16–3/8 gauge machinery to give more open structures. They may or may not require a shock pad, depending on the nature of the base and the requirements of the sport.

Edge fixing and seaming of sand-filled products is less critical than for non-filled products. Once the material has been tensioned, the sand holds it in place. The application of sand is usually at a rate of 20–40 kg/m^2 to leave 2–4 mm of pile exposed; curled pile products may require less sand.

Knitted turf

The original Astroturf was a cut-pile product manufactured by warp knitting and had a nylon monofilament pile. Competitive knitted nylon products are now available. Their principal advantages are that very dense pile may be produced and the backing does not fray as easily as that of a tufted carpet, so that strong seams may be made using heavy duty sewing machines. Backcoating for stability and tuft bind is required, as for tufted products, and the same special procedures are required to create the necessary permeability to rainwater.

Woven turf

High quality surfaces for cricket pitches and golf tees, competitive with tufted products, are manufactured by Axminster, face-to-face or wireloom weaving. They are usually low pile, dense products with pile and all backing yarns composed of polypropylene. Only a light backcoating is required to give adequate tuft bind so that adequate permeability for outdoor installations can be achieved by simpler techniques than those required for tufted products. Shock pads, if required, are supplied as separate items and are not usually laminated to the carpet.

As for knitted products, woven sports surfaces may be seamed to high standards by sewing.

Needled products

Needled products may be intended for indoor or outdoor applications or for use as shock pads for tufted, knitted or woven products. The shock pads may be separate items or, in the case of tufted and needled materials, laminated to the backing.

Quite wide ranges of thicknesses, weights and densities are used. Products 3 mm thick and as light as 600 g/m^2 are made for mini-golf, while dense double layers up to 2000 g/m^2 and 10 mm thick are used for tennis, with intermediate weights being used for sports such as badminton and bowls. Very thick multilayer products are used for golf tees and are supplied with holes drilled for tee pegs (or rubber cups in the case of driving ranges).

A product specially engineered to play like a clay tennis court consists of a needlefelt with a surface structure that holds particulate matter (ie, similar to a sand-filled tufted product) [17]. The filler is selected to be relatively round and the application rate is controlled so that the correct levels of ball/surface and player/surface interaction are achieved.

Needled shock pads are used for both indoor and outdoor sports. Their permeability and resistance to frost damage make them particularly suitable for outdoor use. They are less densely needled than needled wear-surfaces. When the wear surface is needled, a needled shock pad may be integrated with it, simply by needling the two layers together.

Flocked products

Flocked floorcoverings are usually created on a layer of PVC plastisol so they are impermeable and are applied for indoor sports only. The PVC base layer makes them highly suitable for direct stick installations. Line colours are supplied separately in 50 mm width. A typical product has a 2 mm pile and a total thickness of 4.4 mm.

Flocked surfaces are commonly installed in school sports halls.

Combined natural/synthetic turf

Surfaces that combine players' preference for natural turf with the durability of synthetic turf may be achieved by seeding a sand-filled installation. One such example consists of a tufted product with a 28 mm polypropylene pile and a polypropylene backing which is installed on sub-grade, filled with sand and seeded with grass suitable for the locality. The grass grows through and around the synthetic fibres to create a durable matrix of synthetic turf and roots. Both the crown and root system of the plant are protected by the polypropylene.

REFERENCES

1. *Performance Specification for Textile Floorcoverings in National Health Service Patient Areas*, NHS Supplies Authority, UK.
2. L F Story (ed). *Carpets in Hospitals*, Proceedings of a seminar held in Christchurch, 18 February 1976, WRONZ Special Publication, Wool Research Organisation of New Zealand, Christchurch, May 1976.
3. P G H Bakker. Effect of carpets on hospital hygiene – literature survey. *Textil Praxis International*, 1978, 33(8), 941–944; English supplement XXI–XXIV.
4. P L Hauck. Carpets and hygiene, recent key learnings. In: *Proceedings of Tifcon '94 – Technologically Speaking*. The Textile Institute, Manchester, 1994.
5. Antimicrobial issues relating to carpeting. White Paper, DuPont HomePage, www.dupont.com, 1996.
6. Wool carpets – not unhealthy. *Medical Textiles*, 1988, 5(8), 8–9.
7. A Custovic, A Fletcher, A C Pickering, H C Francis, R Green, A Smith, M Chapman and A Woodcock. Domestic allergens in public places. III: House-dust mite, cat, dog, and cockroach allergens in British hospitals. *Clinical and Experimental Allergy*, 1998, 28(1), 53–59.

8. D A Rankin. Unpublished WRONZ report, 1997.
9. L F Allison and S A Hasselbrack. Demands in airplane carpeting. In: *Carpets and Associated Technologies – Proceedings of 10th Annual Conference,* L F Story (ed). The Textile Institute (NZ Section), Auckland, 1982, 117-126.
10. S M Spivak and N Gupta. Carpet cleaning and resoiling. *Textile Horizons*, 1986, 6(2), 45–48.
11. G H Crawshaw. *Textile Sports Surfaces and Artificial Grass*, Elsevier Advanced Technology, Oxford, 1989.
12. D V Waddington, R N Carrow and R C Shearman (eds). *Turfgrass*, No.32 Agronomy Series, American Society of Agronomy, Madison, USA, 1992.
13. D I Rhodes. *The Maintenance of Artificial Sandfilled Turf Used for Sport*, Sports Surface Management, Bolton, UK, 1996.
14. *Handbook of Performance Requirements and Test Procedures for Synthetic Hockey Pitches – Outdoor*. International Hockey Federation, April 1992.
15. *Performance Standard for Artificial Flat Grass Bowls Surfaces*. World Bowls Board, January 1998.
16. *An Initial ITF Study on Performance Standards for Tennis Court Surfaces*. International Tennis Federation, June 1997.
17. I Beswick. Polypropylene in textile surfaces for sports and recreation. In: *Proceedings of World Textile Congress on Polypropylene in Textiles*, University of Huddersfield, 9–11 July 1996.